SCHICK
1/09

The Evolving Universe and the Origin of Life

Perlmutter, Luciano Pietronero, Laura Portinari, Travis Rector, Rami Rekola, Shane D. Ross, John Ruhl, Allan Sandage, Markku Sarimaa, Aimo Sillanpää, Francesco Sylos Labini, Leo Takalo, Gilles Theureau, Malene Thyssen, Luc Viatour, Iiro Vilja, and Petri Väisänen.

We are grateful to Harry Blom, Christopher Coughlin, and Jenny Wolkowicki of Springer-Verlag, New York for very good collaboration and patience during the preparation process of this book.

Similarly, we thank Prasad Sethumadhavan of SPi Technologies India.

August 2008
The authors

Contents

Part II Physical Laws of Nature

List of Tables

Part I
The Widening World View

Chapter 1
When Science Was Born

Thomas Henry Huxley, the eminent British zoologist of the nineteenth century once wrote: "For every man the world is as fresh as it was at the first day." This realization beautifully connects us with ancient minds. It is the same world which puzzles us now, even though we observe it to distances of billions of light years with modern telescopes on Earth and in space, and we penetrate into the incredibly small microworld using microscopes and particle accelerators. These observations and our current knowledge of the workings of the universe are the fruition of a long chain of scientific enquiry extending back into prehistoric times—when the only instrument was the naked eye and the world was fresh.

Prehistoric Astronomy: Science of the Horizon

The Egyptians noted the stars that appeared to attend the birth of the Sun in the eastern morning sky. These were different at different seasons. One star was especially important, Sirius, the brightest star in the sky, in the constellation *Canis Major*, the Great Dog. Around 3000 BC, this "Dog Star" appeared every summer in the eastern sky before dawn. The day of each year when it was viewed the first time, the so-called heliacal rising above the horizon, marked the start of the calendar year in Egypt. This very important event heralded the longed for flood of the Nile, on which agriculture and life depended.

The horizon was a fascinating thing for ancient people. They viewed it as a sort of boundary of the world. "Horizon" comes from the Greek word meaning "to bound." In the Finnish language it is romantically "the coastline of the sky" (taivaanranta). In addition to the Sun's daily motion across the sky, during the year, the places on the horizon where it rises in the morning and sets in the evening shift slowly. As winter progresses to summer, these points on the horizon move from south to north. The Sun remains visible longer and ascends higher in the sky. The day when the sunrise and sunset points are farthest to the north in the horizon and the Sun ascends highest in the sky is the summer solstice (solstice meaning "Sun stand still"

Fig. 1.1 Stonehenge is an impressive monument of Bronze Age interest in celestial events at the horizon (photograph by Harry Lehto)

in Latin). Similarly, there is a day, the winter solstice, when the day is the shortest, and the sunrise happens closest to the south. These and other points on the horizon had both practical and ritual significance. For example, the ancient Hopi people, living in their pueblos in Arizona, used (and still use) the horizon with its sharp peaks and clefts as a convenient agricultural and ceremonial calendar (e.g., the position of the rising Sun indicated when the corn should be planted).

Around the world there are archeological remains dating from thousands of years ago, which seem to have been made to worship, view, and even predict particular celestial events. The pyramids of Egypt may have originally been built to symbolize the Sun god who every morning was reborn in the eastern horizon, a place called "akhet" by the ancient Egyptians. Everybody knows of Stonehenge, one of the wonders of the Bronze Age world in the plain of Salisbury, a hundred kilometers from modern London (Fig. 1.1). It is made of concentric structures of stones and pits, the youngest of which, with the familiar great stones 6.5 m high, dates from about 2000 BC. The rather complex assemblage is surrounded by a ditch that forms a circle 104 m in diameter.

The axis of Stonehenge points at the sunrise direction on midsummer morning. For a person standing in the middle of this monument the disc of the Sun appears just above what is called the "heel stone" 60 m away. Stonehenge may have served other astronomical purposes, too. Its large circles were built first, and may have been directly related to interesting horizon points, while the later structures made of big stones may have had ceremonial significance, perhaps also symbolizing the horizon circle. The great effort needed to make Stonehenge testifies to the status given to horizon phenomena at that time.

A few years ago in Germany, a large circle formation was discovered in a wheat field which archeologists recognized as a Stone Age "observatory of the horizon." When in use, the 75-m circle had three gates, one of which looked to the north (Fig. 1.2). Two southern gates were so directed that on the winter solstice an observer standing at the center of the circle saw the Sun rising and setting at its southernmost horizon points through the gates. This remarkable structure in Goseck is about 7,000 years old. So 2,000 years before the builders started their work at Stonehenge, people in the continent were busy making horizon circles!

Fig. 1.2 A sketch of the large 7,000-year-old circle formation in Goseck, Germany. Two southern gates were so directed that on the winter solstice the observer in the center saw the Sun rising and setting through the gates (credit: Rainer Zenz/Wikipedia)

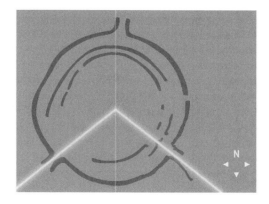

Archeoastronomers have found traces of horizon science all around the world. For example, on Easter Island in the middle of the Pacific Ocean, the famous stone statues standing on great platforms are often directed according to astronomically significant horizon points. For its natives, this island was "the eye that looks at the sky." People everywhere have been fascinated by regularly appearing celestial phenomena, have patiently noted their rhythms, and even have arranged their lives according to them. In this way, our ancestors paved the way for modern astronomy, modern science, and even modern life.

Writing on the Sky Vault and on Clay Tablets

At every point of history, mankind has made the best of what the environment had to offer for living. When the conditions changed, like during the ice ages, human cultures adopted new ways of living as a response to those changes. Sometimes unexpected things resulted. An example is the formation of the fertile delta region between the Euphrates and Tigris rivers flowing into the Persian Gulf. When the surface of the Gulf gradually rose tens of meters after the Ice Age, the flow of the two rivers slowed making the region good for farming. However, when the climate got dryer around 3500 BC, large scale irrigation became important and power became centralized in Sumerian cities. Life was centered on the temple, dedicated to the god of that city. The temples were large administrative and economic centers, headed by the clergy. The polytheistic religion of Sumer was inherited by Babylonia around 1500 BC.

Writing had been invented around 3000 BC by Sumerians. It started a flow of unexpected cultural evolution. The art of cuneiform writing was originally useful for bookkeeping in the economic centers, temples, but it gradually found application in many other fields than business, including sky watching. How celestial bodies move gives us both ancient and modern methods of timekeeping. We know that Sumerian clergy tracked the Moon to build a lunar calendar by recording the information on clay tablets.

However, their direct descendents, the Babylonian priests, were instead curious to learn what signs the divine celestial stage offered about the future of the rulers and the kingdom. The sky formed a huge screen with "texts" that the specialist tried to interpret. Thus, systematic astrology was born, together with a developed state. Interest in the misty future was strong and there were also other methods of prediction, like watching the flight of birds. In contrast to today, at that time astrology was quite a rational undertaking when stars were viewed as gods or their representatives. It was logical to try to find links between celestial phenomena and earthly happenings. Some were indeed known: the seasons are marked by the path of the Sun among the stars and tides obey the Moon. With little artificial light to block their view, the ancients were much more observant of the sky than most people today.

In Mesopotamia, a lunar calendar was based on the phases of the Moon. Each month began on that evening when the thin sickle of the growing Moon was first seen after sunset. Nowadays, the solar calendar (which is consistent with the seasons) dominates everyday life, but the lunar calendar is still important for religious purposes.

Because of the yearly cycle of the Sun, different constellations are visible in the evening at different seasons. The appearance of the sky today is almost the same as thousands of years ago. Many constellations still carry the names that shepherds or seamen once gave them. Certainly the starry patterns initially had real meaning. Various animals, gods, and mythical heroes were permanently etched on the sky. But the constellations also form a map that helps one to identify the place where something happens in the sky. In modern astronomy, there are 88 constellations with definite borders. For instance, when comet Halley last appeared, one could read in the newspaper that in December 1985 the visitor would be in the constellation of Pisces just south of Pegasus. With this information it was easy to spot the famous comet through binoculars. The daily motion of the Earth merely caused the comet and the constellation to move together across the sky, keeping their relative positions.

The Babylonian astrologers were well aware that not all celestial objects move faithfully together with the stars. The Moon shifts about $13°$ (or 26 times its own diameter) eastward relative to the stars every day. It takes a little more than 27 days for the Moon to come back roughly to the same place again among the stars. Also the Sun moves relative to the stars although the glare blots them out. However, during the year, different constellations are visible near the Sun just before sunrise or a little after sunset. Thus it was deduced that the Sun moves around the sky visiting the same constellations through the year. Astrologers divided its route, or the *ecliptic*, into 12 equal parts and the Sun stayed in each for about one month. These constellations came to define the signs of the zodiac. The word ecliptic means the solar path where the eclipses occur.

Constellations and Horoscope Signs

About 2,000 years ago, the signs of the zodiac (familiar from newspaper horoscopes) and the actual constellations corresponded to each other. This is not so any longer. Your horoscope sign may be Aries (the Ram), but this does not mean that the Sun was in the constellation of Aries when you were born! Quite probably the Sun was in Pisces (the Fishes) at the time. The reason for this is that the constellation names and dates in newspaper horoscope columns correspond to those in a book on astrology written by the astronomer Ptolemy nearly 2,000 years ago. The zero point or the start of the sequence of constellations was the vernal equinox, the point where the Sun on March 21 crosses the celestial equator going from the southern to the northern celestial hemisphere. However, this zero point is not fixed but moves slowly relative to the stars and constellations. The time interval from then until now has resulted in a change of about one constellation. This motion makes a full circle every 26,000 years and it was discovered observationally by the Greek astronomer Hipparchus (circa. 190–120 BC). Physically, we now know that the movement of the zero point is due to the Earth's axis slowly wobbling like a top about to tip over due to gravitational effects of the Sun and Moon on the slightly flattened Earth. To read a horoscope corresponding to your "up to date" sign, just read the newspaper entry above the one you would usually consult. Then you can choose the one you like better!

The Babylonians made regular observations of planets that also move close to the ecliptic. They knew Venus, Jupiter, Saturn, Mars, and Mercury, and interpreted their behavior as important signs corresponding to what will happen on the Earth. The various movements of the planets, their encounters with each other and with the Moon, their appearances and disappearances, gradual fading and brightening, all offered information for the interpreter who did not know the real reasons behind such phenomena (Fig. 1.3). The Babylonian astrologers, who were also priests of the great temples, were interested in state affairs, prospects of economy and agriculture, the health of the king, success in war, and such things. It was only later that personal horoscopes based on the time of birth appeared (among the Greeks).

The astrologers noted that the planets followed the same general route as the Sun in the ecliptic, but now and then they slowed down, even stopped altogether and went back a few steps in the sky before again continuing their normal way from east to west. This *retrograde motion* of the planets was a major feature that needed explanation both for the Greeks and later for Copernicus in making mathematical models of planetary motion.

For Babylonian astrologers predicting retrograde motion would be important to predict future events on Earth. Also desired was the ability to foretell the frightening eclipses of the Moon and the Sun. The Assyrians collected accurate statistics of lunar eclipses and found some regularity in their appearances. The Babylonians further developed the art of eclipse prediction. They noted that lunar eclipses had a long period after which they are repeated similarly. This periodicity is governed by the "Saros cycle," a little over 18 years (18 years and $11\frac{1}{3}$ days). It allowed one to calculate tables showing the possible dates of lunar eclipses far in the future. The astrologers found periodicities in the motions of the planets as well and they could predict their future motions and positions by clever arithmetic methods.

Thus ancient sky watchers learned not only to interpret the events in the sky at each moment – but also to predict significant celestial events well in advance. Babylonian astrology/astronomy reached its peak during the centuries before Christ.

Fig. 1.3 The solvogn (the Sun carriage) from the Bronze Age Denmark, expressing the old belief that the Sun was carried across the sky every day. The same idea may be found, e.g., among the Egyptians and the Babylonians, though the vehicles were different. This over 3,000-year-old artifact is at display at the National Museum in Denmark (image: courtesy of Malene Thyssen)

When the "wise men from the east" of the Bible, likely Babylonian astrologers, arrived to worship the newborn after having seen his star, Babylonian culture was already declining. However impressive these predictions were, this systematic gathering of observations was not scientific, in the usual meaning that we today attach to this term. Some key elements were missing. Posing questions and an investigative attitude, which later proved to be a source of real knowledge, were still rare. Modern

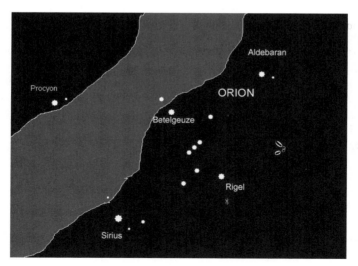

Fig. 1.4 The brightest fixed star in the sky, Sirius in the constellation of the Great Dog (Canis Major) close the Orion, was worshipped in ancient Egypt. The appearance of the "Dog Star" in the morning sky heralded the beginning of the flood of the river Nile. Just across the band of the Milky Way there is Procyon, the brightest star of the Little Dog (Canis Minor). For today's stargazers those brilliant points are material objects in space, and we wonder: How far away are they? What makes them shine?

astronomers observe the sky to understand *what* the celestial bodies are, *how* they are born and evolve (Fig. 1.4).

The Ionian Way of Thinking

The seeds of our science were sowed on the western coast of Asia Minor, where the Ionian Greeks lived in their flourishing colonies. In the seventh century BC Ionian cities, among them Miletus and Ephesus, were centers of Greek culture and economy. In these focal points of trade and exchange of ideas a new mode of thinking was born, characterized by brave individuality, in contrast to the traditional empirical inquiries practiced by the Babylonian priests. Throughout the history of science different modes of scientific activity seem to have been represented in different degrees. In Ionian Greece, thinking and discussions were the primary ways of attempting to understand natural phenomena. Simple, but accurate everyday observations formed the material ("data") for discussion.

We have little first-hand knowledge of the first Ionian philosophers who left no writings. Aristotle, who lived 250 years later, tells how these thinkers began to search for the underlying *principle*, a deep property of the world that ties together apparently different things. It would allow one to understand the great diversity appearing around us and perhaps to predict phenomena that previously were thought to be under the capricious control of the gods. As Aristotle stated: "... this, they say, is the element and this is the principle of things, and therefore they think nothing is either generated or destroyed, since this sort of entity is always conserved ... Yet, they do not all agree as to the number and the nature of these principles. Thales, the founder of this type of philosophy, says the principle is water."

We see that these first philosophers already had in mind the conservation of matter, the predecessor of important conservation laws of modern physics. They also debated about the Aristotle's first element. Thales (624–547 BC) suggested water, while his friend Anaximander (611–546 BC) mused that the first element is something so deep that it even cannot be named from among the known forms of matter. A little later, Anaximenes (585–526 BC) considered that the element is air, taken however in a wider meaning than the blend of gases that we breathe. For him it was a medium that held the whole universe together. It could have different densities, which explained the different forms in which matter exists. His qualitative reasoning was a step toward physics.

These Ionian philosophers did not yet know that the Earth is spherical. Thales and Anaximenes had it flat and floating on the first element (water or air). But Anaximander suggested a remarkable thing. The Earth is at rest in the middle of everything, in the air, and does not move away, because there is no privileged direction where to go! He used in his argument the principle of isotropy, so central in modern cosmology. Aristotle joked that this was as if a hungry man surrounded by food and wine was starving, because he cannot decide from which direction to pick his meal. A Medieval soulmate of the poor fellow was the ass of Buridan, suffering between two huge and delicious haystacks.

Fig. 1.5 It was recognized long ago that stars appear to circle around a point in the sky, the North Celestial Pole. This movement was explained in ancient Greece as the revolution of a giant sphere on whose surface stars are fixed. This photo, with an exposure of a few minutes, shows the Northern Pole, nowadays not far from Polaris, the "Polar Star." In the foreground is the dome of a telescope at Tuorla Observatory, Finland (photo by Aimo Sillanpää & Perttu Keinänen)

Anaximander was also the first, as far as we know, to use models and analogies in science. For example, he explained the daily revolution of the Sun with a mechanical model, a hollow ring. The ring is full of fire that is seen through a round hole. When the huge ring rotates, the glowing hole (the Sun), moves with it. So he already thought that the Sun moves genuinely below the Earth during the night and does not just creep from the west back to the east somewhere just below the horizon.

Anaximenes came up with the idea that the stars are fixed on a spherical vault of the sky (or at least on a hemisphere). This was a splendid example of how the Ionians looked at things. *One* revolving globe could explain the daily rotation of thousands of stars (Fig. 1.5)!

Pythagoras Invents the Cosmos

Pythagoras of the Ionian island of Samos (about 572–500 BC) was an influential but obscure figure in history. It is said that Thales was so surprised by the talents of the young man that he recommended that he should go to Egypt to study under the guidance of priests. An equally uncertain story tells that he received learning while a prisoner in Babylonia. At an age of about 40, Pythagoras moved to southern Italy where he and his wife Theano founded a school in the Greek colony of Crotona. The school was actually a religious fraternity, where mathematics, philosophy, and other topics were practised under the leadership of the master.

To the candidates for the first principle Pythagoras added still another entity, *number*. The cosmos, "ordered universe," is ruled by mathematics. This idea has a far-reaching consequence that we are still feeling in our own science: it is possible for a thinking human being to deduce the structure of the universe, without visiting every corner. The Pythagoreans regarded the Earth as a sphere, as is the starry sky. Planets, among them, the Sun and the Moon, are each attached to their own spheres that revolve around the Earth. Surely there was already evidence for the spheroid of the Earth (e.g., travelers knew that the sky changed when they go from north to south), but likely such empirical aspects just enforced the belief in the primary nature of the complete, beautiful spherical shape.

It is remarkable how one Pythagorean, Philolaus (around 450 BC), taught that the Earth and other cosmic bodies revolve around the fire burning in the center of the world. The fire is *not* the Sun, so this was not a heliocentric system, but it showed that it was possible to imagine the Earth moving in space even though we do not feel anything of the sort under our feet. Philolaus is said to have theorized that we cannot see the central fire, because the Earth always turns with the same half toward it (like the Moon does relative to the Earth).

Pythagoras founded number theory and proved the famous theorem of Pythagoras about the areas of the squares drawn on the sides of a right-angled triangle. Integer numbers were the basis of the Pythagorean worldview. Those thinkers regarded that integer numbers (or their ratios), which were the only type of numbers known at the time, may measure everything in the world. For example, they thought that a line is formed by a large number of points, like atoms put side by side, and hence the ratio of the lengths of any two line segments would always be rational. It was a shock to find, using the very theorem of Pythagoras, that the ratio of the diagonal and the side of a square ($= \sqrt{2}$) cannot be expressed in terms of integers. Along with the old numbers ("rational") one had to accept new ones ("irrational"). In the long run this was necessary for the further development of mathematics.

Irrational numbers served as a healthy reminder that the world is not so simple that first mathematical concepts were sufficient for its description and understanding. Nevertheless, modern scientists view with sympathy the efforts of Pythagoras to grasp the cosmos as a harmonic whole. We also like to believe that the world must be in some deep manner simple and comprehensible.

About 500 BC there was an attack on Crotona, the house of the Pythagoreans was burnt down and several members of the fraternity were killed. Others escaped. Pythagoras himself went to Tarentum (in Italy), but many moved to the mainland of Greece, e.g., to Athens, where the new ideas began to spread.

Chapter 2
Science in Athens

In the fifth century before Christ, the city state of Athens, having defeated the Persian Empire, became the center of Greek culture and science. This city, with a population of at most 300,000, gave birth to an astoundingly rich culture whose influence is strongly present in our western heritage. Sculpture and architecture flourished. The masters of tragedy Aeschylus, Sophocles, and Euripides created drama. Thucydides founded critical historiography. Socrates (469–399 BC) wandered the streets of Athens delighting and angering people with his unusual questions.

Anaxagoras Makes the Celestial Bodies Mundane

Athens was at the focus of new ideas concerning nature. It is regarded that Anaxagoras (ca. 500–428 BC) imported natural philosophy to Athens from Ionia. Perhaps the first scientist in the modern sense of the word, he was born in the city of Clazomenae and had given away his considerable possessions to devote his life to science. When asked why it was that people are born, he replied that it is in order to "investigate sun, moon, and heaven." Around the age of 40, Anaxagoras came to Athens. There he had among his friends the statesman Pericles. The tragedy writer Euripides was one of his pupils.

Anaxagoras still held the view, as did Anaximenes of Miletus, that the Earth is flat and floats in the air. This did not hinder him from making important observations about celestial matters. He suggested that the Moon receives its light from the Sun and he correctly explained solar and lunar eclipses. He taught that celestial phenomena could be understood in terms of the same materials as those down here. So he regarded the Sun as a hot glowing mass or a rock on fire, and the Moon with plains and ravines similar to the Earth. He was impressed by the fall of a meteorite and explained it as a result of an "earthquake" occurring on some celestial body. Ideas like these were not well received by many, as stars and planets were generally viewed as gods. Anaxagoras was accused of impiety. Pericles helped him to escape from Athens to Lampsacus in Ionia. He founded a school and lived there the rest of his life as a very respected person.

Another remarkable thinker of those days was Empedocles (ca. 494–434 BC). We remember this man from Agrigentum (southern Sicily) especially for the four elements. *Fire*, *air*, *water*, and *earth* retained their central role in science for over two millennia. He also made the first steps toward considering the significance of physical forces. In his philosophical poems, he used the allegoric names Love (*philia*) and Hate (*neikos*) for the contrary forces keeping up the balance in natural phenomena – in our more prosaic language these are attractive and repulsive forces. These early views about why the elements behave as they do, forming all those things around us, were in fact qualitative, descriptive physics. But the doctrine of atoms, first formulated at about the same time, did not accept forces into its theoretical arsenal; the atomists had a different way to explain the formation of the various structures in the world.

The Atomic Doctrine

Within Ionian natural philosophy, one of the important ancient systems of thought was created, atomic theory. It can be summarized as "in reality there is nothing else than atoms and the void." Leucippus from Miletus is regarded as the founder of atomic doctrine. It was further developed by Democritus (ca. 460–370 BC), who was born in Abdera (Thrace) but lived a long time in Athens.

According to atomic theory, the ultimate element so eagerly sought by Ionian philosophers was not a continuous substance, but instead, very tiny, indivisible, and extremely hard bodies, atoms (in Greek: indivisible). When taken alone these atoms lack sensible properties like color, smell, and taste, but they may join together to form all kinds of material things. Leucippus suggested that worlds, which are unlimited in number, arise when atoms fall from infinity into the void and meet each other forming a vortex. In our special case, the Earth collected in the center of such a vortex.

Atomic theory seems to us rather familiar and we may be inclined to view ancient atomists as soul mates of today's scientists. But even more important than the superficial similarity is the realization by the early atomists that the phenomena of the sensible "macro" world may be explained by referring to invisible atoms of the "micro" world. The way they inferred from the visible to the invisible was quite similar to what we do in modern science (even though their detailed explanations went often wrong). Clothes hung out to dry offer a good example of how atomists explained visible things. Wet clothes dry in the sun, but we cannot see the moisture leaving them, because it is split up into minute parts.

It was a key element in the worldview of atomists that bodies were formed quite haphazardly from atoms rushing through empty space. There was no purpose or superior intelligence behind all this. Infinite space and endless time guarantee that sooner or later atoms collide to form whole worlds, of which ours is only one example. Since human beings are made of atoms, and so are our souls that fade away when we die – only the eternal atoms remain. On the basis of these materialistic notions, Epicurus (341–270 BC) from the island of Sámos created a view of the world and life which attracted many followers. His ardent Roman admirer, Lucretius (ca. 98–55 BC) later wrote an extensive poem *De Rerum Natura* (On the Nature of Things) where he describes Epicureanism. Its poetic language contains plenty of information on how atoms were thought to explain natural phenomena and the origin of human sensations. At the same time the poem reflects the enthusiasm with which some people accepted rationalistic thinking about nature – it was seen as a way to disperse the fear of the supernatural.

The world view of the atomists differed radically from the views held by Plato and Aristotle which we will encounter below. For the atomists, the random collisions by atoms were the only "law of nature." Similarly to Anaxagoras, the atomists stripped celestial bodies of their divine nature. However, one must say that their achievements in astronomy were not impressive – for example, Democritus still believed that the Earth is flat and Epicurus was not interested in explaining celestial phenomena. It is slightly ironic that an important step in the development of astronomy into an exact science was made by Plato who believed in the divine nature of celestial bodies. The point is that he viewed the regular movements in the sky as controlled by a superior intelligence and therefore being within reach of a rational explanation.

Plato Establishes the Academy

The great thinker Plato (427–347 BC) was from a wealthy Athenian family. In his youth he dreamed of a career in politics, and became a follower of Socrates. He abandoned political plans after Socrates' shocking execution, going abroad for a decade. He spent this time in Egypt and southern Italy, where he became familiar with Pythagorean thinking.

After he returned to Athens, Plato recruited a kind of brotherhood of talented pupils. They gathered outside of Athens in a sacred grove named after the mythical hero Akademos. In this peaceful place, Plato discussed philosophy and science with his pupils. It was here that Plato's Academy was born in 387 BC, the famous seat of learning which operated for nine centuries until the Emperor Justinian closed it in AD 529. Plato's team was very influential indeed. Among his pupils were the philosopher and scientist Aristotle, and the mathematicians Eudoxus, Callippus, and Theaetetos.

Instead of observations, the philosopher Plato emphasized the importance of thinking and reasoning when one attempts to understand what is behind the incomplete and muddy image of our world. For him true reality was the world of concepts. This may reflect the Pythagorean view of reality, number (also an abstract concept). Clearly, these two world views deviated from the material foundation of reality as seen by the Ionians and the atomists.

Plato's approach to the study of nature is revealed in astronomy. In the dialogue *Republic* he introduces an educational program suitable for the philosopher-rulers of his ideal city-state. The aim of the curriculum was to make it easier for the human mind to approach the only true subject of knowledge, the unchangeable world of ideas, not the ever changing phenomena of the world of the senses. In Plato's dialogue, Socrates regards mathematics (arithmetic, geometry) as a way to study unchanging truths. Another recommended field is astronomy, though in a sense that now seems quite alien to us.

Socrates' interlocutor Glaucon eagerly accepts astronomy as useful for farmers and sailors. However, Socrates bluntly condemns this aspect as useless for the philosopher. Glaucon then hopefully asserts that at all events astronomy compels the soul to look upward, away from the lower things. But again Socrates disagrees. For him "upward" is just toward the material heaven, not toward the realm of ideas, as expressed in clear words:... *if any one attempts to learn anything that is perceivable, I do not care whether he looks upwards with mouth gaping or downwards with mouth closed: he will never, as I hold, learn – because no object of sense admits of knowledge – and I maintain that, in that case, his soul is not looking upwards but downwards, even though the learner float face upwards on land or in the sea.*

Glaucon must again admit that he was wrong. But then "what is the way, different from the present method, in which astronomy should be studied for the purposes we have in view?" Socrates admits that "yonder embroideries in the heavens" are more beautiful and perfect than anything else that is visible, yet they are far inferior to that which is true, far inferior to the movements wherewith essential speed and essential slowness, in true number and in all true forms, move in relation to one another and cause that which is essentially in them to move: the true objects which are apprehended by reason and intelligence, not by sight.

And Socrates goes on to clarify what he actually means:

> Then we should use the embroideries in the heaven as illustrations to facilitate the study which aims at those higher objects, just as we might employ ... diagrams drawn and elaborated with exceptional skill by Daedalus or any other artist or draughtsman; for I take it that anyone acquainted with geometry who saw such diagrams would indeed think them most beautifully finished but would regard it as ridiculous to study them seriously in the hope of gathering from them true relations of equality, doubleness, or any other ratio. (Translations of Plato's texts from Heath: *Aristarchus of Samos.*)

Socrates, and Plato, thought that the regular movements of celestial bodies roughly reflect the laws of the ideal world of motions just as hand-drawn geometric pictures offer hints about the mathematical laws governing true geometric figures. However, mere looking or making observations does not lead to genuine confident knowledge about geometry – these must be proved in derivations where visual impressions or measurements of even accurately made drawings do not appear as part of the argument. For example, one might make many scale drawings to approximately verify the theorem of Pythagoras, but one cannot be sure of its complete exactness without a geometric derivation (Fig. 2.1).

Fig. 2.1 Pythagorean Theorem. The area of the square drawn on the hypotenuse of a right-angled triangle is equal to the sum of the areas of the squares on the other two sides. You may try to prove this ancient theorem – there are many ways to do it

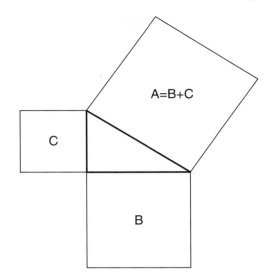

True cosmic motions inhabit the world of ideas as "true velocities" and "true periods," and these make themselves felt in the observed motions of celestial bodies, though only as distorted reflections in the mirror of the senses. By staring at these incomplete phenomena you cannot get genuine knowledge, and "hence we shall pursue astronomy, as we do geometry, by means of problems, and we shall dispense with the starry heavens, if we propose to obtain a real knowledge of astronomy." A modern astronomer studying amazing observational discoveries, would hardly agree with Socrates' assertion. Probably Plato did not hold such an opinion literally. In fact, in his later cosmological work *Timaeus*, Plato thanks our eyesight for having brought the celestial motions within the reach of our senses. "This I declare to be the main blessing due to the eyes."

The strange program of astronomy delineated by Plato is a healthy reminder of how our ideas about science have traveled a long way from those days. We tend to think that laws of nature do not exist independently of natural phenomena even if one can express them using the exact language of mathematics. In any case, we do not imagine that we could discover those laws without observation. Disturbing factors and uncertain observations may affect the accuracy of the inferred regularities, but in principle, this is not fatal at all. Plato aspired to unshakeable knowledge about the world, using the method of pure thinking. We are happy with approximate knowledge that we extract from observations and experiments. Our experience – which the ancients did not have – has shown that this is the fruitful way to gradually increase our knowledge of natural laws, improving the approximation of reality.

It is said that Plato gave his pupils the task of determining what kind of simple and uniform motions could explain the movements of stars and planets. This proposal inspired Eudoxos to devise his famous theory of homocentric spheres (to be discussed in the next chapter). This model initiated construction of planetary mod-

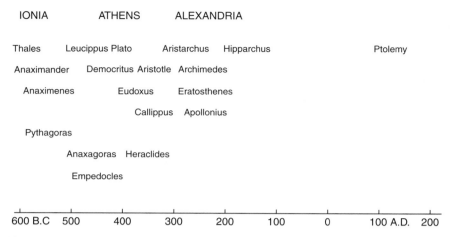

Fig. 2.2 Ancient philosophers and scientists placed above a time axis with their cities given in capital letters. Accurate years of life are often not known

els by others, which had a great significance for the development of science. More important than Plato's concept of good scientific research was the fact that he had around him eager talented disciples who were stimulated by a unique intellectual environment, Plato's Academy. Relationships between these and other central figures if old science are diagrammed in Fig. 2.2 and pictured (with some phantasy) in Fig. 2.3.

The Universe of Aristotle

Aristotle (384–322 BC) was Plato's most famous pupil. He was born in Stagira, Macedonia. Aristotle attended Plato's lectures for two decades till the latter's death, after which he moved first to Asia Minor and then to Pella, the Macedonian capital where he worked for 7 years as tutor to the king's son, the future Alexander the Great. He was already close to 50 years old when he came back to Athens and established his own school. His habit was to stroll with his pupils, teaching and discussing (hence the name "peripatetic school"). Interestingly, not so long ago archaeologists found the place in Athens in which Aristotle's famous school, the Lyceum, was situated.

Aristotle wrote plenty of books, but none of these were preserved in complete form. What remain are "lecture notes" and summaries, and even these were lost for two centuries, before they were found in the cellar of a descendant of one of his pupils. Our link with the past is so weak!

Aristotle was a universal genius who wished to create a system of knowledge covering everything in the world. Among other things, he divided science into different fields of study, investigated the nature of scientific knowledge, and founded logic. As the founder of zoology he was an ardent observer of animal behavior and described about 500 different species. In physics, he was the first to create a doctrine of dynamics, which attempted to explain why the various bodies around us move as they do. His physics was also cosmological in scope. It was closely linked with his view of the universe, which had a great influence on scientific thinking that lasted through the Middle Ages in Europe.

The universe of Aristotle was finite in size, in fact a finite sphere outside of which there was nothing, not even emptiness! He had several arguments in favor of finiteness instead of infinity. For example, he stated "every revolving body is necessarily finite." If an infinite body were revolving, its immense parts would pass in a finite time through an infinite distance, which he thought was impossible. Therefore, as he regarded the daily revolution of the sky as a cosmological property of the universe, the universe must be finite. Also, there was the fact that bodies tend to fall into one point that is situated in the center of the Earth. It was clear for Aristotle that the Earth is a sphere and it seemed that its center was also the central point of the universe. Aristotle reasoned that only a finite universe could have a center.

Aristotle agreed with Empedocles that "down here" there are four elements, one of which is the solid material of which the Earth is made. It was an essential part of Aristotelian dynamics that motions of bodies are governed by their striving toward their *natural place*. The natural position of the element earth is the center of the universe, hence the natural motion "down." Fire was an element opposite of earth and its natural movement was "up." Similarly water and air had their tendencies to settle in different layers, water lower than air.

However, physics is different in the celestial realm. First, celestial bodies are composed of a quite special element, ether. It had been proposed even earlier as a very rarefied substance filling the vacuum, but Aristotle elevated ether into the heavens and gave it the status of the fifth element. Ether is eternal, and stars and planets made of it never decay. Secondly, the universe as a whole is unchanging and eternal, and this is reflected in the regular circular motions of celestial bodies. Circular motion is something special: a body always returns into its previous place, so here apparent change or motion paradoxically is at the service of permanence. In the "sub-lunar" realm of change, natural motions are "down" and "up," but in the heavens the natural motion is circular.

Dynamics developed by Aristotle was based on observations in the terrestrial environment, which may give a misleading picture of what factors govern and maintain motion. Friction and the resistance of air seriously hamper the building of a correct science of motion, and Aristotle did not take these into account. However, even when they were erroneous his ideas gave important impulses for medieval thinkers on the nature of motion.

Aristotle insisted that we understand a phenomenon only if we know its *cause*. This sounds familiar, but Aristotle had in mind a special kind of cause, the final cause (*telos*). It is as if a force comes from the future, influencing what should happen now. We know the final cause when we can tell why the phenomenon happens. For example, a stone falls because its goal is its natural place in the center of the universe. Aristotle was a specialist in biology and there the final or teleological

Fig. 2.3 The philosopher Plato and his most prominent pupil Aristotle during a discussion in the Academy in Athens, as imagined by Raphael in his fresco. Plato points with his finger upward, to the heavens, while Aristotle is more down-to-the-earth in his approach. Hypatia (who lived several hundred years later…), dressed in white on the lower left, stands alone among this gathering of men, turned toward the viewer

reason is at first sight quite a natural way to explain things, then why not elsewhere, too? Aristotle did know other categories of cause, but the final cause was the most fundamental for understanding natural phenomena.

Modern science sees other causes as essential for explaining physical phenomena, with the final cause no longer being fundamental. Causality has replaced finality. Modern science starts its explanations from the past, from certain initial conditions, and follows the chain of cause and effect in an attempt to understand what would happen in the future. When we ask *why* something happens, we have in mind: what are the conditions and natural laws that lead to this phenomenon? We do not ask what its goal is.

Then it is no wonder that in this first ever doctrine of dynamics, the falling motion toward the center of the universe (the Earth) was so important. Now we understand that this phenomenon (falling of a stone), that seems so purposeful, is just one local manifestation of a universal law of gravity. The same happens close to any celestial body. Aristotle knew only one case, that of our Earth.

Aristotle, "the brains of Plato's Academy," was confident that one is able to obtain reliable scientific knowledge about the world. Contrary to what his teacher Plato taught, Aristotle emphasized the importance of observations (Fig. 2.3). By keenly observing the nature, a scientist may intuitively arrive at the fundamental axioms of science, infallible truths. From these initial truths that represent the highest level of knowledge, one may by logical induction infer other true statements about the world, that is, scientific knowledge standing on a firm foundation.

For both Aristotle and Plato, knowledge, to be genuine, had to be really infallible and final, something like mathematical truth. However, experience over the centuries has shown that such a very strict demand makes it impossible to practice science. Maybe science is approaching final truths, but if so, this happens through "partial truths" and temporary assumptions. The growth of scientific knowledge is a more complicated process than was imagined by Aristotle and its reliability is usually restricted and provisional. Nevertheless, in the manner Aristotle thought about science one may see a glimmer of two basic processes which are every modern scientist's basic tools: *induction* or discovering a general law from observations, and *deduction* or inferring logical consequences, for example for predicting what would happen in an experiment.

Chapter 3
Planetary Spheres and the Size of the Universe

Babylonian sky watchers were aware of wandering celestial objects (or *planets*; a wider concept than today). Of these, the Sun always moves eastward among the stars following the ecliptic, its yearly path through the Zodiacal constellations. The Moon stays fairly close to the Sun's path taking about a month to circle the sky with respect to the stars. Most of the time the other planets also move slowly toward the east, keeping close to the ecliptic. It takes a certain time for the planet to make a full circle around the sky from a constellation of the Zodiac back to the same place (its *sidereal period*). However, planets other than the Sun and Moon slow down, stop, and go backward for some time and then resume their normal motion (Fig. 3.1). There is regularity in this odd *retrograde motion*. Each planet has a *synodic period*, the time between successive reversal loops. The synodic period differs from the sidereal, and so successive stops happen at different constellations of the Zodiac. Table 3.1 gives the synodic and sidereal periods of several planets (of which Uranus, Neptune, and Pluto were not known in antiquity).

The Theory of Concentric Spheres

Greek philosophers began a new approach that reached beyond astrology, attempting to explain in a rational manner *why* the planets move as they do. The sphere and circular motion was their preferred ideal for celestial motion (remaining so for two millennia). As geometric forms, the sphere and the circle were much investigated in Greek mathematics. Also, perfect circular motion, always returning back its original point, seemed to be suitable for celestial bodies held to be divine beings or at least eternally existing objects, and in fact the celestial sphere does seem to rotate perfectly uniformly.

Plato asked his pupils what kind of simple motions could explain the complicated movements of planets. Eudoxus (ca. 408–355 BC) took up the challenge. Eudoxus' other achievements include a method to derive formulae for areas and volumes similar to modern integral calculus.

P. Teerikorpi et al., *The Evolving Universe and the Origin of Life*
© Springer Science+Business Media, LLC 2009

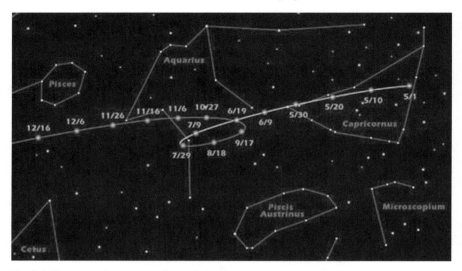

Fig. 3.1 The reversal of motion of Mars in 2003. Its synodic period of 780 days separates any two retrograde loops, which happen in different constellations of the ecliptic. This was a key phenomenon that the ancients and later (more successfully) Copernicus attempted to explain by models of planetary motion. (Credit: NASA/JPL-Caltech)

Eudoxus' theory of spheres concentric on the Earth was the first mathematical model explaining in some detail the motions of the sky, including the puzzling retrograde loops. The model was based on spheres rotating with different but uniform speeds around their axes. These axes connected an inner sphere to the next one and were inclined with fixed angles relative to each other. Beyond all the planets was the celestial sphere of fixed stars revolving uniformly once a day around the immobile Earth. We hope our brief description will not make the reader dizzy! Sets

Table 3.1 Synodic and sidereal periods for the planets (including those discovered in modern times)

Planet	Synodic period	Sidereal period
Mercury	116 days	88 days
Venus	584	245 days
Mars	780	687 days = 1.88 years
Jupiter	399	4,333 days = 11.8 years
Saturn	378	10,744 days = 29.4 years
Uranus	370	30,810 days = 84.4 years
Neptune	368	60,440 days = 165.5 years
Pluto[a]	367	91,750 days = 251.2 years

[a]According to the modern definition Pluto is not a major planet, but a dwarf planet. Note how the synodic period approaches the length of our year for increasingly long sidereal periods (can you figure out why?)

Fig. 3.2 A simplified diagram of Eudoxus' concentric spheres. Spheres rotate with different but uniform speeds around their axes. The axes connect an inner sphere to the next one and were inclined with fixed angles relative to each other. Therefore, the path of the planet as seen from Earth is more complex than a circle

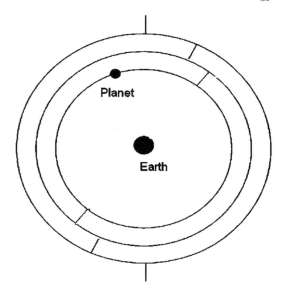

of interconnected spheres provided each planet with its own specific motions. The more uniformly moving Sun and Moon could be dealt with just using three each. The basic idea is shown schematically for the Moon in Fig. 3.2. The first rotated on a north–south axis to create the daily motion. The second was tilted relative to the first to include the tilt of the ecliptic relative to the celestial equator, turning once every sidereal period. Finally, the third turned with the path tilted to include deviations from the ecliptic by the Moon.[1]

With his model, Eudoxus could explain fairly well the planetary motions known at the time. However, Mars proved to be a thorny case whose motions were next to impossible to match with the model. Eudoxus does not seem to have imagined the model as representing a real physical structure but instead as a purely mathematical construction with one planet's set of spheres not affecting the motion of another's even though they were nested concentrically.

Aristotle's planetary model was an expanded version of Eudoxus' using a total of 56 spheres centered on the Earth. Aristotle may have viewed the spheres as physical entities (a sort of celestial crystal). However, he rejected Pythagoras's idea about the music of the spheres. On the contrary, he regarded the silence of the heavens as a proof of the sphere-carriers – noise would be expected if the celestial bodies would rush through some medium. The number of spheres was larger because Aristotle wanted to link together the sets of spheres belonging to each planet with additional spheres so that the fundamental daily motion of the outer sphere of the fixed stars was transferred from up to down.

---

[1] The planets having retrograde loops (Mercury, Venus, Mars, Jupiter, Saturn) required sets of four spheres each to explain their more complicated motions. Hence the number of planetary spheres is $26 \ (= (2 \times 3) + (5 \times 4))$ all nested concentrically.

Fig. 3.3 A schematic illustration of the epicycle model. The planet moves on a small circle (epicycle) whose center moves along a large circle (deferent) centered on the Earth

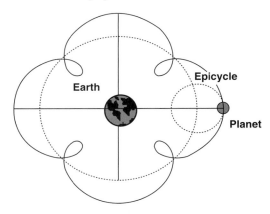

The Epicycle Theory

Eudoxus' planetary model had an observational problem pointed out by Autolycus of Pitane (ca. 360–290 BC). When the planets make their loops westward, they are brighter than at other times meaning that they are then closer to us. In the models based on spheres centered on the Earth, the planets always stay at the same distance from the Earth. This disagreement was reconciled by Apollonius of Perga (ca. 265–176 BC). He worked at the new scientific center of the world, the Museum of Alexandria. Apollonius was a pupil of Euclid and was also known for his studies of geometric curves (the ellipse, the hyperbola, the parabola). Much later, these curves assumed great importance in understanding planet orbits. Apollonius developed a new, but still faithful to circles, way to tackle planetary motions.

In this model, a planet did not stay on its sphere but also traveled on a smaller circle (an *epicycle*) whose center was fixed on the uniformly rotating main sphere. When it is going "backward" along the epicycle, the planet is closest to us on its epicycle, which explains its greater brightness when making a retrograde loop in sky (Fig. 3.3). The motion around a large circle (the *deferent*) has the sidereal period of the planet, while the epicycle is circled once during the synodic period, both at constant speeds. The epicycle described each planet's brightness variations as well as the motions on the sky replacing the two retrograde motion spheres. This scheme was used and elaborated until the end of the Middle Ages.

Hipparchus Discovers the Slow Wobbling of the Celestial Sphere

We know practically nothing about the life of Hipparchus (ca. 190–120 BC) and his writings have almost all vanished, but nevertheless it is evident that he was a great astronomer. Working, among other places, on the island of Rhodes, he developed trigonometry, much needed in astronomy where triangles are used in calculations.

He also made a star catalog containing over 800 stars, their positions in the sky, and their brightness expressed in terms of stellar magnitudes, a quantity still important in astronomy. Hipparchus denoted the magnitude of the brightest stars by number 1. Stars barely visible by the naked eye were assigned the magnitude 6 and all others ranged from 2 to 5.

Later Pliny the Elder (AD 23–79), the Roman writer, expressed his admiration for Hipparchus' catalog: "He made something that would be courageous even for the gods – he counted the stars and constellations, with future generations in mind, and gave them names. For this purpose he built instruments, with which he determined the location and size of each star. Thanks to this it will be easy to learn, not only if stars are born or if they die, but also if they move away from their positions and if their light grows brighter or fainter."

Catalogs of stars and other celestial bodies have been and continue to be very important for our knowledge of the universe. In fact, comparing his stellar catalog with the measurements by two Alexandrian astronomers one and half centuries earlier, Hipparchus discovered a slow motion of the sky. He used *coordinates* to give the location of a star on the sky. These are similar to latitude and longitude on the spherical Earth. To define these two coordinates, one needs a fundamental circle dividing the sphere into halves and on it a fixed zero-point. For the Earth, these are the equator and its intersection with the north–south line (meridian) passing through Greenwich Observatory near London. The longitude of, say, a ship on Earth is the number of degrees from Greenwich along the equator to where a north–south line through the ship crosses the equator. The latitude of the ship is the number of degrees along this circle north or south of the Earth's equator.

In the course of a year, the Sun circles the sky along the ecliptic which is tilted 23° with respect to the celestial equator in the sky directly above the Earth's equator. The Sun thus crosses the celestial equator two times 180° apart, once in the spring at the time of the vernal equinox when it goes from the southern to the northern half of the celestial sphere, and once in the autumn (autumnal equinox) from the north to the south. Hipparchus used the ecliptic as the fundamental circle from which to measure the celestial latitude of a planet north or south. He took the March 21 position of the Sun to define the vernal equinox crossing as zero for the ecliptic, and the angle eastward from zero was the celestial longitude. Comparing the old coordinates with the ones measured by himself, he found that during 150 years the longitudes of the stars had decreased by 2°, while their latitudes had remained the same. Hipparchus realized that the vernal equinox is not a fixed point in the starry sky, but it moves slowly along the ecliptic westward, opposite the yearly movement of the Sun. The points of intersection shift gradually along the zodiac from one constellation to another over thousands of years.

As Copernicus later explained, this quiet but remarkable phenomenon (making the horoscope signs shift, as we mentioned in Chap. 1) reflects a slow spindle-like wobbling of the Earth's axis with a period of 26,000 years, but in old times it was viewed as a mysterious extra motion of the celestial sphere. It has an interesting implication, of which Hipparchus was aware, namely that there are two slightly different concepts of a year (see Box 3.1).

Box 3.1 The sidereal and the tropical years

The *sidereal year* is the interval between the events when the Sun passes by a truly fixed point on the stellar sphere, say a fixed star on the ecliptic. The *tropical year* is the time from one vernal equinox to the next. The tropical year is shorter than the sidereal one, because the Sun encounters the slowly moving vernal equinox (i.e., the equator) about 20 min earlier than "expected." Physically, the sidereal year is the true period of rotation of the Earth around the Sun (about 365.2564 days). The tropical year, as also the name signifies, is the period of about 365.2422 days marked by the seasons (and determined by where the Sun is relative to the equator). In our civil life, we are accustomed to think that there are 365 days in one year (with occasional leap years of 366 days). In fact, the Gregorian year adopted by Pope Gregory XIII in 1582 has in the long run 365.2425 days, while the earlier Julian year decreed by Julius Caesar in 46 BC had $365 + ^1/_4 = 365.25$ days. Note that when you tell that you are so-and-so-many years old, you have in mind the tropical year and not the sidereal year (though it must be said that the difference is insignificant for any practical purposes).

With all such different years at hand, what actually is the *year* in the light-*year* which we often encounter as a distance unit? In fact, as astronomers usually do not use the light year as a basic unit when expressing cosmic distances (they use parsec, see, e.g. Box 8.1), the choice of the length of the year has not been very important. A convenient choice is the Julian year with its exactly 365.25 days (each of exactly 86,400 s). This gives the light-year the length 9,460,730,472,580.8 km (when we take the speed of light which is nowadays defined to be exactly 299,792.458 km/s).

We have here an example of detection of very slowly advancing natural processes requiring long-term, accurate observations (and the ability to write these down!). The shortness of human life and ordinary experience are too limited to reveal the sway of the Earth's axis and many other important phenomena.

Ptolemy

The last great astronomical figure in Greek antiquity was Claudius Ptolemy who lived in Alexandria about AD 100–178. He collected the astronomy of the time into his book best known by its later Arabic name *Almagest* (The Great Book). Adding their own elaborations, Islamic astronomers preserved this work through the Middle Ages, until the time was ripe in Europe for a new start in astronomy. Translations were made from Arabic to Latin, with no translation from the Greek until the fifteenth century.

Ptolemy developed the epicyclic theory. Hipparchus had added to the model the *eccentric circle*: the epicycle moves uniformly along the large deferent circle, whose center is somewhat off-side of the center of the Earth. With this invention he could quite accurately describe the observed variable speed of the Sun during its yearly path. Ptolemy made another innovation: the *equant,* a point inside the eccentric circle. The epicycle center is required to move along the eccentric circle with a variable speed so that when looked at from the equant point, the *apparent* angular speed is constant. This trick further improved the ability of the model to describe planetary motions. However, it meant the abandonment of the traditional uniform circular motion. Later Copernicus, otherwise a great admirer of Ptolemy, could not accept the equant and remained faithful to the idea of uniform circular motion.

The Size of the Spherical Earth

The roots of how to measure astronomical distances go back to Thales who was said to have inferred the height of a pyramid by waiting for the moment when the shadow of a vertical rod was as long as the rod itself Then he measured the length of the shadow cast by the pyramid! This simple, but clever procedure demonstrated how combining observation with mathematics could result in unexpected ways of investigating the world. The foundations of cosmic distance measurements were laid in the land of pyramids, in Alexandria, where Eratosthenes (ca. 275–195 BC), the librarian of its famed Museum measured the size of the Earth, using its spherical form and – once again – the Sun and the shadow.

As geographer he planned a map of the world and needed a scale for its coordinate network. His method was simple: if one knows the distance between two places as measured along the curved surface of the Earth and if one also knows the angular separation of these places, then it is straightforward to calculate the whole circumference of the Earth. For example, the angular separation from the pole to the equator is one-fourth of the complete circle, so that multiplying the corresponding pole-to-equator distance by four gives us the circumference of the Earth.

Eratosthenes considered two places, his location, Alexandria, and Syene (the modern Aswān), which lie roughly on the same longitude (north–south line). He knew that at Syene no shadow was visible on the midday of summer solstice (that is, the Sun was exactly above one's head). At Alexandria, at the same time the Sun was somewhat south of the zenith so a shadow was visible. This angular difference he measured to be about 7° or 1/50 of the full 360° circle. Then by multiplying the ground distance, S, from Alexandria to Syene by 50 he obtained the circumference of the Earth. It is not known how he estimated the distance, but he may have used the time needed for a courier to traverse the distance. In any case, he used the value $S = 5,000$ stadia and hence derived for the Earth's circumference the value 250,000 stadia. This is diagramed in Fig. 3.4a.

The stadium unit was derived from Greek athletic contests, and several such units of different lengths were in use. We do not know for sure which of these

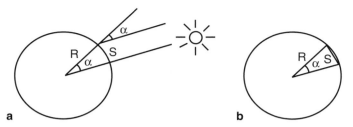

a   b

Fig. 3.4 (**a**) Diagram of Eratosthenes' measurement where R is the radius of the Earth, S is the distance from Alexandria to Syene, and α is the angle of the Sun from overhead at Alexandria as well as the angle at the center of the Earth. The large circle represents the circumference of the Earth. (**b**) Diagram of triangulation where R is distance of object from observer at center, S is the physical size of the object, and α is the apex angular size. The large circle is of radius R, centered on the apex

Eratosthenes refers to when speaking about 5,000 stadia. A short unit 157.5 m (often accepted by historians) would give a slightly too small Earth, a bigger one 185 m would result in an over-sized globe: the circumference would be either 39,375 km or 46,250 km. Modern values are 39,942 km (polar) and 40,075 km (equatorial) or about 24,900 miles. So, remarkably, the spherical shape and size of the Earth was known in antiquity long before Columbus.[2] Eratosthenes showed that it is possible to measure the size of the Earth that you cannot see in totality using local measurements assuming a spherical shape. Even modern cosmologists use a similar process for the universe as a whole.

The way Eratosthenes measured the Earth is a special case of *triangulation*, using an isosceles triangle (two sides equal). As explained in Box 3.2, two typical cases appear in astronomy: the base side of the triangle is the size of a distant object whose distance is desired; or the base side is "down here" and the distant object is at the apex of the triangle.

An example of the first kind of triangulation would be to infer the distance of the Sun from its angular size (about $\frac{1}{2}°$). If its true diameter were known, say, in kilometers, one could easily calculate its distance. But even now, we cannot estimate the true size of the Sun accurately independent of its distance. Neither could astronomers in antiquity. Anaxagoras made the brave guess that the Sun is a glowing rock with the size of Pelopónnisos (about 150 km). The method of triangulation would give the distance of 17,000 km, while the correct value is almost 10,000 times larger (and the Sun is that much larger than Pelopónnisos). The distance of the Sun was a challenge for a long time and could be fairly well measured only in the seventeenth century.

---

[2] From the circumference one can calculate the radius just by dividing it by 2π, giving 6,366 km. Archimedes (whom Eratosthenes knew) had shown that the ratio of the circumference and the diameter was approximately 3.14.

Box 3.2 Triangles and distance

Referring to Fig. 3.4b, if we have an isosceles triangle (two sides, R, equal), then knowing the apex angle size, α, between the two equal sides and the length of the base side, S, we can easily calculate the height of the triangle (base to apex). In astronomical triangulation, the astronomer is usually able to measure α. Also in astronomy, this angle is usually rather small, less than a few degrees making the height nearly equal to R. Drawing an imaginary circle with radius R, centered on the apex, we have the same geometry as Eratosthenes except R is now the distance of the object and S is its physical size. The circumference of the imaginary circle equals S divided by the fraction the apex angle α is of 360°. The distance, R, is just the circumference divided by 2π. Two different cases are typical:

- Suppose the base side is an extended distant object whose distance is desired. Note that the object would be very far away if it is large but much closer if it is smaller (if you hold your finger at arm's length, it has about the same angle as the Earth's Moon but the Moon is a great deal larger and farther away than your finger!) Obviously, to obtain its distance, R, knowing the apex angular size θ that it makes in the sky, one must have the size of the object, S, in kilometers. But how to find its true size without first knowing its distance? This is still a hard problem for astronomers who attempt to find celestial "standard rods" to measure very large distances beyond our Galaxy.
- It may be easier if the base side S is "down here" with the observer and the distant object is at the apex of the triangle. In other words, just like Eratosthenes, we first measure S and the angle α by some other means and then calculate the distance, R, of the object. In effect, Eratosthenes measured the distance to the unreachable center of the Earth using this procedure.

When the base side "down here" is an important quantity itself, naturally appearing in the method, one may be satisfied in using it as a unit, without having to know its exact length in stadia or in meters. From antiquity up to the eighteenth century, the radius of the Earth held this fundamental status when one measured distances within the Solar System. As we will see later, the Earth-to-Sun distance is a natural base side when one measures the distances to nearby stars.

Aristarchus of Samos – The Copernicus of Antiquity Enlarging the Universe

Alongside with the world models based on the central position of the Earth, there were in antiquity "dissident" voices that questioned some basic assumptions of the

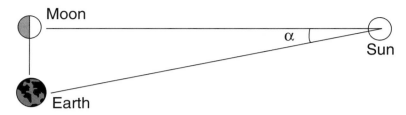

Fig. 3.5 The Earth, the Moon, and the Sun form a right-angled triangle at the moment of the half-full Moon

mainstream cosmology. Heraclides of Pontus (388–315 BC), a pupil of Plato, taught that the Earth rotates around its own axis. The daily motion of the sky is only an apparent phenomenon for the observer on the revolving Earth. Heraclides almost became the head of Academy after the death of Plato's successor Speusippos – he lost the election by a few votes to Xenokrates. It is tempting to think that the question of the Earth's motion may have received more attention in the Academy if Heraclides had been selected.

Aristarchus of Samos (310–230 BC) invented methods to deduce the sizes and distances of the Moon and Sun. He used the Moon as an intermediate step to the more distant Sun. Only one of his works has been preserved, *About the sizes and distances of the Sun and the Moon*. In this book, Aristarchus explains how one may measure (a) the ratio of the distances to the Sun and to the Moon, and (b) the sizes of the Sun and the Moon, using the Earth radius as a unit. The latter method (b) was based on the eclipse of the Moon (using the shadow cast by the Earth). It required the distance ratio derived by the former method (a), based on observations when the Moon is half full.

The Moon's phases (as well as its eclipses) were understood by Anaxagoras two centuries earlier (and completely explained by Aristotle). Aristarchus assumes the Moon to be a sphere which shines only by reflected sunlight. He thus knew that when the Moon appears exactly half full, then the Earth, the Moon, and the Sun form a right-angled triangle with 90° at the Moon (Fig. 3.5). If one now measures the angle between the Moon and the Sun, then one also knows the remaining angle of the triangle and one can calculate the Earth–Sun distance in terms of the Earth–Moon distance. With a modern inexpensive calculator, the calculation is a simple matter, but Aristarchus had to do it the hard way, through a tedious geometrical argument. He took the Moon–Sun angle to be 87° and proved that the ratio of the Sun's and the Moon's distances is greater than 18:1, but less than 20:1. A calculator gives about 19:1.

Next, Aristarchus estimated the size of the Moon compared to the Earth using eclipses of the Moon in a very sophisticated manner. We will describe it in a simplistic manner. Imagine that the Sun is extremely far away so it produces a cylindrical shadow behind the Earth whose diameter equals that of the Earth. One can thus simply observe the Earth's shadow (which is larger than the Moon for it to be eclipsed), timing how long the eclipse lasts, to estimate the relative sizes of the Earth and Moon. Aristarchus found that the Moon was 1/3 the Earth's size. The modern value is closer to 1/4. Both the Sun and the Moon have about the same angular size of $1/2°$. If the Sun is 19 times as far as the Moon, which is 1/3 the size of the Earth, the Sun is roughly $19/3 \approx 6$ times bigger than the Earth. Modern values would give 400 times 1/4 or 100 times the size of the Earth.

It is somewhat strange that Aristarchus did not derive the distances to the Moon and the Sun, as this would have been simple in terms of the Earth radius. Perhaps he did this in some vanished text, in which case he would have obtained with his data: (1) the distance of the Sun = about 1,500 Earth radii, and (2) the distance of the Moon = about 80 Earth radii. The true values are 23,500 and 60 Earth radii, respectively. Aristarchus's mathematics was quite correct, so why the error? The Moon–Sun angle in the half moon triangle is so close to 90° (89.85°!) so that even a tiny inevitable error in its measurement results in a big error in the derived distance ratio.

Both Hipparchus and Ptolemy later obtained via triangulation values for the Moon close to 60 Earth radii. Thus ancient astronomers knew rather well the size and distance of the Moon. The distance to the Sun remained badly underestimated up to the modern times. Even Copernicus still held the opinion that the distance to the Sun was 1,142 Earth radii, 1/20 of the true value.

Aristarchus obtained that the Sun was many times larger than the Earth. Perhaps this led him to propose that the small Earth must orbit the large Sun. His own writings about this have vanished, but we have a reliable younger contemporary source, Archimedes (287–212 BC). After studying in Alexandria, this famous mathematician returned to his native Sicily where he served as advisor to King Hiero II. He was killed as the Romans captured his city. Among other achievements, Archimedes realized that if a heavy body is put into a vessel full of water then the amount of water that flowed over the rim gave the volume of the body. Thus the weight of the body divided by the weight of the over-flowed water expresses the density of the substance out of which the body is made. Without destroying an elaborate crown, he is said to have revealed the fraud of a goldsmith who had used adulterated gold to make it.

Archimedes' book *Sand-reckoner* gives a hint of Aristarchus' lost work on the size of the universe. Here Archimedes presents a new number system intended for dealing with large numbers.[3] In this connection he assumed that the diameter of the universe was less than 10 billion stadia (this happens to be a little more than the size of Jupiter's orbit). Archimedes calculated the greatest imaginable number, that of the grains of sand filling the whole universe. The result was 10^{63}, 1 followed by 63 zeros.

---

[3] Large numbers were a challenge for the clumsy Greek system that used letters to denote numbers. In the traditional system, counting was easy up to 10,000 (let us denote it by M) and with some effort even up to the number $M^2 = 100$ million, but after that the things got difficult. Archimedes took 100 million as a new unit and then the square, cube, and so on of this number were extra units. The largest number in the new system was M^2 raised to the power of M^4. In our notation this respectable number has 80,000,000 billion zeros after 1!

After mentioning the "common account" given by astronomers Archimedes goes on to mention what he considers a really extreme alternative view:

> But Aristarchus brought out a book consisting of certain hypotheses, wherein it appears, as a consequence of the assumptions made, that the universe is many times greater than the "universe" just mentioned. His hypotheses are that the fixed stars and the sun remain unmoved, that the earth revolves about the sun in the circumference of a circle, the sun lying in the middle of the orbit, and that the sphere of the fixed stars, situated about the same center as the sun, is so great that the circle in which he supposes the earth to revolve bears such a proportion to the distance of the fixed stars as the centre of the sphere bears to its surface (translation from Thomas Heath: *Aristarchus of Samos*).

This leaves no doubt that in this lost work Aristarchus had proposed a heliocentric system, even though there are no details. We do not know how Aristarchus dealt with other planets; the above account mentions only the Earth, the Sun, and the fixed stars. It is not known if he used the movement of the Earth to explain the stopping and retrograde motions of the planets in the way Copernicus did (Chap. 5). Archimedes mentions that, according to Aristarchus, the sphere of the fixed stars is hugely greater than the distance of the Sun. This explained why there were no observable annual changes in the directions of the stars (parallax), which were expected if the Earth really revolved around the Sun.

Aristarchus's world model was radical. Now we know that it is true, but at the time it was not yet mature enough to defend itself against the mainstream cosmology. Only one scholar is known to have supported this model – Seleukos who lived in Babylonia a century later. This is no wonder in view of the subtle observations needed to establish beyond doubt that the Earth is really moving. Such effects (aberration of light, stellar parallax) are so small that they were only detected two millennia later.

As to the "size of the universe," i.e., the distance to the outermost stellar sphere, there were no actual ways to measure it. Ptolemy retreated to a minimum estimate where the orbits of the planets were "packed" as tightly as possible, leaving no empty space between them, so that the largest distance achieved by a planet in its epicycle was the same as the smallest distance of the next planet. In this way, he derived for the outermost planet Saturn a distance of 19,865 Earth radii (a modern value is over 200,000). This was also the distance of the enigmatic stellar sphere, beyond which there was nothing.

On the Road Toward the Solar System

The shift from a flat Earth to measuring our spherical globe was a radical step in the view of the world. It is also an example of how local observations combined with mathematical reasoning can literally span the globe of the Earth and measure distance and size of the Moon. We have also seen a first attempt to place the Earth in its true secondary position in the Solar System.

We have seen that some astronomers in antiquity viewed epicycles and deferents as calculating methods, rather than real parts of some cosmic "clockwork." The emphasis was on understanding the apparent planetary motions as combinations of different ideal uniform circular movements rather than their concrete physical nature which may seem strange to us today. But the ancients' scientific inheritance, the available observations, and the conceptual ground from which they approached the world, were very different from ours. Beyond small (and sometimes dangerous) steps such as speculating that the Sun was a glowing ball of rock or that the Moon's appearance came from reflected sunlight, the physical nature of the planets and stars and their real trajectories were something remote and beyond the reach of ancient astronomers.

Chapter 4
Medieval Cosmology

The superb Museum of Alexandria, Egypt, was founded around 300 BC by one of the generals of Alexander the Great. It housed half a million manuscripts (papyrus scrolls) which scholars could use in their studies of literature, mathematics, astronomy, and medicine. Ptolemy was the last of the great scientific figures in Alexandria. He lived when the cultural heritage of Greece was already declining. During the next few centuries creative scientific activity diminished everywhere in the disintegrating Roman Empire.

In AD 312, Constantine the Great embraced Christianity which became the officially sanctioned faith in the Empire. The Church was, during its first centuries of existence, either indifferent to or even against science. There were extremists who opposed classical culture and attacked the Alexandrian library and those working there, murdering the mathematician Hypatia in AD 415. Among other works, she is thought to have assisted her father Theon with a commentary on the *Almagest*. Many scholars found it safer to go to the Academy in Athens and to Constantinople, the capital of the Eastern Roman Empire.

When the library was destroyed and who did it is controversial, partly because there was a main library and an annex in a temple to Serapis in a different part of the city. The Roman historian Plutarch says a city fire started during an attack by Julius Caesar in 48 BC burned the main library. Another possibility is an attack on the city by the emperor Aurelian in the third century AD. Edward Gibbon in *The Decline and Fall of the Roman Empire* cites sources asserting that Theophilus, Patriarch of Alexandria, destroyed the library when the temple of Serapis was converted to a church about AD 391. The last suspect is the Moslem Caliph Omar who is said by Bishop Gregory (writing 600 years later) to have burned the books of the library to heat the numerous city baths after conquering the city in AD 642. So a pagan, a Christian or a Moslem may be to blame. The one certainty is that when ideologies collide, books as well as people have suffered. However, after the wave of expansion, which brought it to the gates of Europe in Spain, the new Islamic empire turned out to be favorable toward classical sciences.

Treasures of the Past

In AD 529, the Emperor Justinianus closed Plato's Academy, after its nine centuries of operation as the longest living institute of higher learning. In Europe, the Roman Empire had collapsed when the Huns invaded Europe, and the Dark Ages had begun. Centuries passed without much interest in science. In the poverty, disorder, and absence of rich cultural centers, scientific work hardly could thrive.

In Christian monasteries, monks copied classical texts, but mostly the thinkers had other ideas. St. Augustine (AD 354–430) was a learned man who valued the achievements of old science. But in his *Confessions* he warned of the "disease of curiosity... which drives us on to try to discover the secrets of nature, which are beyond our understanding. I no longer dream of the stars." The idea was that even if one may be able to partially understand the workings of the physical world, the short human life should be used for the more precious search of God. Today we are in the position to see a historical perspective: while each of us is troubled by the meaning of life and death during our stay on Earth, understanding of nature continues to grow, albeit gradually. Scientific knowledge accumulates and every generation may enjoy and further develop this inheritance, even finding a purpose for life in it.

Fortunately, the empire of Muhammad, which flourished between 700 and 1200, did much to preserve the treasures of the past. Medicine and astronomy were highly esteemed. Scientists, working at the prosperous palaces of Muslim rulers, translated into Arabic language Greek texts some of which had survived the hard times. For instance, in the ninth century the Muslims obtained, as a part of the peace treaty with the Byzantine Empire, Ptolemy's main work, now known by its Arabic title, the *Almagest*.

The Cosmology of the Middle Ages

The Dark Ages was a long period whose "darkness" in various fields of culture in Europe has often been overrated. During the High Middle Ages, in the twelfth century, people started to translate Greek texts into Latin, mostly from Arabic versions. The works of Aristotle and others were received with enthusiasm among European scholars. Astronomers began to study Ptolemy's legacy that had been preserved and developed by Arabic scientists. The words of Bernard of Chartres, a scholar in twelfth century France, give the impression that a treasury had been opened: "We are dwarfs who have been lifted on the shoulders of giants. We thus see more and farther than they do, not because our eyes are sharper or we are taller, but because they hold us in the air, above their gigantic heights..." Centuries later Isaac Newton used similar words to pay tribute to his predecessors.

Aristotle's firm views on the cosmos and natural laws, which seemed to call into question the unlimited power of God, did not at first delight the Church. It was repeatedly prohibited to teach his texts at the University of Paris. But then St. Thomas Aquinas (1225–1274), who taught at the University of Paris, united the

Scriptures and classical ideas. The result was the unique medieval cosmology that held in its paradigmatic grip both the scholar and the layman. This doctrine included God and Man, Heaven and Earth, and made the physics and cosmology of Aristotle the official truth taught in schools and universities. The universe of spheres no longer clashed with the dogma of the Catholic Church. God had made the fixed Earth, and all the rest revolved around Man, sinful, but still the center of Creation.

In his *Divine Comedy,* Dante Alighieri (1265–1321) drew an impressive picture of the medieval cosmology. He wrote the poem while in exile from his home-town Florence due to political reasons. It describes Dante's visit to Hell, Purga-tory, and Paradise. Hell is a cone extending down to the center of the Earth, while Purgatory is a conical mountain on the opposite side. After visiting the less pleas-ant places (where he found his political enemies!) Dante finally rises to Paradise, through its increasingly lovely levels (planetary spheres), ending at Empyrean, the dwelling place of God. Just below this most blessed place was the sphere of *Pri-mum Mobile* or the first mover. This new part had been added to the celestial clockworks by Arabic astronomers to explain the slow wobbling of the eighth

Fig. 4.1 The cosmos of the Middle Ages was bordered by the sphere of Primum Mobile, the first mover. This sphere was just outside the sphere of the fixed stars

sphere of fixed stars (the shift of the vernal equinox as discussed in the previous chapter).

The celestial world differed radically from the Earth so that mortals could not live there for a moment. If, however, one could somehow climb toward the "outer edge," one would see the physical reality change and space and time losing their familiar meaning. Dante imagines that "distance does not decrease nor increase immediately there where God rules; the law of Nature does not exist there."

Dante was not interested in epicycles and other mathematical details studied by astronomers. More important to him was the meaning of the overall structure of the universe for human race. Humanity had two competing natural directions of motion. The balance between a person's material and spiritual sides determined descent after death into the depths of Hell or ascent to the Heavens. This unification of science and faith gave rise to a view where humanity had great cosmic significance, something that was to be lost during the Copernican revolution (Fig. 4.1).

Scholasticism: The Medieval Science

The science of the Middle Ages (scholasticism), was concerned more with thinking and concepts than it was with the physical world. Aristotle had the last word there. In their attempt to understand, people were assisted by logic, also founded by Aristotle. For example, a central question of the time was, whether classes of things, such as cats or stars, are in themselves real things or mere names invented by the human mind (leading to the heated controversy between "realists" and "nominalists").

The notoriously dry scholastic analysis did raise good questions about physical doctrine. At the fourteenth century University of Paris, Jean Buridan and his pupil Nicole Oresme critically examined Aristotle's notion of force, "everything that is in motion must be moved by something." An arrow flies forward, pushed by the air. But there were intriguing problems in all this, and Buridan (ca. 1297–1358) suggested that something that he called *impetus* is added to a body when it is thrown into its trajectory which maintains the body in motion. Impetus is a forerunner of the concept of conservation of momentum, important in modern physics. Impetus theory replaced Aristotle's mechanics, and became a dominant view on the physics of motion in the fifteenth century.

Remarkably, Buridan applied the earthly concept of impetus to the revolving celestial spheres. It was common to think – as in Dante's *Divine Comedy* – that angels were rotating the outer stellar sphere. The giant planetary spheres were kept in rotation by a force from the angel-driven sphere of the fixed stars. However, Buridan reasoned, as the Bible is silent about this, perhaps God gave the spheres their motions at the moment of creation. Having got their impetus, they have rotated until today. This uniform motion happens without friction, allowing us to see impetus at its purest – day and night overhead! This interesting step heralded the coming change where the heavens were found to follow the same physical laws that operate on Earth.

If motion does not require a pushing force, perhaps we could move without being aware of it? Perhaps even the Earth could be in motion? Nicole Oresme (ca. 1320–1382) did not accept Aristotle's proofs for the Earth being at rest. He argued that every motion is relative. The Earth may revolve around its axis (as already Heraclides had proposed), giving the starry sky the appearance of rotation "as a man in the moving ship thinks that it is the trees outside of the ship that move." Aristotle knew this alternative, but had argued against it pointing out that a stone thrown directly upwards falls down on the same spot. In Aristotle's view, if the Earth turned, the point from which it was thrown will have moved aside as the stone falls. But Oresme saw impetus at work: the stone preserves its own share of impetus that it has together with the moving Earth. Thus, both the stone and the surface of the rotating Earth slip the same number of meters aside as the stone returns to the ground.

It may sound curious, but after such reasoning, Buridan and Oresme accepted that the Earth is at rest. As good scholastics they regarded that the truth deserves to be defended by compelling arguments only. In hindsight, their analysis of Aristotle's ideas about physical motion had carried them a little bit toward modern views about rest, uniform motion, and relativity.

Aristotle's writings also inspired thinking about what science is. Remember that in his science one starts from absolutely true axioms and deduces as logical consequences other true facts. But how to find the axioms in the first place? Aristotle said that one should observe nature and use intuition. Robert Grosseteste (ca. 1168–1253) and his pupil Roger Bacon (ca. 1214–1292), who were learned churchmen and philosophers in Oxford, thought about ways to help this process. They suggested that the claims or explanations found by observing nature should be tested by further studies before approval. For instance, there might be two different explanations for a phenomenon, and by an experiment one might exclude the wrong explanation or obtain support for the correct one. One can see here the seeds of modern experimental science, which was to bloom with Galileo four centuries later.

It has been said that the authority of Aristotle slowed down the development of science in medieval Europe. Such a view seems unnecessarily narrow, taking into account that scientific activity as a whole had for centuries been "on the back burner." It was revived by his texts, together with those of other classical masters. Of course, Aristotle is not to be blamed if his followers read his books as the final truth, without realizing that science is a self-correcting activity that will change the content of those books. His ideas, even when incorrect, stimulated independent thinking. Gradually people became prepared to read the "Book of Nature" instead of ancient books.

Infinity Where the Center Is Everywhere...

The finite spherical universe that was popular from antiquity and through the Middle Ages had a central point and, due to the giant outermost sphere, it had some degree of local isotropy (the distance from the center to the sphere is the same in every direction). But in practice, the distance to the sphere could not be measured, so our central position could be deduced only from the apparent revolution of celestial bodies around us. However, remember that Anaximander argued that by taking the Earth at rest, it is at the center of the world without any privileged direction. As we saw in Chap. 2, Aristotle also argued from the rotation that the world is finite in size, otherwise its infinitely distant parts would move at an impossible infinite speed. So the cosmic rotation, the existence of the center, and the finite size of the universe were actually linked.

Back in the third century the "neo-Platonist" Plotinus (205–270), described his spiritual cosmology in a book, *Enneads*. In a section, "The Heavenly Circuit," he wrote, "the heavens, by their nature, will either be motionless or rotate." And then came the astonishing words. "The center of the circle is distinctively a point of rest: if the circumference outside were not in motion, the universe would be no more than one vast center." In other words, if there were no universal rotation, then there were no absolute center, and the universe could be large without limits.

After twelve more centuries around 1440, the German Cardinal Nicholas of Cusa (1401–1464) wrote something similar in his philosophical treatise *Of Learned Ignorance:* "The universe is a sphere of which the center is everywhere and the circumference is nowhere." He came to this cosmological principle when attempting to characterize the incomprehensible infinite God. Interestingly, the context in which this was stated was the relativity of motion, a permanent topic in history of physics. As there cannot be any absolute rest except for God, even the Earth must have some kind of motion, Nicholas of Cusa argued. "Every man, whether he be on Earth, in the Sun, or on another planet, always has the impression that all other things are in movement whilst he himself is in a sort of immovable center!" So "there will be a world-machine whose center, so to speak, is everywhere, whose circumference is nowhere, for God is its circumference and center and He is everywhere and nowhere."

In modern terms, it might be said that various uniformly moving observers in the universe may each think themselves at rest and others to be moving. In this sense, a uniformly moving observer may ascribe to oneself a special status: being at rest, being at the center. However, for Nicholas of Cusa circular motion was the natural one (instead of rectilinear) and even a revolving observer, not feeling the turning, may regard himself resting at the center. This apparent center is defined by the apparent circular movements around that observer.

Saying that the universe is a sphere of which the center is everywhere, one has gone from the finite sphere-world to the world where around every point the observer sees similar landscapes in every direction (isotropy). Nowadays we know, having familiarity with non-Euclidean geometry, that even such a world could be finite and without boundaries. However, the medieval thinker had naturally a nonfinite world in mind (Nicholas of Cusa preserved true absolute infinity to God only and wrote "although the world is not infinite, it cannot be conceived as finite, because it lacks boundaries within which it is enclosed") (Fig. 4.2).

Though Nicholas of Cusa did not present any detailed world model, he liberated the universe from the absolute center. He stated that the number of stars – of which

Fig. 4.2 Nicholas of Cusa (*left*) and Giordano Bruno, who imagined an infinite world and foresaw the modern cosmological principle according to which "Center is everywhere"

the Earth is only one – was unlimited. He also considered it natural that there should be life and inhabitants on other stars, though he admitted that we cannot know what they are like. He comforted those who feared that beings living on stars larger than the Earth were nobler than us by saying that it is the intellectual level that really matters.

... Or Where There Is No Center

Giordano Bruno (1548–1600) went in his youth to a Dominican monastery. His original thinking caused controversies with his superiors who suspected that the young man from Nola (near Naples) supported heretical ideas. At the age of 28, he escaped and spent years wandering around Europe teaching philosophy in universities, all the time stirring up accusations of blasphemy and heresy.

In 1591, Bruno made the fateful decision to return to his native Italy, invited by a young aristocrat who seemed to be eager to learn philosophy, but instead had a shallow hunger for exotics. The disillusioned pupil led Bruno into the hands of the Inquisition. He was arrested and accused of heresy: he had not only claimed that the prevailing view on the universe was erroneous, but more importantly, he viewed God as a pantheistic spirit (roughly meaning that nature and God are the same) and he denied such central doctrines of the Church as transubstantiation and Immaculate Conception. After 7 years in prison, Bruno was burned at the stake in Rome, at the Square of Flowers (Campo dei Fiori) in February 1600.

Though he lived after Copernicus and wholly agreed that the Earth is not the universal center, it is convenient to describe Bruno's ideas here because his spiritual background was in medieval thinking. He knew the writings of Nicholas of Cusa. Bruno's view that God appears as a creating spirit in all things of the universe was accompanied by the parallel idea, for which he is famous, the unlimited power of God corresponds to the infiniteness of the universe.

Giordano Bruno made a huge mental jump toward a new picture of the large-scale cosmos. Copernicus had several decades earlier put the Sun in the center of the universe, but he had thought – as almost everyone did still at the time of Bruno – that the world is bordered by the crystal sphere holding fixed stars. An exception, the English astronomer Thomas Digges (1543–1595) published a map in 1576 of the universe where the stars were detached from their sphere and dispersed in space. But Digges still preserved a special place for the Sun in the center of the infinite stellar universe. It was Bruno who gave stars the physical status of distant suns. It seems that he was the first to imagine and clearly assert that stars, faint points in the sky, actually are as large and bright as our Sun.

Interestingly, the roots of the contemporary debate on how to define a planet (to be discussed in Box 31.1) go back to Bruno who made a clear physical difference between stars and planets: stars shine their own light, while planets reflect the light of their central Sun.

Bruno characterized the universe with his cosmological principle that he expressed as: *In the universe neither center nor circumference exist, but the center is everywhere.* This reminds us of Nicholas of Cusa and means that in the universe all places are alike. This flagrantly contradicted the old cosmology where the center existed, occupied by the Earth. In modern cosmology, the nonexistence of any preferred center is a natural starting point.

Bruno's other cosmological principles were the universality of earthly laws and that the matter in the heavens is similar to the matter on Earth. He wrote that Sun-like stars are scattered in infinite space. "As the universe is infinite... one can assume that there is an infinite number of suns, many of which are seen for us in the form of small bodies; and many may appear for us as small stars." Bruno also concluded that the Earth cannot be the only planet having living beings, as this would make it a preferred place, a kind of center of the universe. This sounds familiar now when modern astrobiology operates on the universality of the laws governing both inanimate and organic Nature.

In his book *About Infinity, Universe, and Worlds,* Bruno longed for the time when there will be means to probe the depths of space: "Open for us the gate, through which we can look at the countless, everywhere similar stellar worlds." Just a few years after Bruno's death Galileo "opened the gate" by pointing his telescope at the starry sky.

Though no astronomer, Bruno was aware of the difficulties that hinder observation of distant celestial bodies. Stars are other suns, but so far away that they look like points of light. Around them are planetary systems, but the planets are too faint for our eyes. Bruno also reasoned that even our Solar System may have other planets that we cannot see because they may be very distant, or they may be small in size, or they may be poor reflectors of sunlight. Having to base his cosmology on scarce observations, Bruno explained the absence of direct evidence as a result of

Fig. 4.3 Leonardo da Vinci described surface features on the Moon and explained correctly why you can see "Earthshine" between the horns of the crescent of the Moon – light from the Earth illuminated by the Sun is faintly reflected back from the surface of the Moon. Note the large contrast between the directly Sun lit side of the Moon and the Earthshine part of the Moon (photo by Harry Lehto)

our limited observing capability. Today similar problems still hamper astronomers trying to look much deeper into space.

But are visions such as those of Nicholas of Cusa and Giordano Bruno, which are not based on new astronomical observations, important for science? Yes, because science lives on both observation and thinking. Sometimes a new idea prompts one to look at old observations in a way that reveals their true significance. Copernicus is a striking example, as we see in the next chapter.

The Middle Ages closed with the birth of the Renaissance that flourished during the fifteenth and sixteenth centuries. Fresh winds were blowing in art and other fields of culture, when people started to look at classical literature, philosophy, and science with new curious eyes. Students of the physical and spiritual nature of humanity and explorers of distant regions were symbols of humanism and the Renaissance. Artists like the universal genius Leonardo da Vinci (1452–1519) began to picture the human being more positively than their predecessors did during the stiff scholastic period. His interests ranged from bird flight to the Moon (Fig. 4.3). *Nature* assumed new significance in the eyes of artists, scientists, and inventors.

Not all scientific progress depends on philosophy and observation. In our discussion up to now, we have lamented the loss of important works. In part this happened because so few copies actually existed even in ancient times. The German printer Johann Gutenberg (ca. 1396–1468) revolutionized the dissemination of knowledge by his invention of movable type. The first book thus printed in Mainz in 1451 was a grammar of Latin. Though book printing was still a slow process (say, 15 pages per hour), it was much speedier compared to months copying by hand a 200 page book. Within a few decades Gutenberg's invention had greatly speeded up the spread of scientific knowledge in Europe.

Chapter 5
The Roots of the Copernican Revolution

The Renaissance burst into scientific life with the work of Nicolaus Copernicus. Mikolaj Kopernik (his name in Polish) was born in 1473 in the town of Toruń in central Poland as the youngest of four children in the family of a merchant. When Copernicus was 9, his father died and he was brought up by his maternal uncle Lucas Watselrode, a churchman who later became the bishop of Remand.

The famed university of Krakow was founded in 1364 and here the young Copernicus started his studies in 1491. The university attracted students from all over Europe where Latin was the common language of teaching and science. The curriculum followed the medieval pattern of seven liberal arts. The *trivium* included Latin grammar, rhetoric, and dialectic, while the more advanced *quadrivium* comprised arithmetic, geometry, astronomy, and music.

Years Under the Italian Sun

After 3 years in Krakow, Copernicus continued his studies in Italy where he spent a few years at the University of Bologna learning canon law (and also Greek and astronomy). In 1501, he returned to his job in Frauenburg (today's Frombork in Poland) as a church administrator, but soon he headed back to Italy, this time to study medicine at the University of Padua. Copernicus finally received his Doctor of Law degree from the University of Ferrara. When he returned to his home country in 1506, at the age of 33, he had been in Italy for 9 years and was a "Renaissance man" trained in many fields.

This peaceful and rather timid servant of the Catholic Church was also decisive and hardworking, writing on various topics, even monetary reform. He also gave medical consultations till the end of his life. However, behind the public face there was ticking a scientific time bomb. It gradually became known outside of Frauenburg, even in nonastronomical circles, that the Canon of Frauenburg had the strange idea that the Earth moves, while the stars and the Sun are at rest.

P. Teerikorpi et al., *The Evolving Universe and the Origin of Life*
© Springer Science+Business Media, LLC 2009

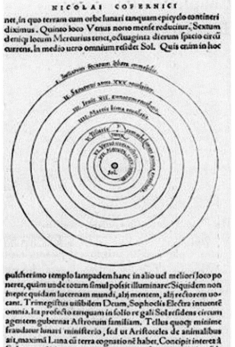

Fig. 5.1 Copernicus and his universe. It was limited outside by the sphere of fixed stars, which was "immobilis," immobile. The drawing is from the book De Revolutionibus

Copernicus was rather silent about the source of his inspiration of a universe centered on the Sun. It is uncertain how much the earlier Sun-centered astronomical ideas influenced Copernicus. He did write "...and though the opinion seemed absurd, yet knowing that others before me had been granted the freedom to imagine such circles as they chose to explain the phenomena of the stars, I considered that I also might easily be allowed to try whether, by assuming some motion of the Earth, sounder explanations [...] for the revolution of the celestial spheres might so be discovered." The absurd ideas turned out to be astronomical treasure (Fig. 5.1).

Copernicus' thoughts about the cosmos, with the Sun standing unmoving in the middle, may have appeared during his Italian years. He seems to have started his great life work, *De Revolutionibus Orbium Coelestium* (On the Revolutions of the Heavenly Spheres), after returning from Italy in 1506. The manuscript may have reached its final form around 1530. Before that Copernicus had written a summary that circulated among astronomers. One of these was the young mathematician Rheticus (1514–1576) from the University of Wittenberg. He visited Copernicus, wishing to persuade him to publish his work in totality. The visit was prolonged to almost 2 years! Thanks to the efforts by Rheticus and another friend, the bishop Tiedeman Giese, Copernicus finally agreed to publish his book. Another representative of the Catholic religious community who, several years earlier, asked him to publish his new theory was Nicolas Schönberg, Cardinal of Capua. In fact, it is thought that Schönberg acted on the insistence of the Pope himself, Clement VII, who was very interested in astronomy.

De Revolutionibus Appears: The Mission Is Complete

It is said that when the 70-year-old Copernicus received the first printed copy of the book, he was already mortally ill in his bed and could not read it. This saved him from seeing the preface that was added without his consent: "To the reader, concerning the hypothesis given in this book." Unsigned, it was written by a friend of Rheticus, the theologian Osiander who took care of the printing while Rheticus was burdened with other matters. Osiander probably was afraid of the controversies the book might give rise to and attempted to veil the true opinion of Copernicus. He emphasized that the Copernican theory is nothing but a new method to calculate positions of planets in the sky and does not allege that the Sun really is in the center of the cosmos. Before condemning Osiander too harshly, we should remember that he may be seen as following the tradition mentioned at the end of Chap. 3 that distinguished between mathematical astronomy and the real physical movements of celestial bodies. Medieval Aristotelians did not attribute a concrete reality to the epicycles. Rheticus was angry about Osiander's intrusion, but Copernicus' own preface to *De Revolutionibus* makes it clear that he presented a new *physical* world model, with the Earth really moving in space.

Why Put Away the Good Old World? Why Copernicus and Why in the Sixteenth Century?

The new system was in some ways not so much simpler than the old one. It was still based on many circles and epicycles, and, in principle, its predictions of planetary positions in the sky were not more accurate than those given by the old geocentric clockwork. However, for a mathematical mind such as Rheticus it was attractive, because it could explain in a simple and natural manner the main celestial movements. Even Ptolemy had written "it is a good principle to explain phenomena by the simplest hypothesis possible, in so far as there is nothing in observations to provide a significant objection to such a procedure." Copernicus gave central importance to the fact that in a Sun-centered system "one motion is sufficient for explaining a large number of apparent irregularities." Let us now discuss the major celestial movements and their relation to how, when, and why the Copernican theory arose:

- The daily rotation of the starry sky
- The yearly wandering of the Sun around the sky and the seasons (Fig. 5.2)
- Most importantly, the regularly repeated retrograde loops of the planets without epicycles (Fig. 5.4)

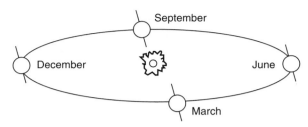

Fig. 5.2 The seasons, together with the varying altitude of the Sun through the year, became understood as due to the inclined axis of revolution of the Earth, keeping its direction fixed in space. This simple but deep explanation is not always remembered – it is not rare to mistakenly believe that in the summer the Earth is closer to the hot Sun (actually when it is summer in the northern latitudes, we are more distant from the Sun!)

As to the daily rotation of everything in the sky, Copernicus emphasized that it is easier to imagine the small Earth rotating around its axis once a day rather than the huge celestial sphere revolving at a breathtaking speed ($9,000 \, \text{km} \, \text{s}^{-1}$ for a star at the equator, if the radius of the celestial sphere is 20,000 Earth radii as assumed by Ptolemy). Such a rapid motion might cause the sphere to fly apart! This is a physical argument beyond any question of the relative accuracy of Sun-centered vs. Earth-centered models. The yearly motion of the Earth around the Sun generates very simply the yearly motion of the Sun around the sky around the ecliptic. This is in place of having the Sun circle the Earth.

The philosopher of science Thomas Kuhn (1922–1996) took the Copernican revolution as a major example of his "paradigm breaking" concept of how science advances via quiescent lengthy periods of "normal science" separated by revolutions. During the revolution the paradigm, roughly the basis of contemporary science, collapses. In Ptolemaic astronomy, the basis was formed by the central position of the Earth and the principle of uniform circular motion leading to increasingly more epicycles. Kuhn thought that by the sixteenth century the old system had been driven into a crisis. It had become an intolerably complex "monster," too clumsy to be viable any longer. At the same time, religious and philosophical movements having the Sun in a position of central importance may have helped. Fig. 5.3 shows the time relations of Copernicus to several other eminent figures of the Renaissance.

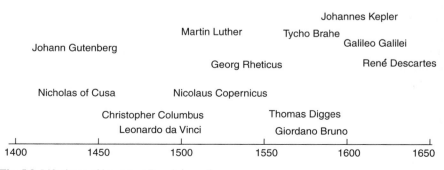

Fig. 5.3 Life times of important Renaissance figures

Fig. 5.4 At regular intervals planets make loops when they move relative to the fixed stars. In the Ptolemaic world model, this dance of planets was described by suitable epicycles, while for Copernicus this key phenomenon naturally followed from the motion of the Earth and the other planets around the Sun

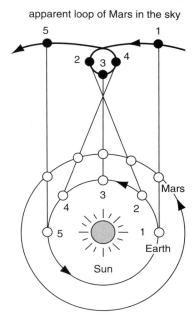

However, the Finnish mathematician and historian of science Raimo Lehti concluded that there was no true 16th century crisis in cosmology. The Ptolemaic system was not viewed as the complicated concept we view it now. Rather, the real key to acceptance of Copernicus' ideas was an interesting feature in the model, which provided a fresh explanation for the annoying retrograde motions of the planets. As described in Ptolemy's *Almagest,* the planets other than the Sun and Moon occasionally show backward (east to west) retrograde motion. Mercury and Venus show retrograde motion when in the same direction as the Sun while Mars and the other planets show it when the Sun is on the opposite side of the sky. Even for the geocentric Earth-centered view, it is as if the Sun were directing the dance of the planets! It is quite possible that Copernicus became convinced of the idea of the central Sun from these regularities that were traditionally viewed as a wonder set by God. In the old geocentric system, retrograde motions each require an individually tuned smaller epicycle attached to each planet's larger deferent. In the heliocentric model, they become simple consequences of the Earth's motion (see Fig. 5.4). Retrograde motion happens when Mercury and Venus pass between the Earth and the Sun. It also occurs when the Earth passes between the Sun and the other planets. The Sun-centered model thus eliminates an epicycle and special "tuning" for each of the planets, a great simplification.

Old and New

Copernicus' world model was still based on the old principle of uniform circular motion and required a complex machinery of deferents and epicycles to explain irregularities other than the basic retrograde motions. It also contained an outermost sphere on which the fixed stars were attached. However, the sphere was now immobile, forming a huge reference frame against which all the motions inside it could be seen.

As mentioned, Copernicus introduced two motions of the Earth: the orbital motion around the Sun and the spin or rotation of the Earth. The seasons result from a 23° tilt of the spin away from a 90° orientation relative to the Earth's orbit. Just like a child's spinning top, during the yearly orbit of the Earth, the spin points in a constant direction. The fact that the rotation axis of the Earth keeps its direction in space is a result of the conservation of angular momentum from Newtonian mechanics. But Copernicus had no access to Newton's laws of motion. In his thinking, the normal situation was that the axis would keep its position relative to the Sun (toward or away, etc.) all the way around the orbit, and there would be no seasons. So Copernicus postulated the *third motion* of the Earth, forcing it to keep the same orientation relative to the orbit plane over one year. From this there was only a short step to also include the movement of the vernal equinox making the third motion just slightly too slow to maintain the direction of the Earth's axis exactly fixed in space![1]

Copernicus was thus compelled to add a rather complicated "too slow" third motion into his model. This was noted and even ridiculed by the opponents of the new system: when previously the Earth was immobile, now it required a total of three motions, one daily and two yearly. A popular poem told about "those clerks who think (think how absurd a jest) that neither heav'ns nor stars do turn at all, [...] who [Copernicus], to save better of the stars th'appearance, unto the earth a three-fold motion warrants."

The followers of Copernicus, Kepler, and Galileo, pointed out that the yearly portion of the third motion is not necessary at all. In his *Dialogue* (1632), Galileo compared the Earth to a ball floating in a water-filled bowl. When you start revolving "on your toe" holding the bowl in your hands, the ball appears to make a rotation counter to the rotation of the bowl. But in fact what happens, Galileo noted, is that the ball remains immobile relative to its surroundings, without any effort of its own. Galileo already saw in the Earth's behavior the Newtonian concept of inertia, unknown to Copernicus.

The orbit of the Earth illustrates the complexity required in Copernicus' model to account for, in this case, observed variations of the motion of the Sun along the ecliptic. The central point of this circular orbit revolves with a uniform speed along a small circle whose center rotates around the Sun. Thus three circular motions are needed to produce the variations in yearly motion of the Sun. To explain all the motions of the Solar System, Copernicus needed over 30 circles, which made his system as complex as that of Ptolemy. Nevertheless, these mathematical complexities, necessitated by the time-honored use of uniform circular motions, did not

---

[1] Previously, the shift of the vernal equinox was interpreted as due to a slow motion of the celestial sphere. In the Middle Ages one more outer sphere was postulated to take care of that extra movement.

change the fact that the model was a breakthrough toward the true laws of planetary motion, which Kepler discovered almost seven decades after *De Revolutionibus*.

The Order and Scale of the Solar System

Astronomy is very much a science of cosmic distances and in this respect the Copernican model had a great practical plus compared to the old model. It became possible to deduce from observations the true order of the planets from the Sun and to determine their relative distances from the Sun. One could give the distances with the Earth–Sun distance as a new natural unit of length (the astronomical unit) to replace the Earth radius.

In the Ptolemaic system, one could choose a planet's distance rather freely; it was only important to fix the size of the epicycle relative to that of the deferent so that the apparent motion of the planet matches that observed. On the contrary, in the heliocentric model, the order of the planets and their distances from the Sun become unequivocally determined. Without going here into details we note that one may determine the planet–Sun distance at the moment when the triangle formed by the planet, the Earth, and the Sun is a right triangle.

Copernicus removed the Moon from among the planets and made it Earth's satellite. He determined the order and distances of the planets as given in Table 5.1 (the unit is the mean Earth–Sun distance, the astronomical unit or AU). We emphasize that after fitting circles and epicycles to match observations, Copernicus did not find that the planets had circular orbits. He calculated the minimum, average, and maximum solar distances of the planets. The table shows the interesting thing that now the maximum distance of a "lower" planet is not equal to the minimum distance of the next "upper" planet. In fact, there was plenty of empty space between the planetary orbits contrary to what Ptolemy had thought. In the Copernican system, the sphere of fixed stars was simply "tremendous," as the Earth's yearly motion did not cause observable shifts in the positions of stars in the sky. It was not until the nineteenth century that the shifts were finally detected. Another noteworthy thing in Table 5.1 is the high maximum/minimum distance ratios for Mercury and Mars. This reflects their quite elongated orbits, which later allowed Kepler to deduce that

Table 5.1 Copernicus' values for the minimum, average, and maximum solar distances of the planets

	Minimum distance (AU)	Average distance (AU)	Maximum distance (AU)
Mercury	0.263	0.376	0.452
Venus	0.701	0.719	0.736
Earth	0.968	1.000	1.032
Mars	1.374	1.520	1.665
Jupiter	4.980	5.219	5.458
Saturn	8.652	9.174	9.696

Mars actually moves along an ellipse. On the contrary, the distances of Venus and Earth from the Sun vary rather little.

We may still emphasize, as did Copernicus himself, that his system had less arbitrary structure than the Ptolemaic one. This may in itself make the Sun-centered model preferable, but more importantly, the predicted unique order and distances of the planets could be checked by other observations later.

The Copernican Principle

The name of Copernicus is attached to two concepts. When speaking about the Copernican Revolution, we may refer to the appearance of the heliocentric world model in 1543. Actually, there was a process lasting a couple of centuries leading to the final establishment of this new astronomical picture of the Solar System. It required many observational and theoretical advances, before the moving Earth became as natural as an immobile globe was for our ancestors.

One concept born with the Copernican Revolution is the cosmological Copernican Principle, referring to the conviction that we are not in a special or preferred place in the universe. In fact, Copernicus thought that the Sun is in the center (or almost so) of the universe, in contrast to the Principle of No Center, as advocated by Bruno. Nevertheless, the abandonment of the immobile central position of the Earth giving it the status of an ordinary planet was such a drastic change that the name Copernican Principle is quite justified. The cosmologist from Copernicus' Alma mater in Cracow, Kondrad Rudnicki, formulated it in a more modern fashion as follows: *The universe as observed from any planet looks much the same.* Nowadays we just replace "any planet" by "any galaxy."

Copernicus did not speculate on the world beyond the distant material sphere of the stars. But he gave a tremendous impetus to look at the stars with fresh eyes. Digges was born in the same year as Copernicus died, Bruno a few years later. They realized that the stars were not on an immobile sphere but were distributed in an infinite space.

De Revolutionibus was not exactly a best seller, and it did not immediately attract much attention. Some enthusiasm was shown by those mathematicians able to go through the difficult text. The Catholic Church was first rather indifferent, perhaps partly due to Osiander's preface, and as we saw some of its officials had supported publishing the new theory. The Orthodox Church did not regard the movements of the physical Earth to be relevant at all. Initial protests came instead from the Lutherans. It took seven decades from the publication of *De Revolutionibus* for the Holy Office to take action in 1616. During that remarkable period, many things happened. Thomas Digges and Giordano Bruno lived and died. Tycho Brahe, Johann Kepler, and Galileo Galilei founded a new astronomy and experimental physics. The telescope was invented. Even the sky seemed to celebrate the Copernican Revolution. The influential comet of 1577 and two supernovae (the last ones observed in historical times in our Milky Way Galaxy) served to demonstrate that the heavens were not unchangeable. In the middle of all this, Shakespeare wrote "There are more things in heaven and earth, Horatio, than are dreamed of in your philosophy."

The universe of Copernicus was still a realm of circles and epicycles. The next step in the Copernican revolution was to replace the overly rigid assumption of circular motions with a more realistic closed orbit. Johann Kepler made this crucial step. For this he needed the very accurate observations by Tycho Brahe. The next chapter is devoted to their work.

Chapter 6
The True Laws of Planetary Motion Revealed

The medieval cosmos formed a tight unity within its spherical boundary, with strict laws of circular motion for its heavenly spheres, while everyday laws and even disorder ruled close to the Earth. Although the geocentric view was deeply rooted in society, this view was bound to erode after Copernicus. Even among astronomers, the heliocentric world system was not accepted immediately. But the search for universal laws of cosmic order, the occupation of rational minds since the Ionian revolution, was reinvigorated.

Tycho Brahe's Nova Lights the Way

Among these searching minds, Tycho Brahe (1546–1601) was a splendid observer of the night sky, who decisively improved the astronomical data available for astronomers. He made careful visual observations of the planets over many years, recording their positions in the sky with an accuracy of 1 arcmin ($1'$), when previously astronomers were content if they had $10'$ accuracy. Tycho achieved the new level of precision by constructing his own large angle-measuring instruments, working every cloudless night, and taking into account various systematic errors affecting the estimated position of a star, including the refraction (change of direction) of the light ray by the Earth's atmosphere (see Fig. 6.2 on p. 60).

Brahe was the oldest son in a noble family in the southernmost part of Sweden (which then belonged to Denmark). His personality may have been affected by the death at a very young age of a twin brother and his being raised by a childless aunt and uncle. The talented boy went to the University of Copenhagen to study rhetoric and philosophy. Here he got interested in stars. When he went to Leipzig in 1562 to read law, he changed to astronomy. Brahe's quick-temper was as much part of his life as astronomy. During his student years, he got into a sword fight with another nobleman, and lost part of his nose. For the rest of his life, he tried to patch up his appearance with an artificial metal nose.

Fig. 6.1 A supernova exploded in 1572 in the constellation of Cassiopeia. Tycho Brahe concluded that this "Stella nova" had to be situated in the stellar sphere, which thus was not unalterable as was previously thought. Modern observations of much more distant supernovae have led to important cosmological conclusions, too

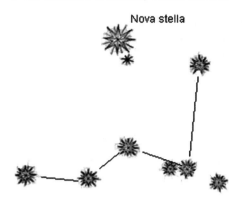

Nova stella

Brahe obtained the island of Hveen from the king of Denmark in 1576. There he built a magnificent observatory, Uraniborg, and received continuous support. Tycho's foster father had died of illness after saving the king from drowning. All this was fairly expensive – a few percent of Denmark's national income went into "The Castle of Heavens," comparable in expense and contemporary technological sophistication to the Hubble Space Telescope project.

The money was a good investment. It raised observations of the sky to a totally new level, even though the observatory was built before the invention of the telescope. It also led to the next phase in the Copernican revolution, when Tycho's accurate observations were used by Johannes Kepler.

Before the days of the Hveen observatory, Tycho made observations of a new bright star that appeared in November 1572. He wrote: "Astounded, as though thunderstruck by this astonishing sight, I stood still and for some time gazed with my eyes fixed intently upon this star. It was near the stars, which have been assigned since antiquity to the asterism of Cassiopeia." The star was at first as bright as Venus and then gradually dimmed until it became invisible after one and half years (Fig. 6.1).

It had previously been observed that the Earth's Moon was close enough so that it shifted among the stars due to the change in position of an observer as the Earth rotated. Brahe's accurate observations showed that the new star, a "nova," did not move at all relative to the stars of Cassiopeia, neither during one day as the Earth rotated nor over longer times. Brahe concluded that (1) the star must be farther away than the Moon and (2) in fact it lies in the sphere of the fixed stars. He wrote a booklet about this phenomenon, where he said that at first he could not believe his own observations, since Aristotelian philosophers agreed that there should not be any changes in the ethereal zone of the heavens. The new star clearly showed that the heavens are not unchangeable after all! This important observation made Tycho Brahe a well-known figure. He continued observations which turned out to be crucial for the Copernican Revolution.

A comet in 1577 further shook the view of the perfect heavens. Brahe's observations convinced him that the comet wandered farther away than the Moon and even followed a trajectory that had to take it directly through the crystal sphere that carried the Sun. All this disagreed with the traditional opinions. The nova and the comet and the conclusions from these showed that rather simple observations, when combined with calculations and reasoning, could give new knowledge about the cosmic realm.

Tycho's World Model

Although Tycho Brahe did not agree with the new world model of Copernicus, it was a sign of the changing times that he proposed his own system which differed from Ptolemy's model. The Earth remained fixed in the center, orbited by the Moon and the Sun. However, the other planets no longer revolved around the Earth, but around the Sun that carried them around the Earth.

Mathematically Tycho's model was equivalent to Copernicus' model. Then why such a peculiar arrangement? It was troublesome for Brahe, a meticulous observer, that in the Copernican system one would expect regular changes in the positions of fixed stars, called parallax shifts, when the Earth moves annually from one point in its circular trajectory to the opposite point on the other side of the Sun. No such changes were seen, which implied that either the stars must be very distant or the Earth does not move. Brahe reasoned that if the stars are so distant, they must be fantastically large (because in this pretelescopic era, he thought the angular size of stars is about $1'$, $1/30$th of the Sun's disc). Keeping the Earth fixed he avoided the puzzle of gigantic stars. In this way there was also no need for the immense "futile" empty space that appeared in the heliocentric universe.[1]

Kepler's Mysterious Universe

Johannes Kepler was a great builder of a world system, perhaps the last one to imagine that Platonic mathematical forms are an ideal reflection of physical reality. His family in Germany seemed to be far from ideal for a future serious scientist. His father was an adventurer and mercenary, who disappeared for good when Johannes was 17 years old. His mother was an erratic character, a kind of sorceress who was threatened with death at the stake for witchcraft – she was freed from prison only thanks to the years-long struggle by her son who was then already a respected astronomer. The family was poor, but Kepler received a grant to go to school – even then there were grants for poor but gifted children. Finally, he entered the University of Tübingen to study theology. There he came to know about the new world system

---

[1] This paradox of immense stars, one objection against Copernicus, vanished when Galileo showed that stars are much smaller than they appear to the naked eye. He stretched a cord against the starry sky and noted at which distance the cord hid the star behind it. He concluded that stars are 5 arcsec wide (i.e. $1/12$ of an arc minute). Actually, the stars are even very much smaller than this; the atmosphere of the Earth smears the sharp images.

Fig. 6.2 Tycho achieved great accuracy in his visual astronomical observations. The picture from a book by Tycho (1598) shows his wall quadrant – the observer is aided by two assistants timing the observations and making notes

from the mathematician Michael Mästlin. He became an ardent supporter of Copernicus, being especially impressed by how the moving Earth explains the retrograde motions of planets.

At the age of 24, Kepler was offered the post of professor of mathematics at the Protestant University of Graz, which had been founded just a few years earlier. After some hesitation, he agreed even though his studies in theology were not yet finished.

Fig. 6.3 Johannes Kepler (1571–1630) in a portrait from year 1610

The theologians in Tübingen may have felt that Kepler was too critical a thinker to preach in the pulpit. In any case, for him this job gave some economic support and time to study cosmology (Fig. 6.3).

The young man's university lectures were not popular. In his first year of teaching, he had several students in his class, but the next year no one. Besides teaching he had the duty of preparing an almanac that included astronomical information and astrological predictions. In his first almanac he predicted an exceptionally cold winter and a Turkish incursion into Austria. These predictions were fulfilled, which brought him considerable fame.

Kepler was most interested in studying the structure of the universe which at that time was thought to be the Solar System surrounded by the sphere of the fixed stars. Guided by Pythagorean tradition he had the feeling that there must be some mathematical law for the particular sequence of distances of the planets from the Sun. Was it a key to cosmic architecture that the number of planets then known, six, was one more than the number of regular solids that Plato knew? At the end of his first year of teaching, Kepler got the exciting idea that the spheres upon which the planets move must be such that they can be drawn inside or outside of the regular solids. That is why there are six planets! He embarked on writing his first book *The Mystery of the Universe* that presented a new model according to which the Great Architect had constructed the universe with the aid of five perfect solids (Fig. 6.4).

Of the five regular solids, the cube is made of six equal squares, while three solids are made of equilateral triangles: the tetrahedron (4 triangles), the octahedron

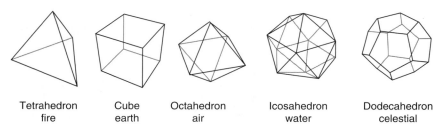

| Tetrahedron | Cube | Octahedron | Icosahedron | Dodecahedron |
| fire | earth | air | water | celestial |

Fig. 6.4 It was proved in Plato's Academy that there are at most five regular solids. For Plato they presented fire, earth, air, water, and celestial matter. Kepler saw in these forms a possible foundation for the architecture of the universe (at the time the Solar System bounded by the sphere of the fixed stars)

(8), and the icosahedron (20). The dodecahedron is formed by 12 pentagons. When one puts two spheres tightly inside and outside of a cube, the ratio of their radii is 0.577. The octahedron results in the same ratio. The dodecahedron and the icosahedron have the spheres with the ratio 0.795 and the tetrahedron produces the ratio of 0.333. These figures have some resemblance to the ratios of the solar distances of neighbouring planets. Even though the agreement was far from perfect, Kepler believed that he was on the right track. Later it became clear that the regular solids hardly have anything to do with the structure of the Solar System. Also, the number of planets has increased. Nevertheless, this first attempt of Kepler to geometrize the cosmos was very important for his career.

The Paths of Brahe and Kepler Intersect

In 1588, Tycho Brahe lost his benefactor when the King Frederic II died. During the following years his relations with the kingdom grew worse. After the successor to the throne Christian was crowned king in 1596, the head of Hveen was deprived of his yearly payment. Brahe could not stay on his island any longer. He left Denmark for good, living first in Hamburg and then in Prague for the remaining few years of his life. He died in 1601 as a sad result of the aftermath of a dinner with heavy eating and drinking. When lying on his death-bed, he repeatedly asked whether his life had been of any use. As a living answer to his desperate inquiry there was a young man beside his bed, Johannes Kepler.

Tycho Brahe had received *The Mystery of the Universe* as a gift from Kepler in 1597. He realized that the writer must be a very talented young man. When Rudolph II, the emperor of Germany, gave Brahe the position of Imperial Mathematician in Prague, Brahe decided to invite Kepler for a visit in 1600. They first met in February in the castle of Benatek close to Prague, just a week or so before Giordano Bruno was burnt at the stake in Rome. Kepler stayed with Brahe till the summer. Then he went back to Graz to find out that he was no longer welcome at the University. He returned to Prague to become the assistant of Brahe. This started an important phase in Kepler's life. In 1602, he became the Imperial Mathematician after Brahe, with half the salary of his predecessor. From a painstaking analysis of Brahe's accurate observations of the planet Mars, Kepler discovered the true mathematical laws of how the planets move around the Sun. This in a sense completed the task Plato had set two millennia earlier.

The New Laws of Cosmic Order

It is a long story how Kepler arrived at his new, revolutionary view on the motions of planets. When visiting Tycho Brahe for the first time he was very excited by the possibility to obtain from Tycho more precise values of the minimum and maximum solar distances of planets in their orbits. He needed these to continue his attempt to make the planetary orbits match the regular solids. After some hesitation Tycho allowed Kepler to collect together all his observations of Mars.

Kepler first tried to understand the motion of Mars following the old principle of circular motion. After years of struggle with circles and epicycles, he finally decided they could not explain the observations of Mars. In fact, it all depended on a small deviation of 8 stubborn arc minutes that Kepler could not explain with the perfect circles. Kepler understood clearly that it is important to test theoretical predictions using accurate observations. Tycho's accuracy of $2'$ was clearly better than the deviation. In Kepler's words, "These 8 minutes of arc that I could not omit, led to the complete reform of astronomy."

He then broke with millennia of tradition and used an elliptical orbit to explain the observations of Mars. Ellipses were known from the time of Apollonius (mentioned in Chap. 3) who studied these curves together with other *conic sections* (the hyperbola and the parabola). Curiously, he was also the inventor of the epicyclic theory of planetary motions. It did not occur to him, or anyone else before Kepler, that planets could move along ellipses. The ellipse is an elongated closed orbit with the circle as a special nonelongated case.

Kepler's life's work is encapsulated in three laws. The first two appeared in his book *Astronomia Nova* (1609) and the third one in his *Harmonices Mundi* (1619). The first, explained above is

I. The planets move round the Sun in a plane along elliptic orbits with the Sun occupying one focus of the ellipse.

Kepler actually found his second law before the first one. He discovered that the Earth moves slower in its orbit when far from the Sun and faster when closer. The speed on the trajectory does not remain constant during the elliptical revolution around the Sun, but the "area velocity" does as follows:

II. The radius vector from the Sun to the planet sweeps out equal areas in equal times.

Table 6.1 The orbital values as calculated by Kepler to check his Third Law

	Mercury	Venus	Earth	Mars	Jupiter	Saturn
P^2	0.058	0.378	1	3.53	140.7	867.7
a^3	0.058	0.379	1	3.53	140.6	860.1

To understand the second law, visualize the areal velocity as a triangle with its apex at the Sun and its base as the short arc along which the planet travels in its orbit during a small unit of time. The triangle will be skinny when the planet is far from the Sun and fat when close to the Sun with the two areas the same (Fig. 6.5).

Kepler's third law compares the orbital sizes and periods of any two planets. The comparison is usually to the Earth, so for the other planet, the time unit is year and the distance unit is Earth–Sun distance (AU). The orbit size, a, is half the long dimension of the ellipse. The sizes of the orbits and the times required for a full circuit, P, are related in a simple manner:

III. The squares of the orbital periods of the planets are proportional to the cubes of the semi-major axes of their orbits.

It is interesting to see how accurately Kepler could check his third law using actual values which appear in *Harmonices Mundi*. In Table 6.1, the upper row shows for each planet the square of its orbital period P, or $P^2 = P \times P$ (with 1 year as the unit). The lower row displays similarly the cube of "a" which is also the mean distance from the Sun, or $a^3 = a \times a \times a$ (with the Earth's mean distance $= 1$ AU). Considering observational errors, the upper and lower rows are virtually identical.

Kepler worked in Prague up to 1612. This was the most fruitful time during his career despite continuous economic problems and personal tragedies (his small

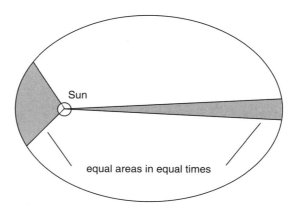

Fig. 6.5 Kepler's First Law: The planets revolve around the Sun on elliptic orbits, where the Sun is situated at one of the two focal points. Kepler's Second Law: The planet moves at a varying speed so that the radius vector sweeps equal areas at equal time intervals (hence, swifter closer to the Sun). Furthermore, Kepler's Third Law states that the period of revolution around the Sun depends accurately on the size of the orbit so that the square of the period is proportional to the cube of the mean distance from the Sun.

son and his wife died). In addition to *Astronomia Nova* he published three books on optics (about one-fourth of his published works are concerned with light and optics).

In 1612, his patron, the Emperor Rudolph II died and Kepler moved to Linz to work as a teacher in a job similar to the one he had previously had in Graz. There he married anew and his young wife gave birth to seven children, two of whom died young. In 1626, Kepler had to leave Linz for religious reasons. Kepler is an example of someone who can solve great riddles of science while beset by many worries. In his late years, Kepler wrote how he has been suffering from a very strange fate, encountering difficulties one after another without being in any way guilty himself.

Kepler and his large family settled at Ulm, where he published his last great work, the *Rudolfine Tables*, which contained astronomical tables based on Brahe's observations, the new laws of planetary motion, and prescriptions for calculating the directions of celestial bodies in the sky at any time.

Kepler's last days were humiliating. He had for years attempted to obtain his unpaid salary from the Emperor Ferdinand II, but without result. He even had worked for 2 years as an astrologer to general Wallenstein, a hero of the Thirty Years' War, thus hoping to get his 11,817 guldens. As a last resort, he mounted a horse and rode to Regensburg where the Diet of the Holy Roman Empire was sitting. It was November 1630 and the long ride in cold weather through Germany ruined by the war was too much for the fragile health of 58-year-old Kepler. He arrived in a poor condition and sold his thin horse for 2 guldens, but then a high fever confined him to bed and within a few days he died. Kepler was buried outside the town in a cemetery for Lutherans. His grave was soon destroyed, together with the cemetery, during the long war that followed.

Orbits and Forces

It was intriguing that the planets could move on closed orbits. How can they find their way back to the same point in space and then repeat the identical elongated trajectory? To explain physically this motion, Kepler had a vision of two forces: one drives the planet along a circle and another, a kind of "magnetism," makes it deviate from the circle. These two forces are somehow so precisely tuned that the result is a perfect ellipse. As we shall see later, 50 years after Kepler's death Newton showed that one force, the universal gravitation, suffices to explain the closed elliptical orbits of planets.

During Kepler's lifetime, his studies did not receive the attention they deserved. He himself never came to know the true significance of his own work. For Kepler, the universe was still finite with the stars sitting on the last sphere. Inside this sphere was our world, subject to the mathematical laws of Nature. This was the message of Kepler, whose one foot was in the past casting horoscopes, while the other one stretched toward modern astrophysics. He no longer believed in material planetary spheres. Planets moved in empty space, supported by forces, and following what we with admiration refer to as Kepler's Laws. In his study of these regularities and in his search for the harmony of the universe Kepler was a predecessor of modern cosmologists and theoretical physicists. When Newton developed his mechanics and theory of gravitation, he said that to accomplish this, he "stood on the shoulders of giants." One was Kepler and the other was Galileo, whom we will discuss next.

Chapter 7
Galileo Galilei and His Successors

Galileo Galilei was born in Pisa into a family of minor nobility. His father, Vincenzio, gave lessons in music (and studied its mathematical theory) and helped his wife's family in their small business. He wished for his son a better than their modest (if not poor) standard of living. However, rather than choosing a career in business as recommended by his father, the 17-year-old Galileo entered the University of Pisa, intending to study medicine. After 4 years he left the university without a degree, but with a basic knowledge of mathematics and Aristotelian physics. Returning home to his parents who now lived in Florence, Galileo began to write mathematical studies and to give private lessons as well as public lectures. He helped his father in musical experiments with strings of different length, thickness, and tension. Interestingly, the founder of experimental physics was occupied with experiments similar to the first known quantitative experiments by the early Pythagoreans, who found that the integer ratios of the strings of the lyre give rise to pleasant harmonies.

Galileo studied texts by Archimedes that were translated into Latin in the sixteenth century. This inspired him to investigate static mechanics topics like the center of gravity of bodies. Thanks to a short paper he wrote on these subjects, he got a temporary position as a professor of mathematics at the University of Pisa. After 3 years, at the age of 28, he went to Padua to teach mathematics and astronomy. He lived there for 18 years carrying out the main part of his famous studies of bodies in motion (Fig. 7.1).

Observation and Experiment

Galileo's writings show a modern approach to nature. In antiquity, observation was appreciated, but the idea of experiments for a particular purpose was unfamiliar. Recall from Chap. 2 that Aristotle insisted that we understand a phenomenon only if we know its cause of a special kind, the final cause. When we know the "motivation," we can tell why something happens. For instance, a stone falls because its goal is to get closer to its natural place, the center of the universe. In Aristotle's approach,

Fig. 7.1 Galileo Galilei (1564–1642), the founder of experimental physics who also started observation of the celestial bodies using telescopes

observing such spontaneous, instead of contrived forced processes, was essential for understanding.

In contrast, modern science considers that if one knows the *initial* state of a system and all the forces present, one can understand the next state without assuming any natural end. This causal relation makes experimentation an efficient way to study nature. By changing the initial state of the experiment, one explores the laws that link cause and effect. An important task for experiments is to test theories that intend to explain phenomena. Experiment and theory also go hand in hand in the sense that a good theory can have practical value since it predicts the course of natural events in different situations. An application, like television, validates the underlying theory every time the "on" button is pushed.

Galileo, the experimenter's, main results in science of dynamics may be stated as a few laws.

I. A free horizontal movement happens at constant speed and without change of direction.

In everyday conditions on Earth, there is always some friction finally stopping any body, e.g., a ball rolling on a plane. However, aided by his experiments and intuition, Galileo could conclude that the ball would never stop if the friction could be totally eliminated, that is, the motion is "free".

II. A freely falling body experiences a constant acceleration.

Acceleration is the change in an object's velocity in a unit interval of time. For a uniformly accelerating object initially at rest, after an interval of time, the velocity v will equal the acceleration a multiplied by the time t ($v = at$). For a falling object at the Earth's surface, the acceleration is $9.8 \, \text{m/s}^2$. After 1 s, the velocity will be 9.8 m/s, after 2 s, 19.6 m/s and so on for progressively larger times. In studies at the Merton College (Oxford) in the fourteenth century, it was already proposed that the distance s, a uniformly accelerating body travels during a time interval, is equal to one half of the product of the acceleration and the time squared ($s = {}^1/_2 \, at^2$). Galileo showed that this formula is valid by studying the gentle acceleration of balls rolling down inclined planes. Extrapolating to the case of a vertical plane, he concluded that freely falling bodies have a constant (but greater) acceleration obeying the same law. Recall the $9.8 \, \text{m/s}^2$ acceleration. After 1 s, the object has fallen 4.4 m. After 2 s, the total distance is 17.6 m, four times that in the first second, and so on.

III. All bodies fall equally fast.

This result, commonly ascribed to Galileo's dropping objects from the leaning tower of Pisa, was actually arrived at earlier by the Dutch–Belgian mathematician Simon Stevinus. He reported in 1586 that bodies with different masses fall with the same acceleration. Galileo was of the same opinion and may have attempted similar experiments with two dense objects of different masses. Indeed, if one could eliminate air friction, a hammer and a feather dropped simultaneously would both hit the ground at the same moment. Apollo astronauts on the airless surface of our Moon found this to be the case!

IV. Galilean principle of relativity: The trajectory and speed of motion of a body depend on the reference frame relative to which it is observed.

One argument against the revolving Earth was that a body released from the top of a tower would not appear to fall to the point directly beneath because the surface of the rotating Earth would move aside during the fall. The validity of the argument may be studied in an analogous situation, by dropping a stone from the top of the mast of a moving ship. Is the stone's trajectory deflected toward the back of the vessel? The French philosopher Pierre Gassendi (1592–1655) made such tests and found that the stone always hit the deck just beside the foot of the mast and there was no deflection! The object shares the uniform motion of the ship, even while falling. The conclusion made by Galileo was that an observer participating in a uniform motion couldn't detect this motion by free-fall experiments. Interestingly, for an observer standing on the shore the falling stone appears to make a parabolic curved trajectory. Which trajectory is the "true" one, the vertical straight line or the curved parabola? Galileo's answer is that both are correct, as the trajectory depends on the reference frame that may be fixed to the shore or to the uniformly moving vessel depending on the location of the observer.

At the time of Galileo, the significance of these laws of motion was twofold. First, they were clearly contrary to the old conceptions based on Aristotelian physics. Secondly, they helped to understand why the Earth could move without any dramatic consequences other than the regular daily rising and setting of the Sun and other heavenly bodies. The atmosphere can move together with the Earth without high winds or escape into space.

The First Steps into Deep Space

It was remarkable that Galileo showed how experiments can be used for testing philosophical claims about matter and motion and how they may unveil new laws of nature on Earth. But this was not all. He also looked at the heavens with a new instrument whose capabilities beyond the eye alone enabled him to discover new phenomena in the universe.

Galileo had heard that in the Netherlands a lens grinder had built a device that made distant objects look close. In the summer of 1609, he succeeded in building such an instrument, now called the telescope, himself. At first he had in mind that the device could have use for naval purposes and thus aid him in obtaining a better paid position. He introduced the instrument to the rulers of Venice, demonstrating to their surprise how one could easily see distant ships in the Gulf of Venice to identify prior to their arrival whether they were friend or foe. Galileo presented the telescope to the supreme ruler of Venice, the Doge. So impressed was the Doge that Galileo's salary was immediately doubled and his tenure as a professor made "for life." Two telescopes made by Galileo are preserved in the Istituto e Museo di Storia della Scienza, in Florence. They have main objective lenses with diameters of 16 and 26 mm. By modern standards, Galileo's magnifying telescope was of course modest. However, it increased radically the ability of the human eye to detect small and faint distant objects, resulting in unexpected discoveries when Galileo aimed his instrument at the sky. His book *Sidereus Nuncius* or *The Starry Messenger* published in 1610 reported his new cosmic discoveries:

- The Moon, thought to be a smooth sphere, actually has a rugged surface, with mountains, holes and valleys, along with wide flatter regions.
- Many new stars invisible to the naked eye appear in the sky inspected with the telescope; especially, the Milky Way is a huge cloud of faint stars.
- There are four moons revolving around Jupiter.

Later in 1610 he made further discoveries

- Venus has phases like the Moon.
- There are spots on the Sun, whose motion over its disk reflects its revolution once in about one month (other astronomers also claimed credit and may be independent discoverers).

Such things were new and radical, and could not be at once accepted by many who had only Galileo's word to believe. And looking through the small telescope did

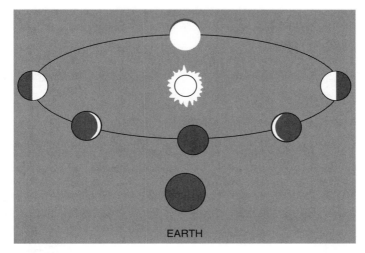

EARTH

Fig. 7.2 The phases of Venus showed clearly that it revolves genuinely around the Sun and does not just wander to-and-fro between the Earth and the Sun as in the old world system (credit: NASA)

not necessarily help – with its blurred, shaky picture the early telescope was not exactly user-friendly. A small modern pair of binoculars gives a better view. You may want to try finding Jupiter in the sky and spotting one of its four big moons with binoculars. You will probably eventually be able to see one or more of the moons but you will also appreciate the need for a sturdy, steady modern mounting like a photographic tripod!

The discoveries by Galileo were sensational news, and his book was a best seller. Its 550 copies were quickly sold out. His fame was not limited to Europe. For example, in 4 years a book was published in China by a Jesuit priest describing the new celestial phenomena discovered in far-away exotic Italy.

Galileo's findings with his telescope supported Copernicus' ideas. Their opponents asserted that if the Earth orbited the Sun, then Earth's Moon would be left behind. Now it was seen that Jupiter's satellites revolved around Jupiter yet were not left behind as Jupiter moved in its orbit. Venus had full and crescent phases like the Moon, which is possible only if it goes from beyond the Sun to between the Earth and the Sun in an orbit around the Sun (Fig. 7.2). Finally, craters on the Moon and spots on the Sun indicated that these bodies were made of material like the "imperfect" Earth (Fig. 7.3).

Kepler and Galileo had quite different personalities, reflected in their approaches to science. Kepler was a quiet and keen theoretician, with fragile health in a slight body. Galileo was physically big and healthy, and was hot-tempered and sharp in mind and words. This tended to drive him into conflicts with other scholars. Though Galileo did not accept Kepler's theories on planetary motion (he viewed circular motion as natural), their work complemented each other during the period that paved the way for new earthly and celestial physics.

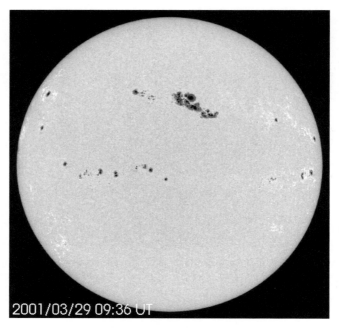

Fig. 7.3 Galileo was among the first to note spots on the Sun. This modern picture shows a huge sunspot group seen in 2001. Sunspots are temporary events which disappear after a time and others appear. Now we know that they are caused by strong magnetic fields emerging from the interior of the Sun and seem to be dark because they are somewhat cooler than other parts of the surface (credit: SOHO (ESA & NASA))

Fighting on Two Fronts

In 1616, the doctrine of the Earth's motion was declared absurd and heretical by the Catholic Church. In fact, this was a result of a complex chain of events, with jealous lay professors, disputes between the fiery-natured Galileo and university officials, and a plan to draw Galileo into controversy about the world system and the statements in the Bible. As a result, the book of Copernicus and another book were "suspended until they are corrected."[1]

One valid argument by both the religious and science community was that the Earth's motion had not yet been proved. The far-reaching theory had to fight on two intertwined fronts, in science and in society. In 1632–33, the famous trial of Galileo before the tribunal of the Inquisition in Rome took place. The specific reason for the trial was the *Dialogue Concerning the Two Chief World Systems*. Pope Urban VIII, who showed interest in cosmology, had encouraged Galileo, his old friend, to write

---

[1] A book by Foscarini was totally forbidden – the Carmelite Father had tried to show that the moving Earth is in accordance with the Bible. In 1620, "all other books teaching the same thing" were forbidden. It was not until the 1835 edition of the *Index* that Copernican ideas were no longer suppressed.

the new book. But he said that the Copernican system should be given only as a hypothesis (as allowed by the Decree of 1616), to which Galileo consented. However, when the book came out, it clearly tried to prove the Earth's motion. It did not help matters that a not-so-witty Earth-center supporter, Simplicio, could be viewed as a caricature of the Pope. The verdict forced Galileo to declare in public that after all the Earth did not move. Fortunately, the 70-year-old scholar was treated rather well during the process, was never put into a cell and was not tortured.

Galileo's trial, like those of Socrates and Bruno, has come to symbolize the struggle for freedom of thought. But it would be too simple to describe it as just a collision between science and religion. The scientific revolutionaries Copernicus, Kepler, and Galileo, as well as Newton, were believers in God, like their contemporaries in general in Europe, and they did not view the Bible as contradictory to science. The new ideas caused hostility among religious leaders who had adopted the world system of Ptolemy among their doctrines, later referred to as an "illegal marriage of science and religion."

Cartesian Physics

The trial was a part of the Copernican revolution, giving extra motivation for scientists to search for additional evidence for the new world system. However, in the short term it also must have prevented some the open discussion of this topic. One man who was alarmed by the news from Rome in 1633 was René Descartes (1596–1650), the French philosopher and mathematician, who was just finishing his work *Le Monde* (the World). The book contained his physical world system, including heliocentrism. He decided to put the manuscript aside and it was published only after his death.

However, Descartes did many other things that influenced philosophy, physics, and mathematics during his life. A starting point in "Cartesian physics" was the law of inertia. This had been discussed by Galileo, but only Descartes formulated it for an idealized particle residing in infinite space. If the particle has no contact with other particles, it either would keep its initial state of rest or would move with a constant speed along a straight line. Descartes' law of inertial motion for a free particle is quite similar to Newton's first law of motion, to be discussed later. But, in contrast to gravitational attraction across empty space, in Descartes' physics nothing else happens until a particle is deflected by a collision with another particle, i.e., changes in our world are caused by impacts. There is no mysterious action-at-distance and bodies are all the time in contact with other bodies. The space between stars is not empty, but filled by ethereal particles.

From such considerations, Descartes interpreted various phenomena, including the motions of planets: instead of gravity their motions are forced by an ethereal particle vortex around the Sun. Similar whirlpools exist around other stars. The Solar vortex may have swallowed dead stars that happened to pass by and thus the planets, including the Earth, were born.

In describing planetary motions, Cartesian physics could offer just qualitative, vague explanations of phenomena. With his other laws of motion, including gravitational attraction across empty space, Newton would build quantitative mathematical physics that replaced Descartes'. Nevertheless, the investigative attitude of Descartes influenced scientific thinking during the period when the Copernican revolution was still in process. Descartes is often called the father of modern mathematics. He combined geometry and algebra when he invented analytic geometry where the positions of points on a mathematical plane are given by two coordinates, x and y. A story tells that the roots of this idea go to his childhood when he watched a fly crawling on the ceiling above his bed. How to describe the path of the fly? This could be done by labeling each point of the ceiling by an (x, y) pair of numbers! An example is the rectangular coordinate system. Then the distance between any two points is obtained simply from the coordinate differences: $(\text{distance})^2 = (x\text{-difference})^2 + (y\text{-difference})^2$.

Introducing Accurate Time

Galileo brought time into physics in a modern sense. In his experiments on balls rolling down an inclined plane he used the pulse of his own heart as a clock. He also measured time by weighing how much water came out from a vessel with a hole in it, but he realized the possible value of the pendulum in this respect. It is said that when he was 20 years old, he attended a mass in a cathedral, and his attention was drawn to heavy chandeliers hanging on long chains from the ceiling and swinging majestically. The chandeliers had equally long chains, but were of different weight. Interestingly, they were swinging at the same rate. This led to experiments showing that indeed the period of swinging does not depend on the weight of the bob of a pendulum, but on the length of the cord. Galileo had the idea that one could construct a clock-work utilizing the regular swinging of the pendulum, if only one could keep the swinging going and count mechanically the number of swings. If the cord is shorter, the period becomes shorter, too, and one could measure short time intervals with accuracy.

This idea of a pendulum clock was brought to reality by Christiaan Huygens (1629–1695), a physicist from the Netherlands. His pendulum clocks solved the problem of keeping the swing going, measuring time with an accuracy of about ten seconds a day, compared to a quarter of hour a day previously achieved by previous mechanical clocks.

Also related to whether the Earth moves and Newton's later work on gravity, Huygens, in 1659, calculated the acceleration toward the center required for an object to move along a circular path. He demonstrated how to calculate the central acceleration: just divide the circular speed squared by the radius of the circle. For example, at the equator of the Earth the speed is 464 m/s and the radius is 6.380×10^6 m. Hence the required acceleration inward to make air stay on the Earth's surface is $(464 \times 464)/6{,}380{,}000 = 0.0337 \, \text{m/s}^2$. On the other hand, the

Earth's gravity gives to masses an inward acceleration of $9.8\,\mathrm{m/s^2}$, much larger than the required value. Earlier there was the concern that the rotating Earth might cause winds and even result in the air flying into space. The above calculation shows that the acceleration of gravity is much greater than the acceleration needed to keep air on the surface of the rotating Earth. There is thus no risk of the Earth's atmosphere flying into space.

The Developing Telescope

Galileo's first astronomical observations demonstrated how even a small telescope can exceed the capabilities of the human eye in many respects. The telescope *collects much more light* than the eye. This makes it possible to see much fainter objects than by the naked eye. For example, Galileo saw in the direction of the Pleiades 36 stars instead of the usual 6. Photographs by modern telescopes show hundreds of stars in this stellar group. The big lens also makes *resolution* much better. This means that while two close-by stars are seen as one dot of light by the naked eye, the telescope shows them as separate. The ability to collect more light than the eye and the improved resolution allow one to see much more structure and fainter objects in the starry sky. The improved resolution also makes the measurements of stellar positions (their coordinates) more accurate. This proved crucial for the determination of stellar distances as we will discuss in Chap. 8.[2]

The first telescopes suffered from poor image quality. Simple lenses are hampered by a color error (chromatic aberration), which means that rays of light of different colors do not focus onto the same point and hence the image of a star is an indistinct spot surrounded by colored circles. The lens acts a bit like a prism. This problem was greatly improved in the eighteenth century with the invention of achromatic lenses. Before that a remedy was to make very long telescopes. When the ratio between the diameter of the objective lens and the focal length is small, the rays of light are only slightly refracted, the color error is smaller and the image sharper. Figure 7.4 shows such long telescopes in Paris Observatory.

Christiaan Huygens also built telescopes, the biggest of which had a length of 123 ft. or 37 m. It was not possible to make such gigantic solid tubes and one had to put the objective lens on the top of a pole or on the edge of a roof and to control its movements with a long rope, while standing oneself on the ground and keeping the ocular before one's eye. It must have been quite inconvenient to follow the revolving starry sky with such instruments, but nevertheless interesting observations were made. For example, Huygens found that the curious appendages of Saturn, which Galileo had noticed, were actually a thin flat disc around the planet in the equatorial plane.

---

[2] Kepler improved Galileo's telescope with a design still used today. In the "Keplerian" telescope, a large objective lens forms an image of a celestial object at a large distance from the objective. The detail and brightness of this image are then examined by a magnifying convex eyepiece lens.

Fig. 7.4 "Aerial telescopes" of Paris Observatory in the seventeenth century. Even though inconvenient to use, such instruments led to new astronomical discoveries (photo credit: Georges Paturel)

Another famous observer during the era of the long lens telescopes was the Polish Johann Hevelius (1611–1687) who had his own observatory in Danzig, the first one in the world complete with a telescope. His wife Elisabeth made observations too. Hevelius' record-sized instrument was 150 ft. or 45 m long. Its complicated system of ropes and long rods reminded one of the rigging of a sailing boat and certainly required seaman's skills to handle! With his telescopes Hevelius studied the surface of the Moon and drew fine maps of it. Our habit of speaking about the "seas" on the Moon goes back to Hevelius. We now know these to be depressions filled with solidified lava.

The development in the eighteenth century of achromatic lens telescopes in which color fringes are greatly reduced ended the era of the long lens telescopes. Large diameter objective lens telescopes up to about a meter in diameter were built through the 1800s but another kind of telescope was developed that gradually came to dominate the research field today. In 1671, Isaac Newton built the first *reflecting* telescope where a concave curved mirror gathers the light, instead of the lens as in the refractors. His experiments with glass prisms and refracted colors had led him to the conclusion that the color error in refracting telescopes is here to stay. And this led him to consider an alternative way to focus rays of light into one point by reflection which is the same for light of all colors. The image formed at the focus of the mirror does not show color fringes. The concave mirror surface must be a parabola so that all the rays, close to the center of the mirror as well as near its edge, will converge into the common focus. Newton's telescope, built with his own hands,

Fig. 7.5 The 3.5-m mirror made by the Finnish optical firm Opteon for the European Herschel Space Telescope, together with the team of specialists. The mirror surface had to be polished to make it so extremely smooth that its "bumps" are smaller than a few thousandths of a millimeter. This is the largest space telescope built up to now. From *left to right* A. Sillanpää, T. Lappalainen, D. Pierrot (Astrium), T. Korhonen (the director of Opteon), M. Pasanen, P. Keinänen (Credit: Opteon)

has survived. Its mirror, made from metal, had a diameter of 3.5 cm. Newton used a small flat secondary mirror to direct the light to side through a hole in the tube of the telescope where the eyepiece then magnified it.

Large modern reflecting telescopes often have a hole in the center of the main mirror, through which the light reflected from the secondary mirror goes into a detector. The detector is nowadays, instead of the eye or a photographic plate, a highly light-sensitive CCD camera or a spectrograph. This so-called Cassegrain type reflector was invented by a Frenchman, G. Cassegrain (of whom little is known) shortly after Newton's reflector.[3]

An important plus for the reflecting telescope is that its main mirror can be made much larger than the glass lens of the refractor, allowing a large light gathering power and observations of very faint and distant objects. The mirror can be supported over its entire back while the objective lens can only be supported at the edges. Once mirror silvering and later aluminizing was developed, glass could be

---

[3] In fact, Cassegrain's telescope was an improvement over one suggested by James Gregory before Newton. Gregory did not actually build his version. In the Cassegrain telescope, the secondary mirror is convex which results in a short telescope.

used rather than the metal Newton used. The glass does not even have to be transparent. Overall a color-free and larger reflecting telescope can be built for the same price as a smaller lens telescope.

Even though the reflecting telescope started to dominate astronomy in the nineteenth century there were still many important tasks left for the lens telescope. It was better for accurate measurements of the positions of stars, once the problem of the chromatic aberration was reduced. This finally made possible the dream of measuring the distances to stars.

Today, telescopes are still more sophisticated. Besides visual light, they operate at x-ray, ultraviolet, radio, and infrared wavelengths invisible to the eye. Some orbit in space, thus leaving behind the atmosphere which blurs optical images and absorbs radiation at most wavelengths (excepting visual light and radio waves). Figure 7.5 shows a big mirror made for a space telescope. For radio telescopes, one has a concave reflecting dish rather than a mirror as part of the telescope with a radio receiver at the focus. The long wavelengths of radio waves make their resolution much worse than that for the same size visual telescope, so the radio dishes are typically much larger, perhaps 100 m in diameter or more, much larger than the 10 m size of the largest visual telescopes today. Radio astronomers have learned to combine signals from separate dishes simulating a single dish comparable to the size of the Earth. These are called interferometers. With modern electronics even optical astronomers are doing this with several telescopes at the same observatory.

Finally, some modern telescopes are hardly recognizable as such. Devices have been constructed which have detected subatomic neutrino emissions from the Sun and a supernova. Gravitational wave detectors have been built to detect field variations from orbiting black holes or their formation in supernovas.

Indeed, this explorative spirit is strong in astronomy – one wants to look deeper and deeper in space, to see what nobody else has seen before. The discovery and further study of all those unexpected celestial bodies and cosmic phenomena require larger and larger telescopes.

Chapter 8
How Far Away Are the Stars?

According to Ptolemy, the distance to the stellar sphere was about 20,000 Earth radii. However, for Copernicus the distance was simply "immense," because the stars did not show any swings when the Earth makes its journey around the Sun. The absence of "annual parallax" was already noted by Ptolemy, who used it as an argument for the immovable Earth. For Aristarchus as well as for Copernicus, the absence was an indication of the immensity of the universe.

The Copernican Revolution did more than just remove the Earth from the center of the universe and put it in motion. It shattered into dust the old crystal sphere that had carried the stars since antiquity. Copernicus and Kepler still believed in this outermost sphere, but in fact it became obsolete when left without its original function. This new world order is clearly expressed by the ardent supporter of Copernicus, Bruno: "As soon as we realize that the apparent celestial rotation is caused by the real daily motion of the Earth ... then there is no reason to make us think that the stars are at equal distance from us." Even earlier, as noted in Chap. 4, Digges had detached the stars from their sphere and dispersed them in space: "This orbe of stares fixed infinitely up extendeth hit self in altitude sphericallye ... with perpetuall shininge glorious lightes innumerable, farr exellinge our sonne both in quantitye and qualitye."

Galileo and the Annual Parallax

The detection of the small annual parallax was important to prove the Copernican system. It also offered the possibility of measuring the distances to stars. A star's parallax means the angle in which the radius of the Earth's orbit would cover or subtend at the distance of the star. It is also one half of the total variation of the star's direction during 1 year. If the parallax angle is 1 arcsecond, it is said that the star's distance is 1 parsec. This definition is cleverly hidden in the name of the unit (**par**allax = 1 arc**sec**). One parsec equals 206,265 radii of the Earth's orbit. Box 8.1 explains the origin of this special number. It is also good to remember that

1 parsec is 3.26 light years. One light year is the distance covered by light in 1 year $(9.46 \times 10^{12}\,\text{km})$.

Box 8.1 The length of 1 parsec

At what distance r does the radius of the Earth's orbit R subtend an angle of 1 arcsec? Considering R as the length of the small segment of the circle with radius r, then $R/(2\pi r) = 1\,\text{arcsec}/360°$.

As the whole circle contains $360 \times 60 \times 60$ arcsec then $R/r = 2\pi/(360 \times 60 \times 60) = 1/206{,}265$. Hence 1 parsec is 206,265 times the Earth–Sun distance, or $3.8057 \times 10^{13}\,\text{km}$. Its relation to the popular light year unit is 1 parsec = 3.26 light years or 1 light year = 0.307 pc.

In his *Dialogue*, Galileo devotes much attention to the question of how to detect and prove the motion of the Earth. Similarly just as on board a ship we do not feel its motion, we do not feel the Earth's steady motion, unless the Earth encounters some obstacle stopping it and throwing us toward the stars, as Salvia, Galileo's alter ego paints in a nightmare scene. But we can look at the stars, and get some hints of the Earth's motion. Such hints had not yet been detected in his time. Salvatio first inspects the case where a star lies exactly on the ecliptic. As looked at from the moving Earth, such a star should make over 1 year a saw-like to-and-fro motion along the ecliptic just like the planets make their loops relative to the fixed stars (Fig. 8.1). But Salvatio explains that such motions are difficult to observe for the stars, because one would need points of comparison that are farther than the stars! And there would be none if the stars were on a celestial sphere.

It might be simpler to consider a star lying away from the plane of the ecliptic. Then the star would change its angular distance from the ecliptic during the year, being either "lower" or "upper." Now the unmovable plane of ecliptic would be the reference frame relative to which to measure the angle.

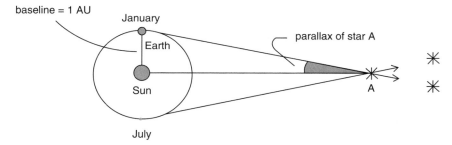

Fig. 8.1 The direction of a star changes when the Earth goes around the Sun. This gives the possibility to measure by triangulation the distance of the star. However, this fundamental method is suitable only for relatively nearby stars

Galileo also mentioned the possibility of observing the *relative parallax* of two stars at different distances, which would be another proof of the Earth's yearly motion. He expresses his opinion that the stars are not at the same distance from us, but some could be "two or three" times farther than others. If now two such stars lie in the sky close to each other, then the more nearby star would move relative to the more distant one and the astronomer might be able to measure those small changes. Yes, such a measurement was made, but two centuries later! In the meanwhile, people tried to detect the "up-and-downs" in Galileo's ecliptic method. The attempts failed (it is hard to measure accurately the angles from the ecliptic to see the shifts) but another very important phenomenon that also changes a star's direction was found in the process. The unexpected optical phenomenon was the aberration of light.

Before the hunt for the parallax could properly start, a real animal had to intervene. Namely, around 1640, a spider had built its web inside the telescope of a British amateur William Gascoigne. The telescope was Kepler's type, so that the location of the image formed by the objective was inside the telescope in front of the eyepiece. Those parts of the web that happened to hang just at the focal plane were seen sharply when the telescope's owner (not its occupant!) looked through it next time. This gave Gascoigne the bright idea to construct a measuring device for his telescope; he arranged two parallel thin fibers on the focal plane so that he could change their mutual distance by turning a screw. This screw micrometer developed over the years into an accurate measurer of small angles. It proved to be incredibly useful for measuring the tiny motions of stars.

Bradley Discovers the Aberration of Light

The star called Eltanin in the constellation of Draco between the Small Dipper and Lyra is rather faint and ordinary. However, its position in the sky happens to be such that when looked at from the latitude of London, its daily trajectory takes it close to the zenith. This makes it a quite suitable object to observe by a zenith telescope that measures the angle of the star from the zenith when it crosses the north–south meridian. Already Robert Hooke (1635–1703), the well-known English physicist, tried to detect Eltanin's annual parallax and reported the results in 1674 in a booklet *An Attempt to Prove the Motion of the Earth from Observations* (the title tells that Hooke had in mind a crucial cosmological test – to prove that the Earth really moves as it does in the Copernican world model). He believed to have observed changes in Eltanin's position, up to 24 arcsec, but the observations were small in number and the accuracy of his instrument was poor.

Decades later an enthusiastic amateur Samuel Molyneux started observations of Eltanin with his zenith telescope that was longer and better than that used by Hooke. His friend, the Royal Astronomer James Bradley (1693–1762) also inspected the movements of Eltanin with him. To their surprise, the star's position did change from its average position, but not in the way expected from parallax. During 3 months it

Fig. 8.2 A simplified analogue of the aberration of light: a man hurrying in the rain has to turn his umbrella in the direction of the motion as if the raindrops were coming down at an angle. (Drawing by Georges Paturel)

went to the south by 20 arcsec from its average position and then in 6 months rose 40 arcsec from the southern point to its northernmost position after which it went down again and so on. The ±20 arcsec change was real, but it could not be due to the annual parallax, as the behavior was 3 months out of step from what the parallax would cause. The displacement was always in the direction of the orbital velocity of the Earth relative to the star.

After 3 years' further observations and a lot of thought Bradley realized the reason for the peculiar changes of Eltanin's direction in the sky. It is said that the idea came to him out of the blue when he was sailing on a ship on the river Thames in September 1728. He noticed that when the ship made a turn, the weathervane on the top of the mast also turned. Bradley reasoned that the direction of the weathervane gives the sense of the wind relative to the moving ship and not its true direction. This prompted him to ponder what happens to the apparent direction of light traversing space, when watched from a moving observing site, that is, the Earth. He concluded the following by assuming that the velocity of light is finite:

The apparent position of a light source, when looked by a moving eye, is generally not the same as when looked by a stationary eye. It remains the same only if the eye moves exactly along the line of sight to the object (toward it or away from it). But if there is some motion perpendicular to the line of sight, then the object is seen at a different position, slightly shifted in the direction of the motion.

The reason for this *aberration of light* may be seen every rainy day. When you sit in a stationary car, the raindrops fall vertically. However, when you are driving the drops seem to come down along an oblique trajectory from a direction shifted toward the direction of motion of the car. This is because you are still using the car as a reference frame but it is moving. When the observer moves much slower than light, as is the case for the Earth's motion around the Sun, one may well understand the aberration in a simple way and derive for it a mathematical formula (Fig. 8.2).

The angular shift of the images depends on the ratio of the velocity of the observer and the velocity of light (V/c). The shift depends also on the angle that direction of the object (say, a star) and the direction of the motion make. If the angle is zero, there is no shift. The shift is maximal, when the motion is perpendicular to the star (the angle is 90°). For example, the Earth's orbital speed is about 30 km/s. The ratio of this speed to that of light is 1/10,000, making it about 20 arcsec ($= (1/10,000) \times 360 \times 60 \times 60$). It is no coincidence that the observed swings of Eltanin from its average position were 20 arcsec – Molyneux and Bradley were simply seeing the aberration of light.[1]

Today we know that Eltanin is so far away that its annual parallax is only about 0.02 arcsec, much less than 20 arcsec. There was no way to detect such a tiny parallax effect with Molyneux's instrument and method amidst the large aberration.

Fifty Years Earlier: Rømer and the Speed of Light

The discovery of the aberration was remarkable in several ways. It was crucial information for astronomers measuring positions of stars and attempting to derive their distances. But it also caught two fat flies at one blow. The aberration showed that the Earth is really moving in space relative to the stars, i.e., going around the Sun (for this purpose this was as good or even better test than the much smaller annual parallax). The moving Earth was now an observed fact. Also, it confirmed that light has a finite, albeit high, velocity. Before Bradley's discovery, the speed of light was still a debated issue, even though in 1676 the Danish astronomer Ole Rømer (1644–1710), working at the time in Paris, had published a report that, in effect, contained the first measurement of the velocity of light.

He studied Jupiter's innermost moon Io as a "clock" that could be used at sea for determination of geographic longitude, as originally suggested by Galileo. But this clock was less precise than expected. Sometimes it was "slow" sometimes it was "fast" depending on whether Jupiter was on the other side of the Sun relative to the Earth or on the same side as the Earth. Rømer was convinced that this variation of 22 min was not due to imperfections in the cosmic clockwork, but caused by the finite velocity of light, being the time that it takes for the light to traverse the diameter of Earth's orbit. His report did not contain an explicit calculation of the speed. It rather tells how he detected the apparent variation in Io's motion and it conveyed his conviction that the reason is the finite velocity of light. If calculated, in modern units the result would be a speed of about 227,000 km/s, in comparison with the modern value:

$$c = 299,792.458 \, \text{km/s}$$

The difference was due to problems with timing Io's motion. In any case, the speed of light is huge compared with familiar motions on the Earth. To measure it, it was necessary to go to the "cosmic laboratory" where even light takes a noticeable time to cover large distances.

---

[1] "Aberration" derives from Latin verb *ab erro* (to turn aside, to deviate). It seems to have been first used to signify tiny apparent shifts of stars by Eustachio Manfredi in the same year 1629 when Bradley reported his discovery. The Italian astronomer did not yet know the reason for the shifts.

DEMONSTRATION TOVCHANT LE
mouvement de la lumiere trouvé par M. Rømer de
l'Academie Royale des Sciences.

IL y a long-temps que les Philofophes font en peine de decider par quelque experience, fi l'action de fa lumiere fe porte dans un inftant à quelque diftance que ce foit, ou fi elle demande du temps. Mr Rømer de l'Academie Royale des Sciences s'eft avifé d'un moyen tiré des obferva-tions du premier fatellite de Jupiter, par lequel il démontre que pour une diftance d'environ 3000 lieuës, telle qu'eft à peu prés la grandeur du dia-metre de la terre, la lumiere n'a pas befoin d'u-ne feconde de temps.

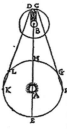

Soit A le Soleil, B Jupiter, C le premier Satellite qui entre dans l'ombre de Jupiter pour en fortir en D, & foit EFGHKL la Terre placée à diverfes di-ftances de Jupiter.

Or fuppofé que la terre eftant en L vers la feconde Quadra-ture de Jupiter, ait veu le pre-mier Satellite, lors de fon é-merfion ou fortie de l'ombre en D ; & qu'en fuite envi-ron 42. heures & demie a-

Terre, il s'enfuit que fi pour la valeur de chaque dia-metre de la Terre, il faloit une feconde de temps, la lumiere employeroit 3½ min. pour chacü des inter-valles GF, KL, ce qui cauferoit une differéce de prés d'un demy quart d'heure entre deux revolutions du premier Satellite, dont l'une auroit efté ob-fervée en FG, & l'autre en KL, au lieu qu'on n'y remarque aucune difference fenfible.

Il ne s'enfuit pas pourtant que la lumiere ne demande aucun temps : car apres avoir examiné la chofe de plus prés, il a trouvé que ce qui n'é-toit pas fenfible en deux revolutions, devenoit tres-confiderable à l'égard de plufieurs prifes en-femble, & que par exemple 40 revolutions ob-fervées du cofté F, eftoient fenfiblemént plus courtes, que 40. autres obfervées de l'autre cô-té en quelque endroit du Zodiaque que Jupiter fe foit rencontré; & ce à raifon de 22. pour tout l'intervalle H E, qui eft le double de celuy qu'il y a d'icy au foleil.

La neceffité de cette nouvelle Equation du re-tardement de la lumiere, eft établie par toutes les obfervations qui ont efté faites à l'Academie Royale, & à l'Obfervatoire depuis 8. ans, & nou-vellement elle a efté confirmée par l'Emerfion du premier Satellite obfervée à Paris le 9. Novembre dernier à 5. h. 35.' 45." du foir, 10. minutes plus tard

Fig. 8.3 Part of the communication to the French Academy where Rømer reported the detection of the finite speed of light

The conclusion was received with little enthusiasm, because it was generally thought that light rays travel instantaneously. Such people as Kepler and Descartes had shared this opinion, but Galileo had already considered an empirical way to test the idea, using two sharp-eyed and quick-fingered men with lanterns. A decade later Newton, in his *Principia* already declared that the speed of light is finite, "as measured by astronomers." In Paris, the things were not made easier by the fact that Rømer's boss, Giovanni Cassini, had earlier proposed a similar interpretation for Io's peculiar behavior, but had withdrawn it, likely as too speculative for this careful observer of planets. So it happened that during Rømer's lifetime the French Academy in Paris could not decide how fast light goes, finite or infinite (Fig. 8.3).

Bradley's discovery of the aberration of light resolved the question for good. From the speed of the Earth in its orbit and the observed changes of Eltanin's ap-parent position Bradley could calculate the speed of light. The result was roughly what Rømer had measured! These two quite different observations convinced the scientific community of the finite velocity of light. If the velocity were infinite, the aberration would be zero.

Instrumental Advances

Newton was pessimistic about the possibility to make a lens without the color er-ror. However, in the eighteenth century opticians succeeded in making such a lens, among them John Dollond, a scientist from London, who was the first to patent the *achromatic* lens around 1757. This was made of two parts, but later his son Peter arrived at the solution where the object lens was triple. The outermost ones were ordinary lenses, while between them there was squeezed a concave lens made of strongly refracting silicon glass. With such an arrangement the light rays of different colors were focused approximately onto the same point at the focal plane.

The early achromatic lenses were small, less than 10 cm in diameter. It was only in 1799 when a Swiss–French artisan and amateur optician Pierre Luis Guinand learned how to make large good-quality silicon glass disks and achromatic lenses out of these. His largest lenses were 35 cm in diameter. At first Guinand kept his methods secret. Then in 1805 he moved to Munich, where he began collaborating with Joseph von Fraunhofer. Now Guinand's art of lens-making could influence science by using Fraunhofer's instruments.

When 11 years old, Joseph von Fraunhofer (1787–1826) was orphaned and had to go to work. He got a job as an apprentice of a mirror-maker, who unfortunately died 3 years later at an accident in the workshop. Also Fraunhofer got hurt in the accident, but this did not end his career. He luckily could go to work for Joseph von Utzschneider (1763–1840) who had a firm making optical instruments. The unschooled but talented young man advanced to become a business associate of von Utzschneider and a versatile student of light and optics. With more than 50 workers, their firm became a world's leader in the field of precision instruments for geodesy, navigation, and astronomy.

Along with optics, other parts of telescopes developed. We already mentioned the filar micrometer (that useful spider) needed for accurate position measurements. Another tool needed by astronomers was the clock. As discussed in Chap. 7, Huygens built the first pendulum clock. It revolutionized time-keeping both in everyday life and in science, and had an immediate application in astronomy.

The starry sky revolves at a steady rate and to know where a star is to be seen one should know the time. Or vice versa, if one observes the star when it exactly crosses the south meridian, the accurate time of transit gives the star's longitude coordinate ("right ascension") in the sky. In fact, the time here is the *sidereal time* that differs from our usual solar time, because the starry sky rotates a bit quicker than the Sun. The reason is that in addition to its daily rotation the Earth goes around the Sun, which gives the sky one extra daily rotation every year and makes the sidereal time go faster than the solar time by 4 min every day ($1/365 \times 24$ h = 4 min). Using meridian instruments pointing at the south, together with precision clocks, astronomers later measured accurate values of the coordinates for thousands of stars, thus laying the groundwork for first successful determinations of stellar parallaxes.

Rebirth of Galileo's Method

The measurement of the aberration made it clear that the yearly parallax of stars must be a significantly smaller effect than aberration and stars are at much larger distances than had been thought. Astronomers were compelled to develop better methods of observation and figure out how to find promising stars, i.e., candidates for nearby stars whose parallaxes were large enough to be measurable.

It was William Herschel (discussed elsewhere in this book) who first tried to apply Galileo's relative parallax method to real stars. He made a list of hundreds of star pairs in the sky and selected for measurements such pairs where one star was much fainter than the other. If the fainter star is much more distant than the bright one, it might act as a comparison star relative to which the parallax shifts of the bright and nearby star could be measured. Note also that both stars in a pair are influenced by the same amount of aberration which thus automatically cancels off from the measurement.

When Herschel attempted to use Galileo's method to detect the parallax, he saw with his telescope that there are surprisingly many pairs of stars in the sky. He had assumed that all pairs are made of stars at different distances, which by chance happen to be almost at the same direction when looked from the Earth. But their great numbers made him suspect that many of these could be true, physical pairs. Later he became quite convinced by his observations of the star Castor in the constellation of Gemini. Castor has two component stars and Herschel could establish that they revolve slowly around each other. Yes, he had initially searched for the parallax, but discovered *binary stars* instead! This discovery of true binaries can be seen similarly important as was Galileo's discovery of Jupiter's moons: gravitation is a universal phenomenon as Newton had assumed.

The Race Toward Stellar Distances

During his short life, Joseph von Fraunhofer made important advances to telescopes. He constructed a stand on which the telescope could rotate equatorially, with the axis of rotation pointing at the North Pole. It had a clock mechanism keeping the correct rate of rotation so that the desired star remained in the field of view and its position could be carefully measured by an astronomer. He also manufactured a special kind of refracting telescope, so-called heliometer, which was very suitable for precision measurements of angles between two stars.

Fraunhofer's skill of making instruments led to the first reliable measurement of a star's parallax by Friedrich Bessel (1784–1846). This director of the Observatory of Königsberg was a self-made man, whose teenage dream had been to go on a trading expedition to China and the East Indies. In preparation for this trip he desired to add some acquaintance with the art of taking observations at sea. He was thus led from navigation to astronomy and from astronomy to mathematics.

Fraunhofer built the first heliometer for Bessel's observatory. However, it was completed only after the death of the master optician and was mounted in 1829. Bessel was well aware of the high quality of the instrument but only in 1837 did he find time to make a serious attack on the problem of parallax. Unlike Herschel, he did not use stellar brightness as a criterion of closeness; rather he reasoned that a star with a rapid motion across the sky should be nearby. One century earlier the British astronomer Edmond Halley (1656–1742) had shown that stars are not fixed on the celestial sphere, but move slowly. For example, since the time of Ptolemy, Sirius had shifted its position by half a degree (the diameter of the full Moon). These *proper motions* reflect the motion of both our Sun in space and the intrinsic motion of the star itself. In any case, it is expected that a distant star has a small proper motion, while nearby stars would appear to move more swiftly (similarly when sitting in a moving train the things close to you seem to move quickly while the distant landscape is crawling slowly). Bessel's criterion explains why he chose a rather inconspicuous star, 61 Cygni, at the back edge of the "wing" of the Swan (the constellation Cygnus). This star is actually a "sprinter" among stars, as it moves more than three diameters of the full Moon during one thousand years (the record-holder is Barnard's star in Ophiuchus, running across one Moon diameter in 180 years; in fact, it is the second nearest star).

Bessel measured for over one year the angular distance of 61 Cygni from three faint comparison stars. His careful analysis of the measurements revealed that the star had a parallax of 0.3136 ± 0.0202 arcsec. A parallax of one second of arc corresponds to a distance of 206,265 radii of the Earth's orbit (Box 8.1), Bessel's result put 61 Cygni at a distance of about 650,000 times the Sun–Earth distance.[2]

The first measurement of a star's distance aroused much attention, being an important breakthrough in astronomy. The tiny effect, to which Ptolemy and Galileo had referred to, was finally observed, and determination of cosmic distances had moved from the Solar System to the realm of the stars (Fig. 8.4).

Just 2 months after Bessel had communicated his result, the Scottish Thomas Henderson (1798–1844) informed the astronomy community that he had measured the parallax of the bright southern star alpha Centauri. The result, based on his observations several years earlier at the Observatory of the Cape of Good Hope in South Africa, was 0.98 ± 0.09 arcsec. The modern value for this nearest of all stars (excepting our Sun, of course) is 0.75 arcsec. In fact, alpha Centauri is made of three stars revolving around each other, of which Proxima Centauri is the nearest one. Its distance is 1.3 parsec.

In fact, the question of stellar distances was much "in the air." The head of Dorpat (Tartu) observatory Friedrich Struve (1793–1864) had ordered from Utzschneider and Fraunhofer a high-class telescope. Its 24-cm objective lens made it the largest refractor in the world, when it started operating in 1824. Among other objects, Struve focused the telescope on the brightest star of the northern heavens, Vega. Observations in 1835–36 put its parallax into the range 0.10–0.18 arcsec, as he reported to St. Petersburg Academy of Science in 1837. His notice was read in the meeting of the Academy, but it got buried in the archives. The modern value for Vega's parallax

---

[2] The modern looking "plus/minus" error estimate in Bessel's result was calculated by the new recipe by mathematician Carl Friedrich Gauss, who had found out how one can derive from observations not only an average value for the result, but also an estimate for its accuracy. Modern measurements have given for the parallax of 61 Cygni the value 0.299 ± 0.0045 arcsec, so Bessel's measurement was not far from the true value.

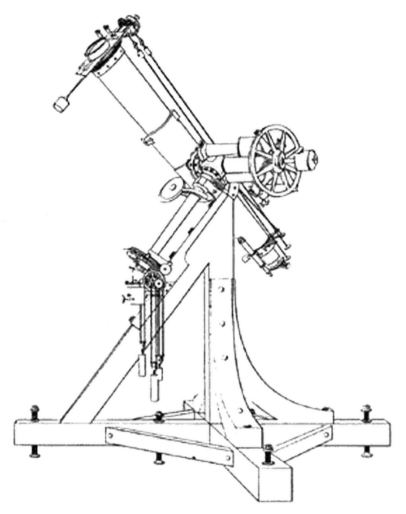

Fig. 8.4 Fraunhofer heliometer at the Royal Observatory of Königsberg which was used to make the first measurement of the parallax (distance) of a star. In 1838, Bessel determined that the distance of 61 Cygni is about 650,000 times the distance to the Sun

is 0.12 arcsec (distance = 8 pc), so Struve was on the right track. However, he was not yet satisfied by the result and continued with the observations. When he finally published, in 1840, his final results, he derived the parallax 0.26 ± 0.03 arcsec. For some reason he had got twice the true value, or the distance 50% too short.

After these pioneering efforts by three astronomers, parallax measurement was a demonstrated technique of obtaining star distances and became an important specialty in astronomy. The large distances proved that stars are so remote that to be visible in our sky, they must be pouring out as much or even more light than the Sun. If one gives stellar distances in kilometers, cumbersomely large numbers appear, since 1 pc is about 3×10^{13} km. Even the nearest star is 3.9×10^{13} km away, an immense distance. If stars were squeezed down into the size of an apple, they would still be separated by some 20,000 km. Stars are really very sparsely scattered in space and collisions between them are extremely rare!

The parsec unit is comparable to the huge distances between stars and is directly related to the method of measuring star distances. Astronomers usually express cosmic distances in parsecs. In this book we also use the light year (remember that 1 parsec is about 3.3 light years).

At first the number of stars having their parallax measured grew quite slowly. At the end of the 1870s only around 20 parallaxes were known, because visual measurements through the telescope were tedious. But when astronomical photography matured in the 1880s, astronomers started to make also parallax measurements from photographs, which speeded up the process. By the present day more than 7,000 parallaxes have been measured with ground-based telescopes.

All known stars are more distant than 1 parsec, so the parallax shift in the sky is always less than one second of arc. Such very small shifts are difficult to detect even with the widely separated astronomer's "eyes" (the diameter of the Earth's orbit). The restless air spreads the image of a star into a fuzzy dot, which limits ground-based parallax measurements to stars closer than 50 parsec.

A Three-Dimensional Look at the Winter Sky: Sirius, Stars of Orion, and Aldebaran

The beautiful winter constellation Orion and the nearby brightest fixed star Sirius are well known. On the opposite side of Orion is the bright Aldebaran in the constellation of Taurus, or the Bull. Only two centuries ago, the distances to these stars were unknown. A watcher of this section of the starry sky would think of it as two dimensional. But now, when admiring this region, we have the extra bonus of knowing how far away they are. We indicate in Fig. 8.5 this part of the sky, together with distances for most of the stars. Here the nearest ones are Sirius at 2.7 pc and Aldebaran at 20 pc (or 65 light years). The other bright stars are all farther than 100 pc and generally at such large distances, parallaxes cannot be measured from the ground, and the distances have been estimated by other means.

Today, the measurement of parallaxes is a fundamental step in the cosmic distance ladder. Stars beyond 50 pc may be reached from above the Earth's atmosphere, where the stars look sharp. The European Hipparcos satellite in the 1990s measured stars several times more distant. A total of 100,000 were measured, nevertheless reaching only a small fraction of the size of our Milky Way. In the 2010s, the space observatory Gaia will measure out to at least 20,000 pc almost covering the whole Milky Way!

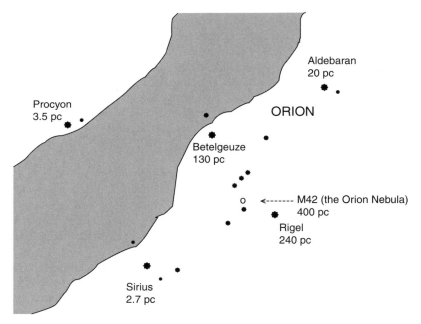

Fig. 8.5 Sirius, the stars of Orion and Aldebaran of the Bull (Taurus) make a beautiful view in winter evenings. The stars are located at very different distances in space, here given in parsecs (1 pc = 3.26 light years)

What If All Stars Were Like the Sun?

It may sound strange, but Newton knew roughly how tremendously distant stars are. How was this possible before the era of parallaxes? In 1668, the Scottish mathematician James Gregory (1638–1675) proposed a novel method to measure stellar distances: the *standard candle*. If all stars were as luminous as our Sun, then by comparing the apparent brightness of a star to the Sun one could infer the distance of the star in terms of the Earth–Sun distance. Faintness tells us the distance!

Naturally, it is difficult to compare the light of the dazzling Sun to that of a faint star. As a clever way to do this, Gregory's method used a planet as an intermediate step (the planet's brightness that can be compared with that of the star depends on the reflected light of the Sun). Thus Newton could calculate the distance to Sirius, with the help of Saturn. It turned out to be about 1 million times the distance to the Sun. This is twice the true distance, but delivered a good idea of the enormous remoteness.

The method of the standard candle is based on the important law, stated by Kepler, that the flux of light from a star diminishes inversely proportionally to the square of the distance (see Box 8.2). This photometric method is still the most important way to measure large cosmic distances, when the parallax method fails.

Instead of the Sun, many other classes of stars and galaxies are used as standard candles.

Box 8.2 Distance, luminosity, and observed flux of light

Suppose that a star radiates with the luminosity L, i.e., this amount of light energy in every direction per one second. At a distance R from the star, the light energy flying from the star is evenly distributed on the surface of a sphere with radius R. As the area of the surface is $4\pi R^2$, the flux of light f falling on every unit area is

$$f = L/4\pi R^2$$

inversely proportional to the square of the distance R. If one measures the flux f and knows the luminosity L, then this formula gives the distance R. Or if one knows the distance R, one can calculate the star's luminosity L. This is a very important formula in astronomy.

In real life, stars are not identical. In luminosity, i.e., their light emitting power, they may differ very much from the Sun. Some giant stars radiate as much as one million suns, while some dwarfs ten thousand times less. A nearby example is Sirius which is actually a binary star. Sirius A has a luminosity equal to 23 suns, while its faint companion Sirius B emits only 1/500 of the light of the Sun. If one just compares the Sun and a star, assuming the star is like the Sun, one is likely to make a large error in the distance estimate. Clearly, it is very desirable to be able to classify celestial bodies into classes having a narrow range of luminosity. The fact that Sirius differs from the Sun only by a factor of 20 in luminosity, explains why the early attempt by Newton led to a reasonable value of its distance.

We have seen that the Earth–Sun distance has appeared as a natural unit of length when one measures distances to the stars using the parallax method (and even in the attempts to use the Sun as a standard candle). But what is the value of this unit in our usual units of length? In other words, how large is the Solar System? We will see in the next chapter that it has not been so easy to measure the distance of the Sun, even though this is the closest of all stars and so bright in the sky.

Chapter 9
The Scale of the Solar System

In antiquity, the radius of the Earth was a basic unit of length for attempts to infer distances to the Moon and the Sun. Measurement of the distance to the Sun was attempted by Aristarchus, Hipparchus, and Ptolemy, but they failed, since the Sun is so far away. Copernicus' heliocentric system gave the Earth–Sun distance special importance; it could be used as a measuring ruler within the Solar System (Table 5.1). Kepler's Third Law emphasized the same thing: the times of revolution around the Sun, obtained from observations, determine the relative sizes of the planetary orbits in Earth–Sun distance units. When astronomers started determining distances (parallaxes) of stars, the Earth–Sun distance finally replaced our planet's radius as the natural unit.

However, one would also like to know cosmic distances in the earthly units of length used by physicists in their experiments. For example, to know the total radiation power in watts (J/s) of a star, as inferred from its radiation flux measured on Earth in W/m^2, one must know its distance from Earth in meters. To derive this distance from the star's annual parallax, one must know the distance of the Sun in meters! But it is not obvious how to measure this Earth–Sun unit.

A Hint from the Cathedral of San Petronio

The solar distance was still poorly known to Copernicus and Kepler and the size of the stellar sphere was simply unknown (Table 9.1). From the seventeenth to the nineteenth centuries the Earth–Sun distance was a central astronomical problem. A variety of methods were invented and tried and expensive expeditions were sent to remote regions of the Earth. One important permanent result was – along with the increasingly accurate solar distance – the beginning of international scientific collaboration.

Giovanni Cassini (1625–1712), the young astronomy professor at the University of Bologna in northern Italy, used a measuring device that he had constructed in

P. Teerikorpi et al., *The Evolving Universe and the Origin of Life*
© Springer Science+Business Media, LLC 2009

Table 9.1 Derived values of the solar distance

Author	Solar distance/Earth radius	The stellar sphere
Aristarchus	1,520	"Much more distant than Sun"
Ptolemy	1,210	19,865 Earth radii
Copernicus	1,142	"Immense"
Kepler	3,469	About 60,000,000 Earth radii
Cassini and Flamsteed	21,000	–
Modern	23,500	No actual sphere

the Cathedral of San Petronio to determine the elevation angle of the Sun when it was directly in the south. It was actually a giant "camera obscura," throwing the round image of the Sun onto the floor of the Cathedral. He had been entrusted with determining a new meridian line for the Cathedral.

Though his original goal was not the solar distance, Cassini's careful measurements over the year led him to an unexpected conclusion: to understand the variation of the Sun's elevation angle one had to shift the Sun much farther away than the generally accepted value at that time, 3,469 Earth radii, recommended by Kepler. We may understand why the variation of the Sun's elevation angle depends on the solar distance. Namely, the daily rotation of the Earth moves the observer, relative to the Earth center, a distance of the order of the Earth's size. This motion is reflected in the direction of the Sun, and the effect is larger if the Sun is nearby. From his measurements, Cassini was compelled to assign to the Sun the unheard-of large distance of at least 17,000 Earth radii to explain his observations.

Invited by Louis XIV, Cassini moved to Paris in 1669 to head the brand-new Paris Observatory. The solar distance had a high priority in his research program there. Since the value suggested by the measurements in Bologna could have been influenced by variations in atmospheric refraction, it was important to use some other method to prove, or disprove, the longer Earth–Sun distance scale.

Using Mars as an Intermediary

As mentioned above, and in Table 5.1, Copernicus determined the relative distances within the Solar System. Especially important, the Sun–Mars distance was known to be 1.52 times the Sun–Earth distance. If only one knew any *difference* between these distances, one could by simple arithmetic derive the Earth to Sun distance. One useful difference is the distance between the Earth and Mars, when Mars is in its opposite position relative to the Sun (in other words, when the three bodies lie on the same line: Sun–Earth–Mars). Every 16 years there is an especially close opposition, when Mars is closest to the Earth, when its distance is easiest to measure.

Such a good opposition was predicted to occur in 1672, and Cassini quickly organized an expedition to Guyana in South America. This was a French colony

Fig. 9.1 Illustrating the scale of the Solar System relative to distances to nearby stars and galaxies

and regular ship connections existed. The goal was to use the Paris–Guyana line as the base line of the cosmic triangle with its apex at Mars, and see if simultaneous measurements of Mars in Guyana and Paris revealed a difference in its direction relative to fixed stars. Alas, no difference was found!

However, even a "null result" can be valuable. Cassini interpreted the situation so that Mars is so distant that its parallax shift was hidden behind observational errors. This finally led to the conclusion that the solar distance must be at least 21,000 Earth radii, which confirmed the suspicion about the old distance scale raised by the image of the Sun wandering on the floor of the Cathedral of San Petronio.[1]

Note that the increase of the distance to the Sun immediately expanded the size of the Solar System as a whole. So the farthest planet Saturn was now at a distance of 200,000 Earth radii from the Sun – beyond the sphere of the fixed stars as imagined a century before! (Fig. 9.1)

Transits of Venus

Seventeenth century astronomers had obtained a lower limit for the Earth–Sun distance. A new method used during the next two centuries was the passing of Venus across the Sun's disk. This method has since been replaced by more accurate techniques, but it has an important place in the history of astronomy and as the first extensive international collaborative research project.

When it goes around the Sun inside the Earth orbit, Venus sometimes passes through the Earth–Sun line and is then seen on the disk of the Sun as a small dark spot. Such transits are quite rare, but then they happen in pairs separated by 8 years as given below:

6 December 1631	6 June 1761	9 December 1874	8 June 2004
4 December 1639	3 June 1769	6 December 1882	6 June 2012

---

[1] Support for the large distance also came from across the English Channel. James Flamsteed (1646–1719) used a method suggested by Tycho Brahe. He followed the movements of Mars in the starry sky for several hours. Its apparent motion reflects the orbital motions of both Mars and the Earth. The daily rotation of the Earth also causes a shift which is smaller the more distant Mars is. Thus Flamsteed also concluded that the solar distance must be "at least 21,000 Earth radii."

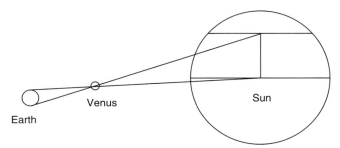

Fig. 9.2 The transit method. Venus crosses the Sun along different lines depending on the geographic latitude of the observer on the Earth. As the ratio of the Earth–Sun and Venus–Sun distances is 7:5, the apparent trajectories can differ no more than 5 Earth radii, meaning about 44 s of arc on the Solar disk. We greatly exaggerate this difference in the drawing (remember that the diameter of the Sun is actually half a degree, 40 times the maximum difference). In essence, the Sun is used as a background to measure accurately the parallax angle of Venus, after which the distance of the Sun is obtained from the 7:5 ratio

Transits are seen either in June or December, when the Earth passes those points in its orbit where the slightly inclined orbit plane of Venus cuts the plane of the Earth's orbit. Edmond Halley realized the possibility of measuring the Earth–Sun distance with transits in 1716 when he observed a similar event for Mercury. However, he did not live to see its first application to Venus in 1761. The idea was that observers stationed in widely separated geographic latitudes would observe the transit and measure accurately the interval of time it takes Venus to cross the Sun's disk. Observers at southern latitudes will see Venus cross the Sun closer to its northern pole than those observing at northern latitudes. The time intervals give the precise positions of the trajectories of Venus on the solar disk. Combining that knowledge with the known geographic latitudes of the observers and the known ratio of the orbit sizes of the Earth and Venus leads to the distance of the Sun (Fig. 9.2).

The measurement procedure is surprisingly simple, requiring only a telescope and a good clock. However, the observers were not happy to find that the timing of the moment when Venus within the solar disk touches the edge of the Sun could not be made as accurately as was hoped, since the image of the touching point becomes fuzzy. Related optical phenomena gave the first indication that Venus has an atmosphere (see Fig. 9.3. and Chap. 31). Since the accuracy of timing is critical for the method, the results of the 1761 and 1769 transits did not quite reach the level of precision hoped.

The observations of the second transit were carefully planned. There were 77 observing stations all around the Earth and 151 observers. It took decades to analyze and combine all the observations. The end result was that the distance to the Sun is 24,200 (±250) Earth radii. More modern determinations with different methods give the more accurate result 23,494 Earth radii. One needs the size of the Earth in meters to complete the calculation of the Earth–Sun distance.

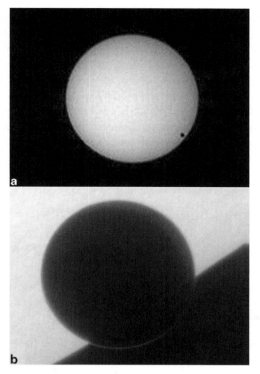

Fig. 9.3 (**a**) The 2004 June transit of Venus, 8 June, 11 UT (Credit: USNO). (**b**) Venus on the limb of the Sun. The brightening of the edge of Venus' disk against the black sky is caused by refraction in the thick Venus atmosphere (credit: Dutch Open Telescope at La Palma, Utrecht University)

The Size of the Earth 2,200 Years After Eratosthenes

Recall that Eratosthenes determined the approximate size of the Earth. He measured the angle of the Sun from overhead (the zenith) to measure the difference in latitude between Alexandria and a point on Syene, a known distance directly south. It is more accurate to use stars, especially those close to the zenith, and measure their angular distance from the vertical defining the zenith, when they are directly in the south. The French astronomer Jean Picard (1620–1682) was the first to make such measurements using a telescope equipped with the newly invented screw microme-ter (Chap. 8). He reached an accuracy of 5 arcsec in the measured zenith distances. In other words, he could measure the circumference of the Earth with an accuracy of about 50 km.

It became also possible to investigate whether the form of the Earth is exactly spherical. Christiaan Huygens and Isaac Newton had concluded that the Earth's ro-tation around its axis should cause it to be slightly flattened at its poles, bulging at the equator. On the other hand, Jacques Cassini made measurements of the length of the arc in different places in France and inferred that the polar radius of the Earth is somewhat longer than the radius at the equator, contrary to the predictions by Huygens and Newton. However, his measurements were made on a rather short arc of the meridian (9°). To resolve the riddle of the Earth's form once and for all, the French Academy organized in the 1730s two expeditions, one to the south, close to the equator (Peru) and another far to the north (Lapland). The measurements clearly demonstrated that the arc of 1° in the north was longer than close to the equator, as expected for a flattened Earth. Modern measurements using satellites have given the following values for the size of the spheroid best describing the form of the Earth:

The radius at the equator $= 6,378$ km

The radius at the poles $= 6,357$ km

The Modern View of the Scale of the Solar System

Expressed in kilometers the currently accepted value of the astronomical unit AU is

The mean Earth–Sun distance $= 1$ AU $= 149,597,870$ km

This value relies on several measurements, among them the radar-based distance to Mars, together with Kepler's Third Law. As we already mentioned, once the Earth–Sun distance is known, all other distances in the Solar System become fixed. Table 9.2 gives data on the orbits of planets, including Pluto that lost its status as a major planet in 2006.

From the table one may see several interesting things. The orbit of Venus is closest to a circle, its distance from the Sun varying only by 1%. Mercury has a very elongated orbit (not to speak of Pluto!). Also the orbit of Mars is quite elliptical, which helped Kepler to derive its correct form. The table also shows that the Earth's distance from the Sun varies by five million kilometers. It is closest to the Sun when it is winter in the northern hemisphere!

Table 9.2 Data on the orbits of planets (plus the dwarf planet Pluto)

Planet	a (million km)	a (AU)	Distance variation (million km)	% Distance variation
Mercury	57.9	0.387	23.8	41
Venus	108.2	0.723	1.5	1
Earth	149.6	1.000	5.0	3
Mars	228.0	1.524	42.6	19
Jupiter	778.4	5.203	75.5	10
Saturn	1,427.0	9.539	158.7	11
Uranus	2,869.6	19.182	270.9	9
Neptune	4,496.5	30.057	77.3	2
Pluto	5,946.5	39.750	3,013.7	51

It may be helpful to visualize the proportions of the Solar System with a minia-ture model (following an early attempt by Christiaan Huygens). Let us place in the center a sphere of the size of a large apple, say, 10 cm. This is the Sun. The Earth, a 1-mm-sized grain, revolves around it at a distance of 11 m. Saturn is orbiting at a distance of 103 m. The Pluto–Sun distance would be typically 425 m, though it would vary a lot. If we add to this map the nearest star, it would be at the respectable distance of 3,000 km! To be exact, it would be the triple system of Alpha Centauri, with its two major stars A (perhaps like a large grapefruit) and B (a small apple) circling each other at a distance of some 300 m, while the small C (Proxima), with a size of a blueberry would wander very slowly at a distance of about 100 km from those two.

We have come a long way from the mid-summer Sun lighting up Stonehenge to the nearest stars four light-years away. It is time to come back for a moment, to look at the secrets of our backyard, and ask, together with Isaac Newton, what makes an apple fall – and the Earth go around the Sun.

Part II
Physical Laws of Nature

Chapter 10
Newton

Sir Isaac Newton (1642–1727) is one of the most influential scientists who ever lived. He completed the revolution started by Copernicus, Kepler, and Galileo, allowing us to understand why planets move as they do. The orbits of spacecrafts are safely calculated from Newton's laws. He also created a new scientific method pivotal for future researchers; experimental testing became a partner to induction and deduction. Newton emphasized observation and experiment. The idea is to infer mathematical theories from empirical data and to compare values obtained from the theories with new measurements. A good theory not only explains the original observations, but in addition, it predicts phenomena allowing it to be tested. If there is a contradiction, one must adjust or perhaps even discard the theory. As the president of Royal Society, Newton wrote "Natural Philosophy consists of discovering the frame and operations of Nature, and reducing them, as far as may be, to general Rules or Laws – establishing these rules by observations and experiments, and thence deducing the causes and effects of things. . . . "

From Woolsthorpe to *Principia*

Newton had a rather unsettled childhood. He was born at his parents' Woolsthorpe manor house in Lincolnshire 3 months after his father's death. His mother remarried when he was 3 years old, and left him with his grandmother. When Newton was 11, his mother's second husband died, and Newton moved in with his mother along with two half-brothers and a half-sister. His mother wanted Isaac to become a farmer, but he was not interested. Instead he showed talent in constructing mechanical toys and was successful at secondary school. The local priest persuaded his mother to send her son to the University of Cambridge after he graduated from the secondary school. Isaac was admitted to Cambridge in 1661 when he was 18, rather old for a university student in those days (Fig. 10.1).

Newton's studies progressed in an average way, but he read a lot on his own. This was noticed by the professor of mathematics, Isaac Barrow, who lent Newton

Fig. 10.1 Isaac Newton when he was 46 years old. Godfrey Kneller's 1689 portrait

books from his own library. Therefore, when Newton passed his final examinations 4 years later, he had excellent knowledge in astronomy, mathematics, physics, and chemistry. He was ready to start creating modern physical science.

But there was also another Newton beyond our usual conception of a scientist. He studied alchemy, which became his lifelong serious hobby, and the Bible, which he knew better than many theologians. These interests remained with Newton all his life. In words of Lord Keynes, "he was the last of the magicians, the last of the Babylonians and Sumerians, the last great mind which looked out on the visible and intellectual world with the same eyes as those who began to build our intellectual inheritance rather less than 10,000 years ago."

In 1665, a plague spread through England, and the university closed down. Newton returned to his home at Woolsthorpe. Later, Newton described how he passed his time there, first finding "the Method of approximating series & the Rules for reducing any dignity of any Binomial into such series." And:

> The same year in May I found the method of Tangents of Gregory & Slusius & in November had the direct method of fluxions & the next year in January had the Theory of Colours & in May following I had entrance into ye inverse method of fluxions. And the same year I began to think of gravity extending to ye orb of the Moon & (having found out how to estimate the force with w^{ch} [a] globe revolving within a sphere presses the surface of the sphere) from Kepler's rule of the periodical times of the Planets being in sesquialterate proportions of their distances from the center of their Orbs, I deduced that the forces wch keep the Planets in their Orbs must [be] reciprocally as the squares of their distances from the centers about wch they revolve: & thereby compared the force requisite to keep the Moon in her Orb with the force of gravity at the surface of the earth & found them answer pretty nearly. All this

was in the two plague years of 1665–1666. For in those days I was in the prime of my age
for invention & minded Mathematicks & Philosophy more then at any time since.

This is impressive for a newly graduated physicist. Actually, historians have
shown that in his old age Newton exaggerated the achievements of his youth. Probably Newton thought about all the matters he mentions during the plague years, but
certainly many of his works were completed years later. Newton was retiring by
nature and did not like sharing all his knowledge. When another researcher started
to trace the same tracks as Newton, he would hurry to publish his research and to
claim priority. Afterwards, disputes ensued as to who got there first. By shifting all
of his most important inventions to the plague years, Newton may have solved the
priority questions in his own mind all at once.

When Newton returned to Cambridge in 1667, he started to lay the foundation for
several fields of science. His method of fluxions is known today as differential and
integral calculus. In the theory of light, he was especially interested in the nature of
color, and using mechanics he solved the ancient riddle of the motion of planets. The
results appeared in final form much later: *Philosophiae Naturalis Principia Mathe-
matica* (the *Principia*) appeared in 1687 and *Opticks* appeared in 1704 (Fig. 10.2).

The *Principia* has been called the most important work in history of science.
The Royal Society, founded in 1662, and especially its members Christopher Wren
(1632–1723), Robert Hooke, and Edmond Halley can take some credit in getting
this work started. When Wren gave his inaugural speech as professor of astronomy
at Oxford University, he declared that the most important physical problem of the
time was the explanation of Kepler's laws. He prophesied that the man who would
explain them was already born. Not a bad conjecture: Newton was 15 years of age at
the time. Wren and Hooke were experimenting with pendulums which lead Hooke
to a hypothesis that planetary motions are compounded of a tangential motion and
"an attractive motion towards the centrall body."

After becoming the secretary of the Royal Society in 1677, Hooke tried to start a
correspondence with Newton, well known for his mathematical skills. Hooke suggested the above hypothesis as a topic of the correspondence, writing: "It remains
to know the properties of the path that a body would follow when attracted by an
inverse square force. I doubt not but that by your excellent method you will easily
find out what that Curve must be, and its proprietys, and suggest a physicall Reason
of this proportion."

Hooke did not get an answer to his question. Perhaps inspired by Hooke's challenge, Newton arrived at his universal law of gravity in the early 1680s together with
the derivation of Kepler's laws. For some time, scientists had discussed the possibility that the attraction between the Sun and the planets could weaken as the distance
squared (the so-called inverse square law). One could reach this conclusion by combining the formula for centripetal acceleration by Huygens with Kepler's third law.
Robert Hooke was aware of this, but he could not say if this force law resulted in
orbits in accordance with Kepler's first and second laws (ellipses, equal areas).

Since Hooke was not able to initiate conversation on this problem with Newton,
in August 1684, he sent young Edmond Halley to meet Newton. As Newton later
described to Abraham DeMoivre: "After conversing some time, Halley asked Newton 'what he thought the Curve would be that would be described by the Planets

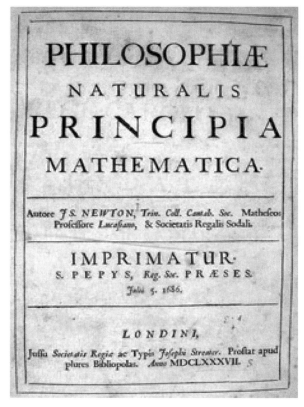

Fig. 10.2 The title page of the first edition of *Principia*

supposing the force of attraction towards the Sun to be reciprocal to the square of
their distance from it.' Sir Isaac replied immediately that it would be an Ellipsis,
the Doctor struck with joy & amazement asked him how he knew it, whereupon
Dr Halley asked him for his calculation without any further delay, Sir Isaac Looked
among his papers but could not find it, but he promised him to renew it, & then to
send it him ... "

Newton decided to dedicate his lectures in the following term "On of the Motion
of Bodies in an Orbit" which he wrote as a nine page treatise ("De motu"), and
delivered to Halley in November. Prompted by Halley, he continued writing. This
led to the *Principia* (partly financed by Halley) 2 years later.

Newton's Physics

One of the most significant concepts in the *Principia* was universal gravity. Of
course, gravity holds us to the ground on the Earth. Something forces the distant

Moon to circle the Earth and the planets to circle the Sun. Is it the same force? We discussed earlier how Huygens found that the acceleration of an object in a circular path toward the center is its speed squared divided by the radius of the path. To establish that the universal law of gravity follows the inverse square law, Newton compared the acceleration toward the center of the Earth at its surface with the acceleration caused by the Earth further away, at the distance of 60 Earth radii on the Moon. The gravitational acceleration should be lowered by 60^2 when we go from the Earth to the orbit of the Moon. It should equal the circular acceleration of the Moon toward the Earth. Newton carried out the comparison by using the value of the radius of the Earth and was able to confirm the inverse square law. A splendid result! Because of the greatly reduced acceleration, the Moon falls in one minute as far as an apple (on Earth) falls in one second.

Newton summarized his research on motion in three laws of mechanics. The first rule of Galileo (and also used by Descartes) was presented as Newton's first law,

I. Every body continues in its state of rest, or of uniform motion in a right line, unless it is compelled to change that state by forces impressed upon it.

Under the influence of an external force the state of motion changes, in other words, the body experiences acceleration. In his second law, Newton concluded that

II. The change of motion is proportional to the motive force impressed and inversely proportional to the mass of the body; and is made in the direction of the right line in which that force is impressed.

We may state this more briefly: acceleration = force/mass or as is often written,

$$\text{force} = \text{mass} \times \text{acceleration}.$$

The law of reaction (Newton's third law), completes the basic rules of mechanics:

III. To every action there is always opposed an equal reaction; or, the mutual actions of two bodies upon each other are always equal, and directed to contrary parts.

In other words, for the force exerted by one body (an "agent") on another, the other exerts an equal and opposite force on the "agent." Newton could thus write down the mass dependence of the law of gravity. Recall that the acceleration caused by gravity follows the inverse square law. According to Newton's second law, the force has to be proportional to the mass of the affected body. For example, the force by which the Earth pulls the Moon must be proportional to mass of the Moon. But from Newton's third law, consider the case from the point of view of the Moon, the force by which the Moon pulls the Earth must be equal and opposite, and also proportional to the mass of the Earth. Thus in all, the gravitational attraction between two bodies has to be proportional to the *product of the masses* of the two bodies as well as being *inversely proportional to the distance* between them.

It should be noted that rocket flight is based on Newton's third law of action and reaction. Two centuries after Newton's times, the theoretical basis of astronautics

Fig. 10.3 Konstantin Tsiolkovski (1857–1935) was the father of astronautics, who clearly understood how Newton's third law of action and reaction makes space travel possible. On the right his plan for a rocket

was built by the Russian mathematics teacher and visionary Konstantin Tsiolkovski who said: "The planet is the cradle of intelligence, but you do not live in the cradle forever" (Fig. 10.3).

Nature of Gravitation

The *Principia* was not accepted at once. First, the work was mathematical and hard to read. It was said that Newton wanted to make the text so difficult that his rival Hooke would not be able to understand it. But even among other readers there were doubts. Huygens wrote in 1690: "Newton's theory has the problem that in space there is only very rarefied matter, as is shown by the motion of planets and comets through it. Then it appears difficult to explain the propagation of gravitation or light, at least in the way I am used to it." Huygens was used to the ideas of Descartes (Chap. 7). It also bothered Newton that he was not able to explain how the force was transmitted between astronomical bodies. He could only describe the force mathematically.[1] In his correspondence with Richard Bentley on cosmological

---

[1] Newton wrote in the second edition of *Principia* (1713) his famous words about not making assumptions, also containing in a nutshell his scientific method: "I have not as yet been able to discover the reason for these properties of gravity from phenomena, and I do not feign hypotheses. For whatever is not deduced from the phenomena must be called a hypothesis; and hypotheses, whether metaphysical or physical, or based on occult qualities, or mechanical, have no place in experimental philosophy. In this philosophy particular propositions are inferred from the phenomena, and afterwards rendered general by induction."

matters (we discuss it in Chaps. 23 and 28), Newton wrote in 1693: "Gravity must be caused by an agent acting constantly according to certain laws; but whether this agent be material or immaterial I have left to the consideration of my readers."

This explanation did not satisfy everybody. For example, Fontenelle in France argued in the eighteenth century: "Pull of gravity and vacuum which Descartes seems to have expelled from physics for ever, have now been brought back by Sir Isaac Newton with new vitality which I could not have believed possible, and in a somewhat disguised form." For similar reasons, Newton's ideas about the nature of light were difficult for Huygens to accept. If light consists of waves, e.g., like sound waves, it is necessary to have a medium penetrating everywhere to propagate it. Newton gave up the concept of a medium; he viewed light as particles speeding through space.

In 1669, Newton succeeded Isaac Barrow as the professor of mathematics at Cambridge. In 1689, he was elected to the Parliament as the representative of the university. According to an anecdote the introverted professor spoke before the house on only one occasion: he stated that a window had been left open and was causing a draft. Then he promptly sat down...

Newton's interest in science declined. In 1696, he was appointed Warden of the Royal Mint, the second highest position in the hierarchy of the Mint; he was appointed to its highest post, the Master of the Mint, 3 years later. It was an important position; the coinage system of the British Empire was under revision. Newton carried out the revision with enthusiasm and success. He was knighted in 1705, acting as the President of the prestigious Royal Society during the last decades of his life but with his scientific work far behind him. Late in life, Sir Isaac commented on his achievements as follows: "If I have seen further it is by standing on ye shoulders of Giants." Other touching words are:

> I do not know what I may appear to the world; but to myself I seem to have been only like a boy playing on the seashore, and diverting myself in now and then finding a smoother pebble or a prettier shell than ordinary, whilst the great ocean of truth lay all undiscovered before me.

Armed with the mathematical methods and natural laws discovered by Newton, we move back to the Solar System, the test ground of the new science of mechanics. We left it knowing its scale and with the six planets of the seventeenth century.

Chapter 11
Celestial Mechanics

The new branch of mathematics called fluxions by Newton[1] allowed astronomers to calculate orbits of celestial bodies and led to a flowering of physics in the following century. Joseph Louis Lagrange summarized this success story in his *Mécanique analytique* (1788) in which he developed a method for casting different mechanical problems into mathematical form. He was proud that he did not need a single figure in his famous book which was not exactly "reader-friendly." Everything could be represented by formulae and algebraic operations (Fig. 11.1).

Discovery of Uranus

Over practically all of recorded history, only seven special objects (the Sun, the Moon and five planets) were known to move among the fixed stars in the same band of constellations. Their number was heavily enshrined in culture. The seven were named after gods and goddesses and even used to name the days of the week. It is interesting that, even up to the late eighteenth century, when their physical nature was understood, the possibility of additional planets in our Solar System was not taken seriously.

This all changed when William Herschel (1738–1822) discovered a new slowly moving object in 1781, which he first regarded as a comet. Subsequently, the Finnish astronomer Anders Johan Lexell (1740–1784; working in St. Petersburg, Russia) and later Pierre-Simon de Laplace calculated its orbit and found it to be circular making it obvious that the object was a planet. The names Georgium Sidus (after George III in England), Herschel, and Uranus were suggested for this new object with the last one becoming accepted. Thus, not only a new planet but also the notion of even more unknown objects beyond the orbit of Saturn was introduced. For his momentous discovery Herschel received a permanent salary from the English Crown. We return to Herschel's other achievements in Chap. 20.

---

[1] The new mathematics is for us better known in the calculus notation developed independently by Gottfried Wilhelm von Leibnitz (1646–1716).

P. Teerikorpi et al., *The Evolving Universe and the Origin of Life*
© Springer Science+Business Media, LLC 2009

Fig. 11.1 Joseph Louis
Lagrange (1736–1813),
a great mathematician,
developed Newtonian
mechanics

His sister, Caroline Herschel, was a faithful assistant from lens grinding to making observations. She was an astronomer in her own right discovering at least eight comets as well as several nebulae and clusters of stars. She received a gold medal in 1828 from the Royal Society for her publication of a catalog of the star clusters and nebulae observed by her brother. This and other catalogs published by her form the basis of modern catalogs. After the discovery of Uranus, she also received a government salary as perhaps the first woman in England to hold such a position.

The Race to Discover Neptune

An important astronomical problem of the eighteenth century was the calculation of orbits when more than two bodies were influencing each other. For example, the orbit of the Moon around the Earth is influenced, in addition to the gravitation between the Moon and the Earth, by the gravitational pull of the Sun upon both. Not only is there a resulting motion of the Moon–Earth system around the Sun, but the motion of the Moon relative to the Earth is not exactly an ideal ellipse. Similarly, planets perturb each other's elliptical motions around the Sun.

A famous case is the orbit of Uranus which had been calculated with great precision in the 1820s. The English astronomer Mary Somerville (1780–1872) predicted that orbit perturbations could be used to discover new objects. Uranus was not observed to keep the expected path, so that in 1830 it was 20 arcsec off the predicted course, by 1840, 1.5 arcmin off, and by 1845, reaching 2 arcmin away from its expected position. This was in addition to any perturbations by the known planets; thus there must be an unknown planet whose gravity perturbs the motion of Uranus.

In 1843, John Couch Adams (1819–1892), a student at Cambridge University, began calculating where the unknown planet should be situated to cause the observed misbehavior of Uranus. The calculations were complex; to simplify them, Adams assumed that the unknown mass orbits the Sun beyond Uranus at a distance from the Sun as given by the Titius–Bode law. This "law" derives its name from a remark in a work by Johann Titius von Wittenberg in 1766, pointing out that the distances of the planets from the Sun can be expressed using a simple rule. Six years later Johann Bode, the director of Berlin observatory, saw the remark, and added it to the latest edition of his astronomy text (see Box 11.1). By October 1845, Adams had calculated the current orbital position of the unknown planet, and informed his professor of astronomy Challis who showed the coordinates to the Astronomer Royal Airy. Neither of them took the student's calculation very seriously, and no observational search was conducted.

Box 11.1 The Titius–Bode law

The Titius–Bode law is an empirical formula which gives the distances d of the planets from the Sun. It can be expressed in terms of the Earth–Sun distance (AU) as

$$d = (4 + 3 \times 2^n)/10$$

Here $n = -\infty$ for Mercury (or $d = 0.44$), and $n = 0, 1, 2$, etc., for Venus, Earth, Mars, etc. It may help one to recall the formula by noting the order of numbers in it (43210). The formula gives a fair result for all the planets known up to 1845, and applies also to the biggest of the minor planets Ceres, known in 1845. The calculated and measured distances are shown in the following table:

	Distance from the Sun	
	Titius–Bode law	Observed value
Mercury	$4/10 = 0.4$	0.4
Venus	$(4 + 3 \times 1)/10 = 0.7$	0.7
Earth	$(4 + 3 \times 2)/10 = 1.0$	1.0
Mars	$(4 + 3 \times 4)/10 = 1.6$	1.5
Ceres	$(4 + 3 \times 8)/10 = 2.8$	2.8
Jupiter	$(4 + 3 \times 16)/10 = 5.2$	5.2
Saturn	$(4 + 3 \times 32)/10 = 10.0$	9.2
Uranus	$(4 + 3 \times 64)/10 = 19.6$	19.2
Neptune	$(4 + 3 \times 128)/10 = 38.8$	30.1

In the same year, French astronomer Urbain Le Verrier (1811–1877) started similar calculations, not knowing that Adams had already finished them. In the spring of 1846, he had a result consistent with Adams' coordinates. Le Verrier wrote Johann Galle, director of Berlin observatory, and asked him to search for the planet. Right away, on September 23, Galle's telescope had enough magnification to tell him that one of the stars in the predicted area was not point-like but a disk as a planet should appear in the sky. Moreover, the next night it had moved relative to the stars. From various suggested names, Neptune was deemed to agree best with the names of the other planets and became official.

The finding of Neptune so close to the expected position, within 1° of the calculated coordinates, was considered a great victory for mechanistic view of the world based on Newton's theory. When one wanted to emphasize the supremacy of modern science over earlier beliefs, one would mention that with modern science one could even predict and discover new planets. Actually, there was also plenty of good luck needed for the discovery. Neptune is actually quite a bit closer to the Sun than what the Titius–Bode law gives. With bad luck, the calculations of Adams and Le Verrier could have gone way off the mark.

There was – naturally! – a dispute in British and French press about which nation deserved credit the discovery of Neptune. The English knew the predicted position of the planet first. However, the planet was discovered by a German astronomer making use of the calculation of a French astronomer. Generally the greatest honor in the discovery of Neptune is given to Le Verrier. John Adams and Urbain Le Verrier retained mutual respect – later the former as President of the Royal Astronomical Society presented the latter, Director of the Paris Observatory, with a gold medal.

More Planetary Perturbations

The discovery of Neptune encouraged researchers to watch for other unexplained effects in the orbits of planets. Small perturbations were thought to exist in the orbit of Neptune. Percival Lowell explained these by an unknown planet further out than Neptune and seven times more massive than the Earth. Inspired by this prediction, a search was conducted for decades until the young assistant Clyde Tombaugh (1906–1997) identified it in a photograph in 1930. The planet was named Pluto, but at about 1/500 Earth's mass it is now known not to be big enough to cause an observable perturbation in the orbit of Neptune. Thus, the discovery of Pluto only 6° away from the expected place is entirely due to good luck and persistence![2]

In 1993, Jet Propulsion Lab astronomer Myles Standish found using new space probe data for the masses of the planets that any remaining irregularities in the positions of Uranus and Neptune were nonexistent. Thus, there is no dynamical evidence for an additional planet beyond Pluto. Recent discoveries of small "Kuiper Belt" objects near and beyond Pluto have been found by observational searches. None are large enough to cause perturbations of planets. However, one rather distant object has been found which is larger than Pluto. Since more such objects probably exist, in 2006, Pluto was demoted to the status of a "dwarf planet" along with the newly discovered larger Kuiper belt object and the largest asteroid Ceres (more about the new definition in Box 31.1).

---

[2] If Tombaugh had missed Pluto, it would have been found later by Yrjö Väisälä of Turku University during minor planet searches in 1935–1945 (about Väisälä in Chap. 22).

Within the inner Solar System, the motion of the planet Mercury has irregularities which cannot be entirely explained by Newtonian gravitational perturbation by other planets. Le Verrier calculated that the orbital ellipse of Mercury turns (or precesses) in one century 35 arcsec more than known planetary perturbations can explain. Simon Newcomb improved the calculation and found that the unexplained precession is 43 arcsec per century. The precession could be explained by a small unknown planet orbiting closer to the Sun than Mercury. It would thus be difficult to observe. This planet (tentatively called Vulcan) was not seen, in spite of many searches. As an alternative, Newcomb proposed in 1895 that Newton's inverse square law is not exactly valid. In a way Newcomb was correct. The excess precession of Mercury was one motivation for Einstein in his search for an improved theory of gravity which successfully explained the precession.

We have seen how the calculation of orbits using Newton's law of gravity started a new science called Celestial Mechanics where the accuracy of calculations and the related observations were unprecedented. A deviation of the orbit of Mercury by only 43 arcsec per century from the prediction was considered significant and worth closer study. In one year, the unexplained deviation is 0.43 arcsec. Compare this with Tyco Brahe's observations of Mars where the deviation was 500 arcsec from the predictions of Ptolemy or Copernicus. This shows a huge increase in accuracy, by a factor of 1000, both in the theory and observations of planetary motions during the previous three hundred years. The deviation in the motion of Mars was just barely big enough to motivate Kepler in his search for a new planetary theory.

Laplace's World View

The triumphs of Newton's theory promoted the mechanistic view of the world. The foremost proponent of this view was Pierre-Simon, Marquise de Laplace (1749–1827) whose five volume *Mécanique céleste* was not only a translation of the *Principia* into differential calculus, but contained additional details. Laplace presented the universe as a kind of huge clockwork. He said: *"If an intelligent being knows at some moment of time all forces of nature, and the positions of all particles of nature, he would be able to write a single formula which describes the states of motion of all particles in the universe, from the biggest ones down to the smallest atoms, supposing that the being is able to handle all the information; there would be nothing unclear to this being, but both the past and the future would unfold in front of his eyes."*

Laplace thought that the evolution of any system, be it even the whole universe, is fully determined when the initial states of all of its particles are given: "All the effects of Nature are only the mathematical consequences of a small number of immutable laws." If nature is this simple, then according to Thomas Huxley (1825–1895) "science is nothing but trained and organized common sense." However, physical reality turned out to be much more complicated than the ideal clockwork.

A very significant result of Laplace's research is the calculation of the long-term perturbation of the orbits of planets. One might worry about the fate of the life on Earth if the effects of other planets cause it to spiral into or away from the Sun. Happily Laplace showed that such effects do not systematically shift planetary orbits to any particular direction toward or away from the Sun. The perturbations are only cyclic. Therefore the Earth remains at the same mean distance from the Sun for billions of years in spite of Mercury and Venus giving it nudges toward and the outer planets away from the Sun.

Laplace also discussed the origin of the Solar System in his book *Exposition du système du monde* (1796). He hypothesizes based on Newton's theory that the Solar System was originally a rotating cloud of gas which shrank, and as a result of shrinking, rotated faster and faster. At last the rotation was so rapid that the cloud started to shed rings off its equator. Every ring formed a planet at a later time, what was left in the center made the Sun. Originally the planets were rotating gas clouds which also contracted and shed off rings from their equators. Subsequently the moons of the planets condensed out of these rings. Somewhat similar ideas had been presented by Emanuel Swedenborg (1688–1772) and Immanuel Kant. These old theories have elements still shared with modern views of the origin of the Solar System (Chap. 30), even if the physical processes turned out to be more complicated.

Exposition included a prophetic sentence that "the gravity of a celestial body can be so great that light cannot flow out of it." Such a body is now called a black hole. This same idea was presented earlier in 1784 by John Michell; the two may have arrived at the concept independently (Chap. 15).

The Three Body Problem

The calculation of the perturbed orbit of the Moon is a difficult problem, said to be the only problem that ever made Sir Isaac's head ache. Partly this is because one has to consider the attraction of the Sun as well as of the Earth on the Moon. After Newton, the great practitioners of celestial mechanics, the French mathematician Jean le Rond d'Alembert (1717–1783) and the Swiss astronomer Leonhard Euler (1707–1783) who worked most of his career in St. Petersburg, both attempted to explain the complications in the motion of the Moon and the related variations of the axis of the Earth. The Earth's axis precesses with a 26,000-year period, but in addition, it has a nodding motion with an 18-year period which is the cause of the Saros period of eclipses mentioned in Chap. 1. The nodding, or nutation as it is called, was discovered by James Bradley in 1748. Only a year later d'Alembert published a theory of nutation based on Newtonian mechanics. He communicated the result to Euler who found it hard to read. Euler produced a simpler version of d'Alembert's work. However, for unknown reasons, Euler failed to mention d'Alembert. This led to a complete breakdown of relations between these two great scientists of the time. A late apology by Euler did not help the matters.

Tidal friction phenomena further complicate the problem. The tides cause a gradual slowing down of the Earth's rotation. Through tides, the Moon attempts to force

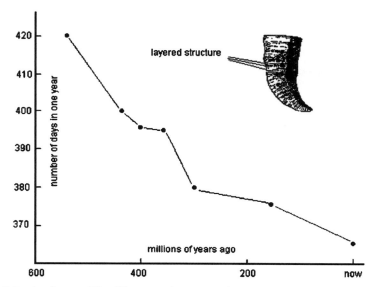

Fig. 11.2 During the past 600 million years the number of days in a year has decreased from about 420 to 365 days, as shown by counting the layers in fossil clam shells and corals, i.e., the day was shorter than now

the Earth's rotation to its own orbital cycle but in turn the Moon's orbital period lengthens. Eventually, the Earth's day and the Moon's month will be of equal length, 55 (present) days each. At this time, the Moon will also be further away from the Earth than today. However, the change is slow. During the past 400 million years the day has increased from 22 h to 24 h. The change has been verified in the layered structures of fossil clam shells and corals which can be used to calculate the number of days and months in the year when they were alive, like finding the age of a tree by counting rings in its trunk. Corals deposit a single, very thin layer of lime once a day. It is possible to count these diurnal (day–night) growth lines. One can also count yearly growth variations. So, given the right piece of coral, one can measure how many days there are in a year (Fig. 11.2).

Aside from long-term effects of tidal friction, the Earth–Moon–Sun system is an example of a relatively simple *three body problem*, with the very massive Sun at a very large distance from the other two. When a spacecraft is launched toward the Moon, we have to solve a more complicated three body problem with comparable distances between the bodies: at what direction and at what speed do we have to launch the low mass spacecraft from the vicinity of the Earth in order that it arrives at the Moon in a suitable orbit. In the *General Three Body Problem* where there are three celestial bodies of comparable masses at similar distances from each other, the orbits can become even more complex (Fig. 11.3). The motion is composed of close encounters between two bodies while the third body looks on at a distance. The encounters happen again and again while the pairs are exchanged. This takes place until the system breaks up by throwing away one of the three bodies. After

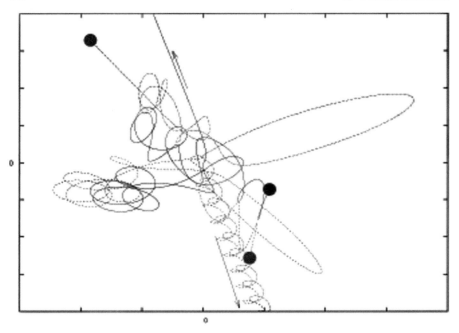

Fig. 11.3 Orbits in a system of three bodies. The orbits continue in a complex fashion, until one body is thrown away with a binary of the other two bodies remaining dancing around each other. This orbit came from a computer experiment made by Seppo Mikkola at Tuorla Observatory (University of Turku)

that the orbits are simple: there is a binary system with elliptical orbits and a third body escaping from the binary. The shapes and the sizes of the final orbits can be predicted in a statistical sense, but what happens in each individual case can be found out only after long and exact calculations. Often the statistical description is all we need to know. For example in a star cluster, three body encounters take place frequently, and only their statistical effect is of general interest.

Only a hundred years ago the solution of the three body problem was not at all clear. There were two schools of thought. Following the clockwork tradition of Laplace, one could always describe the orbits of three bodies if the initial values are known. The prime proponent of this view was the Finnish astronomer Karl Sundman (1873–1949) who presented in 1912 a mathematical formula as a solution of three body orbits. The French mathematician Henri Poincaré (1854–1912) noticed that "it may happen that small differences in the initial conditions produce very great ones in the final phenomena." For the three body problem it means that there is *deterministic chaos*: the small changes in initial condition produce such big differences in the final state that the outcome is chaotic, unpredictable.

At the end of the nineteenth century, the question of the solution of the three body problem was posed by King Oscar II of Sweden with a cash prize promised to whoever answered it definitively. Poincaré won the prize with his publication "On

The Problem of Three Bodies and the Equations of Equilibrium." Through this investigation Poincaré came to understand that infinitely complicated behavior could arise in simple nonlinear systems.[3] Without the benefit of computers, only through his mathematical insight, he was able to describe many of the basic properties of deterministic chaos. The term "chaos" came to use much later, and it is now understood as an essential part of the description of complex systems in nature (limiting, for example, the forecasting powers of meteorologists).

But it is only fair to note that also Sundman was partially correct. If one of the three bodies always stays far away from the other two, then one may describe the orbits in a predictable way, and even write mathematical formulae to describe them. Thus the three body problem shows the two sides of natural phenomena: if the initial state is known, at some level and in some circumstances the phenomena are predictable, just as Laplace claimed, but at another level and in other circumstances the phenomena are unpredictable.

The three body problem becomes considerably easier if one of the bodies is negligibly small in comparison with the other two. Then the two primary bodies follow elliptical orbits about each other and are hardly influenced by the third body. Then only the description of the orbit of the small body remains to be found. An even simpler case arises when the two primary bodies are in a circular orbit (the *Restricted Three Body Problem*). Karl Gustav Jacob Jacobi (1804–1851) made good progress in the study of this problem. His work allows one to decide right away what kinds of small body orbits are possible and what are not. Since the orbit of the Moon around the Earth is practically circular, one may apply the restricted three body theory to the motion of a spacecraft to be sent to the Moon. In case of travel to another planet, the planet and the Sun are the main bodies, and the spacecraft is the third body.

Orbits of Comets

Another important application of the restricted three body problem is the orbits of comets. The icy bodies of comets are much smaller than planets, only a few kilometers across. When comets pass near a planet, their mass is too small to affect the practically circular orbit of the planet. On the other hand, the orbits of comets are far from circular. In most cases the orbits are so elongated that the orbit is almost a parabola. Contrary to planets, which are close to the common plane of the Solar System, comet orbits are oriented more or less randomly with respect to this plane.

The present orbit of a comet is unlikely to be the original one. In a typical orbit, a comet goes 1000 times further from the Sun than Pluto. But when it comes to the region of the planets, in particular to the mighty Jupiter's gravitational field, its orbit is easily perturbed. The orbit may shrink so much that the comet is captured into a

---

[3] In a nonlinear system, a change in the state of the system depends on its present state. For example, $y = kx + b$ is a linear deterministic law for which the derivative $dy/dx = k$ does not depend on x. But the simple quadratic law $y = kx^2 + b$ is nonlinear; its derivative $dy/dx = 2kx$ depends on the value of x.

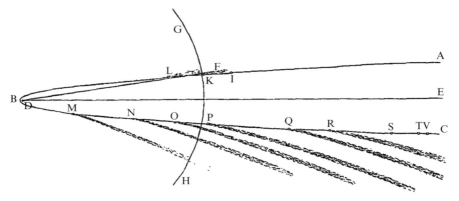

Fig. 11.4 The orbit of the great comet of 1680 was a very elongated ellipse as shown in an illustration from Newton's *Principia*

smaller orbit for a long period of time. Alternatively, the comet may gather speed from perturbations and escape from the Solar System altogether. Even if the comet orbits had been initially in the plane of the Solar System, the planetary perturbations would have thrown them off the plane, to orbits similar to where they are observed today.

A good example of a comet captured into the planetary region is Halley's comet. Its discovery history goes back to Newton, who showed how to calculate a comet's orbit after it has been observed in the sky on several nights. Using this method Edmond Halley started to calculate the orbits of comets which had been discovered during previous centuries. He got especially keen on the comets of 1531, 1607, and 1682 whose orbits appeared practically identical. He concluded in 1705 that it was one and same comet which, at the intervals of 76 years, closely approaches to the Sun in its elongated orbit. Also, information on the comets of 1305, 1380, and 1456 agreed with the orbit of the same comet. Thus, Halley predicted that the same comet would be seen again in 1758 (Fig. 11.4).

When the time of return of the comet came near, it occurred to the French astronomer Alexis Claude Clairaut (1713–1765) that the planetary perturbations may change the orbit so much that the comet does not return as predicted. Thus he started, with two assistants, a quick calculation of the effects of the planets. Clairaut worried that the comet would return before he had finished his calculations, but he was lucky. The calculation (finished in the fall of 1758) predicted that the comet should be late in appearance by more than a year, and that it should not be at its closest point to the Sun until the following March. The comet was discovered toward the end of 1758, and it reached the closest point to the Sun in March, as calculated by Clairaut. Halley's successful prediction, complemented by Clairaut's calculation, was regarded as a triumph of Newton's theory.

The comet was named after Halley, and its subsequent visits in 1835, 1910, and 1986 to the neighborhood of the Sun have been followed with keen interest. The methods of calculation of orbits have improved in 200 years to the extent that during its last visit in 1986 the arrival time of the comet was known in advance with the accuracy of 5 hours. If there were no other forces affecting the orbit than gravity, the arrival time would be known much more accurately. Gases are vaporized from the comet to form its extensive tail (see Fig. 11.6). These gas outflows act as small rockets which tend to take the comet off course in somewhat unpredictable way.

Interesting changes in orbits of comets can be caused by the perturbation of Jupiter. In 1770, Charles Messier discovered a new comet which came almost straight at the Earth and passed by us within just over 2 million kilometers. Anders Lexell calculated the orbit of the comet and found that its orbital period is only 5.6 years. This comet was the first example of a class of comets called short period comets. When the comet was not seen again for 10 years, Lexell started to look for a reason. According to his calculations, in 1779 the comet had passed close to Jupiter and it had obtained a new orbit which does not bring it close to Earth any more. The comet was rediscovered in its new orbit and is now called comet Lexell.

Lexell was probably the first scientist to realize the extreme sensitivity of the three body problem on the initial conditions, i.e., the deterministic chaos mentioned

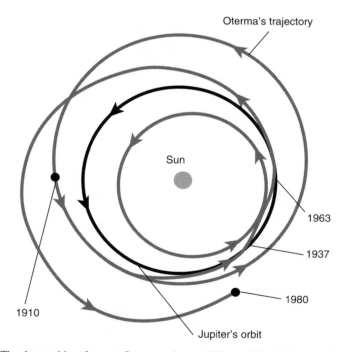

Fig. 11.5 The three orbits of comet Oterma: prior to 1937, in 1939–1962, and after 1964. The orbit of Jupiter is shown for comparison (adapted from a drawing by Shane D. Ross based on his calculations; by his permission)

above. This is apparent in unpublished comments which Lexell wrote in connection of the calculation of the orbit of Lexell's comet. It is interesting that in late eighteenth century the nondeterministic nature of Newtonian mechanics was already known even though it was totally overshadowed by deterministic successes of d'Alembert, Clairot, and others.

In another example of Jupiter's perturbation, in 1943, Liisi Oterma (1915–2001) of Turku University, Finland, discovered a faint comet and calculated its orbit. Surprisingly, the orbit turned out to be almost circular, contrary to the highly elliptic orbits of other comets. There is only one other comet known with similarly circular orbit. According to Oterma's calculations, this orbit was only temporary. Prior to 1937, the comet had been far from the Earth, outside the orbit of Jupiter. An encounter with Jupiter brought the comet inside Jupiter's orbit where it was discovered. Oterma predicted that the comet will return to its distant orbit past Jupiter in 1963, and this is what happened. Since then comet Oterma can be observed only by the biggest telescopes (Fig. 11.5).

Fig. 11.6 The Hale–Bopp comet as photographed at Tuorla Observatory in April 1997. At this time the comet's tail was double. The straight tail is the ionic tail pointing directly away from the Sun and the curved tail is the dust tail and trails the ionic tail. This comet has a long period of about 4,000 years after which one expects the next visit (photo by Harry Lehto)

Finally, the famous comet, Shoemaker–Levy was captured by Jupiter from a Sun-centered orbit into a bound Jupiter-centered orbit. In the close encounter, the head of the comet was torn into at least 21 fragments. In an event observed all over the world and from space, the fragments crashed into the atmosphere of Jupiter in 1994. Although the largest fragments were only a few kilometers in size, the impact sites could be seen even with small telescopes on Earth (see the color supplement).

Chapter 12
Nature of Light

What is light, that wonderful swift carrier of information without which we cannot investigate the depths of the universe nor the secrets of the microworld? For Newton, light was made of particles while Huygens regarded light as waves in a hypothetical medium, the ether. Thomas Young solved the question once and for all, or at least this is how it appeared.

Young started his career in medicine which he studied in London, Edinburgh, and Göttingen; however, he also eventually graduated from Cambridge University. Before graduation Young received an inheritance from a grand-uncle which was enough to secure his finances for the rest of his life (Fig. 12.1). Young practiced medicine in London but at the same time he became interested in matters connected with light, such as the eye, the origin of the rainbow, and others. He carried out experiments where a beam of light was split in two, and then the rays were brought together again.

Light as a Wave Phenomenon

What happens when two beams of light are combined? If light is made of particles, the intensity of light should increase: light + light = more light. But if light consists of waves, there is also another possibility: light + light = dark. Imagine waves like those on a lake with high crests above the surface and troughs below. Waves can destroy each other if the trough of one beam's wave strikes the surface at the same time as another wave's opposite phase crest. Young observed this phenomenon which is known as destructive interference (Fig. 12.2). Interference is characteristic of waves. As a very useful application of his experiment, Young was able to determine the tiny distance from one wave crest to the next, the wavelength of light. It varies between $0.4\,\mu m(1\,\mu m = 0.001\,mm)$ for violet light and $0.7\,\mu m$ for red light.

If light is a wave, what is vibrating? The surface of our lake example vibrates up and down, perpendicular to the direction of motion; it is a transverse wave. A sound wave propagates through air as a compression wave which takes place in

P. Teerikorpi et al., *The Evolving Universe and the Origin of Life*
© Springer Science+Business Media, LLC 2009

Fig. 12.1 Thomas Young (1773–1829) demonstrated the wave nature of light

the direction motion like a compression traveling along a spring. Young showed that light waves are transverse like the lake waves, as was independently found by Augustin Jean Fresnel (1788–1827) a little later. Namely, light can be polarized (as seen in today's familiar Polaroid sun glasses), which is not possible for compression

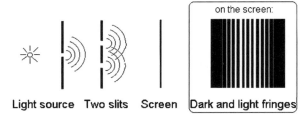

Light source Two slits Screen Dark and light fringes

Fig. 12.2 Young's interference experiment. Light arrives at the screen from two vertical slits. Instead of having two bright lines in the screen, we see a whole lot of alternately bright and dark fringes. At the bright fringes, the waves arriving from the two different slits reenforce each other. The path length difference from the slits to the location is either zero or one whole wavelength, so the two crests match. At the dark fringes, they cancel each other because the path difference is one-half wavelength, so a crest from one matches a trough from the other. The experiment shows that light is a wave phenomenon

Fig. 12.3 Newton decomposed the sunlight into the colors of the rainbow by using a prism, shown on the right. He then used a second prism like the one on the left to demonstrate that separate colors could not be broken down any more and concluded that light is a heterogenous substance essentially made of different components, colors. Illustration from Newton's *Opticks*

waves. As an argument against the wave theory, scientists of the time pointed out that the nature of the medium for light waves, the ether, as proposed by Young and Fresnel, was unknown.

As Newton had noted, when a beam of sunlight from a hole passes through a prism, the beam is spread out in all colors of the rainbow which apparently form a continuous band of colors, the spectrum of the Sun (Fig. 12.3). As sketched in the figure, light of a given color could not be broken down further by a second prism. From his clever experiment, Newton concluded that light is a heterogenous substance made of separate components, each having its own color.

Because the different colors of the wide beam overlap, Joseph Fraunhofer (mentioned earlier) used a careful arrangement of a very narrow slit and lenses as well as the prism to avoid overlap of the colors (this device is now called a spectroscope). Examining sunlight, Fraunhofer found that there are apparently missing colors in the spectrum of the Sun! The missing colors show up as dark lines; no image of the narrow slit is formed in those positions or wavelengths of the Sun's spectrum.

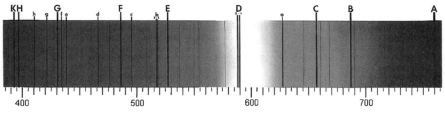

K & H = ionized calcium   A and B = molecular oxygen (our atmosphere)
f and F = Balmer lines of hydrogen (434 and 486 nm)
E = iron   D = natrium   C = Balmer line of hydrogen (656 nm)

Fig. 12.4 Positions of main Fraunhofer's lines on the spectrum of the Sun. Note that "A" and "B" have nothing to do with the Sun itself – they are due to oxygen molecules in our own atmosphere. The unit of the wavelength scale is $1\,\text{nm} = 10^{-9}\,\text{m}$

William Wollaston (1766–1828) had discovered the same phenomenon earlier, in 1802. He observed only a few lines and regarded them as natural borders between major colors. But Fraunhofer saw and measured as many as 600 dark lines; these missing parts of the spectrum became known as Fraunhofer lines. He noticed also that in sparks and in the flame of fire the spectrum of certain elements shows bright spectral lines which appear exactly at the same colors as certain dark lines in the spectrum of the Sun. For example, sodium produces a bright yellow color at the same wavelength as Fraunhofer's solar "D" dark line. Other Fraunhofer's lines are indicated in Fig. 12.4.

Spectral Analysis – Toward the Physics of Stars

The true value of Fraunhofer's findings was not appreciated for decades. Finally, around 1860, Robert Wilhelm Bunsen (1811–1899) and Gustav Robert Kirchhoff demonstrated the significance of the spectral lines in chemical analysis. Kirchhoff studied in Königsberg obtaining a professorship in Breslau University at the young age of 26. There Kirchhoff met Bunsen, and they became friends. When Bunsen moved to Heidelberg, he managed to secure a position there for Kirchhoff also. In 1871, Kirchhoff became professor of theoretical physics in Berlin. It was said that Kirchhoff had the ability to put his students to sleep rather than make them enthusiastic, but his students included Heinrich Hertz and Max Planck, both of whom became famous physicists (Fig. 12.5).

Kirchhoff's longest lasting achievements were in collaboration with Bunsen. Bunsen had started analyzing the chemical composition of samples on the basis of the color which they gave to the colorless flame in his famous burner. Kirchhoff suggested that it would be worth while to use a spectroscope so that the wavelength (or color) could be measured precisely. When this was done Fraunhofer lines were finally understood.

It turned out that the characteristic color of a flame was due to bright spectral lines which are at different wavelengths for different elements. Every element has its own fingerprint, a pattern of spectral lines, which appears when the sample is heated hot enough to make it into a thin gas. From the spectral lines, one can identify the chemical composition of the sample. In a letter dated 1859, Bunsen wrote: "At present I carry out research with Kirchhoff which does not allow us to sleep. Kirchhoff has made a completely unexpected discovery. He has found the reason for the dark lines in the spectrum of the Sun and he is able to reproduce these lines... in the continuous spectrum of a flame at identical positions with the Fraunhofer lines. This has opened up a path to the determination of the chemical composition of the Sun and of the fixed stars..."[1]

---

[1] As a matter of fact, Jean Foucault (1819–1868) had discovered in Paris the coincidence of the laboratory spectral lines and the spectral lines of the Sun already in 1849. But for some reason his discovery was forgotten, and unaware of Foucault's work, Bunsen and Kirchhoff repeated and improved his experiments.

Fig. 12.5 Gustav Robert Kirchhoff (1824–1887) identified dark lines in the spectrum of the Sun with spectral lines of earthly elements

Kirchhoff summarized the results of his researches in what are called Kirchhoff's laws (see also Fig. 12.6).

Kirchhoff's I law: Hot dense gases and solid bodies radiate a *continuous spectrum*. A spectrum is said to be continuous if all colors of the rainbow are represented in it so it does not possess dark spectral lines.

Kirchhoff's II law: A rare (low density) hot gas radiates a spectrum of bright spectral lines. Bright lines at only certain wavelengths are also called *emission lines*.

As mentioned, an emission line spectrum arises from heated thin gas in the flame of a Bunsen burner seen against a dark background. However, if an intense beam of light is shone through the gas in the flame from behind the burner, one might think that the light from the burner and the light coming from behind would add up. If the light coming from behind the burner has a continuous spectrum, one would expect that the bright lines from the burner would be superposed on the continuous spectrum. But this is not what Kirchhoff saw. Instead, the continuous spectrum is now seen with dark lines where each of the emission lines was previously located! He recorded this in his third law.

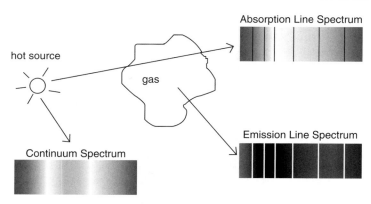

Fig. 12.6 The light of a hot source coming directly to the spectroscope shows a continuous spectrum, while the light that has traveled through gas has dark absorption lines. When we view the gas from the side against a dark background, its spectrum has bright emission lines. Investigating the spectra of stars and galaxies, astronomers measure their temperature and chemical composition and even their masses, speeds, and distances

Kirchhoff's III law: When a continuous spectrum passes through rare gas, dark lines are generated in the spectrum.

The dark lines are called *absorption lines*. In the spectrum of the Sun, the continuous spectrum comes from the lower, relatively hot (about $5,500°C$) and dense layers of the Sun's surface. On the way out from the Sun, light then passes through cooler and more rarefied layers of the solar atmosphere which imprint the dark Fraunhofer lines.

With the help of spectral analysis it became possible to analyze the chemical composition of the Sun and even of stars. For example, the two nearby dark "D" spectral lines in the solar spectrum are also seen as bright lines in the spectrum of hot sodium gas. From this Kirchhoff and Bunsen concluded that the Sun has plenty of sodium gas. In addition, they recognized in the solar spectrum signs of iron, magnesium, calcium, chrome, copper, zinc, barium, and nickel. By the end of the century, the discoveries included hydrogen, carbon, silicon, and an unknown element which was named helium after the Greek name for the Sun. In 1895, helium was discovered also on the Earth. Among all the elements, hydrogen had the simplest spectrum. Spectral lines appear in such a neat order that a Swiss high school teacher Johann Jakob Balmer (1825–1898) discovered a simple formula which gives their wavelengths. These spectral lines of hydrogen are called Balmer lines.

One cannot deduce the abundances of elements in the Sun entirely on the basis of the strengths of the spectral lines of each element. Through complicated calculations, including the temperature, it has become clear that hydrogen is by far the most common element in the Sun (though its spectral lines are not strong), and an equally clear second place goes to helium. Other elements constitute less than 2% of the Sun (see Table 12.1, where we also show the abundances of the most common elements in the Earth and in the human body). We return to the spectra of stars in Chap. 19.

Table 12.1 Relative proportions (by mass) of chemical elements in the Sun, the Earth, and the human body

The Sun		The Earth		The human body	
Hydrogen	71	Iron	34.6	Oxygen	65
Helium	27	Oxygen	29.5	Carbon	18
Oxygen	0.97	Silicon	15.2	Hydrogen	10
Carbon	0.40	Magnesium	12.7	Nitrogen	3
Iron	0.14	Nickel	2.4	Calcium	1.5
Silicon	0.10	Sulphur	1.9	Phosphorus	1.2
Nitrogen	0.10	Calcium	1.1	Potassium	0.2
Magnesium	0.08	Aluminium	1.1	Sulphur	0.2
Neon	0.06	Sodium	0.57	Chlorine	0.2
All other elements	<0.2		<1		<1

At present, the chemical analysis shows that other stars are not very different from the Sun. In particular, hydrogen is the most common element; its share of the mass of a star is typically about 72%. The mass fraction of helium is about 26% which leaves only about 2% for the rest of the elements. However, the latter fraction varies greatly from the surface of one star to another.

More Information from a Spectrum

Besides chemical composition, the spectrum of a star carries much more information. One of the most important is the motion of the star relative to the observer. The measurement is based on a principle proposed in 1842 by the Austrian scientist Christian Doppler (1803–1853). According to the Doppler principle, the wavelength of light changes in proportion to the speed of the emitting body. The phenomenon is well known in case of sound waves; for example, the siren of an emergency vehicle is heard at a higher pitch (or shorter wavelength) when the vehicle approaches us, and it turns to lower pitch (or longer wavelength) as soon as the vehicle has passed us (Fig. 12.7). In the same way, spectral lines in starlight have shifted toward the blue end of the spectrum, i.e., wavelengths have shortened, when the star approaches us. Oppositely, when a star moves away from us, the spectral lines shift toward the red end of the spectrum. The relative amount of shift, called redshift, tells the recession speed of the star.

In fact, Doppler thought that one might be able to infer a star's speed from its color. However, for typical stellar speeds the change of the color is undetectably small. A few years later, the French physicist Hippolyte Fizeau suggested, without knowing about Doppler's work, that one could use the narrow spectral lines as sensitive indicators of the tiny shifts of spectra of moving stars.

The proportion of energy in different wavelengths in the spectrum is very much the same independent of nature of the radiating body, be it, for example, a piece

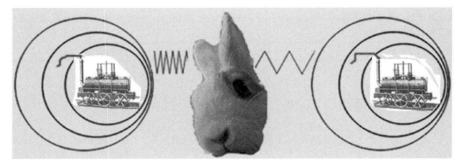

Fig. 12.7 Doppler effect: the sources emitting waves are moving relative to the sharp-eared observer who detects systematic differences in the wavelengths of the waves from the approaching and receding emitters

of iron or a distant star. This weighting of colors seems to depend only on the temperature of the body. Porcelain manufacturer Thomas Wedgewood noticed this already in 1792 while heating up different materials. About one hundred years later German physicist Wilhelm Wien (1864–1928) expressed the same idea more exactly in what is now called Wien's displacement law: The maximum energy wavelength of radiation is inversely proportional to the temperature of the body in Kelvin degrees (Box 12.1).

To be exact, Wien's displacement law is valid only for ideal bodies which both emit and absorb light with 100% efficiency. The name "blackbody" given to such ideal bodies emphasizes the absorption; if the body was not emitting any light, it would appear black. A hole in a laboratory oven makes a good blackbody since it is not possible that light can be reflected from the hole. Thus the thermal light coming through the hole can be regarded as blackbody radiation. Stars are also rather good examples of blackbodies. This concept originates from Gustav Kirchhoff.[2]

We remind that Kelvin degrees (K) are obtained from Celsius degrees by adding 273. The zero point in the Kelvin scale is the lowest possible temperature, the so-called absolute zero = −273°C. The absolute zero point of the temperature scale goes back to William Thomson (1824–1907). Thomson's father, a mathematics professor at Glasgow University, took his son to listen to his lectures at an early age. At the age of 10, William became an official student of the university, and at 15 he was reading the works of leading physicists. Two years later he enrolled at Cambridge University. At Cambridge he placed only second in the mathematics tripos, a great disappointment to him. When his father died in 1846, William followed him as professor of mathematics at Glasgow. He was in this position for 53 years.

---

[2] When a body is heated, its maximum emission color not only changes from red to blue, but also its total radiation (energy per second) increases. Austrian physicist Josef Stefan (1835–1893) proposed a formula for the increase of brightness (Stefan's law): the brightness of a body is proportional to the fourth power of its temperature (in K degrees).

Box 12.1 Wien's displacement law

The wavelength of the radiation maximum λ_{max} depends on the temperature T in Kelvin (K) as follows:

$$\lambda_{max} = 0.2898/T \text{ cm}$$

Temperature (K)	λ_{max}	Nature of radiation
3	0.97 mm	Radio waves
300	9660.0 nm	Infrared radiation
4,000	724.5 nm	Red light
6,000	483.0 nm	Yellow light
8,000	362.3 nm	Violet light
24,000	120.7 nm	Ultraviolet light
300,000	9.7 nm	X-rays
3 billion	0.1 nm	Gamma rays

Most celestial bodies shine because they are hot. Then the wavelength range where the radiation is peaked roughly tells the temperature of the body. Optical (visual) radiation comes from stars like the Sun (around 6,000 K), while very hot stars (say, 30,000 K) emit much ultraviolet light. Infrared light is emitted by much cooler planets and interstellar dust. X-rays come, e.g., from the corona of the Sun or from the several million-degree gas in clusters of galaxies.

The thermodynamical studies of Thomson led him to propose an absolute scale of temperature in 1848. At the absolute zero of this scale, thermal motions of molecules would theoretically cease. The Kelvin absolute temperature scale, as now known, derives its name from the title, Baron Kelvin of Largs, that Thomson received from the British government in 1892. Kelvin is the river flowing near his university.

We have come a good way toward understanding light: we have described its properties as a wave phenomenon and have told about some applications allowing us to measure a remote star's speed, chemical composition, and surface temperature from the spectrum of its light. However, to better understand the role of light in physical reality we must now turn to seemingly quite other phenomena, electricity and magnetism.

Chapter 13
Electricity and Magnetism

By the first decades of the eighteenth century, mechanics alone of all the branches of physics had obtained a somewhat modern form. When Newton died in 1727, another major branch of physics, the study of electricity and magnetism, was still rather elementary. The most important discoveries in this area were made in the following hundred years, finally leading, unexpectedly as it often happens in science, to a new unified view of electromagnetism, light, and other kinds of radiation.

Naturally magnetized iron ore or magnetite was known in antiquity. Also the electrostatic attraction of amber was mentioned by Plato, among others. However, we may regard William Gilbert (1544–1603), the private doctor of Queen Elizabeth I, as a pioneer of the scientific study of electric and magnetic phenomena. Gilbert studied medicine and mathematics at Cambridge, and practiced medicine in London. He was also an early supporter of Copernicus and the moving Earth. His studies in physics (which were a hobby) appeared in the book *De Magnete* in 1600.

Nature of Electricity

Gilbert regarded electricity as a liquid which is created or transported by rubbing, for example, when amber is rubbed by fur. He called this liquid elektrica after the Greek word for amber (many related words derive from this term, for example electron). He also showed that the Earth is a huge magnet, and studied it by using a miniature model of magnetite (Fig. 13.1). This helped him to explain why the compass needle points roughly in north–south direction. The actual magnetic pole of the Earth is at a latitude of 83° in northern Canada and is slowly moving to the north, about 40 km per year. By definition, a magnet's north pole is the end that points roughly to north. As we mentioned, Kepler contemplated a role for magnetism in planetary motion which was, of course, on the wrong track.

Another Englishman, Stephen Gray (1666–1736) announced in 1729 that electricity from rubbing can be conducted from place to place. He divided materials into conductors (e.g., copper) and insulators (e.g., glass) on the basis of this

P. Teerikorpi et al., *The Evolving Universe and the Origin of Life*
© Springer Science+Business Media, LLC 2009

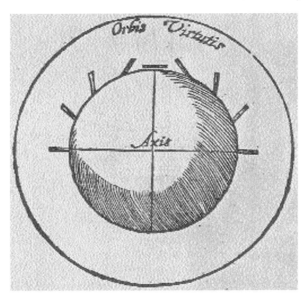

Fig. 13.1 An illustration from William Gilbert's book *De Magnete* (On Magnets). Gilbert knew that the compass needle was affected by Earth's magnetic field which he called Orbis virtutis. The compass had been in use in China since the early centuries AD, and it was known in Europe in the thirteenth century. The north and south magnetic poles are to the left and right

property. Frenchman Charles Du Fay (1698–1739) heard about Gray's work and started his own research. He concluded that there are two kinds of electricity: glass electricity and amber electricity. The former is generated, for example, by rubbing glass by a silk cloth while the latter arises in amber when it is rubbed by a piece of fur. He made this distinction by noting that bodies charged with like electricity repel each other while bodies of opposite electricity attract one another.

Du Fay's discovery was interpreted in many ways: there could be truly two kinds of electric fluid, or the fluid is only of one kind, but there can be an excess or a deficit of it, as suggested by Benjamin Franklin, among others. He regarded glass electricity as real, positive electricity, while the amber electricity would imply a shortage, or negative electricity. In his view, rubbing or any other operation does not create or destroy electricity, but it only leads to a transfer of electricity from one body to another. Thus he anticipated the law of conservation of electric charge, one of the cornerstones of current physics. The same idea had been proposed earlier by William Watson (1715–1787).

Franklin not only was one of the "founding fathers" in the American Revolution but he also invented the efficient Franklin stove, bifocal glasses, and the lightning rod. Franklin started as an apprentice of a bookbinder, and later became a book seller and publisher. At the age of 40, Franklin was a well-to-do man who could do whatever he liked. By accident, he happened to see an exhibition of the miracles

Fig. 13.2 Benjamin Franklin (1706–1790) was a polymath whose versatile interests, in addition to diplomacy included electricity

of electricity in Boston, and was so captivated that he spent the following 10 years studying electricity. However, he had to share his time with diplomatic duties, like helping write the Declaration of Independence and the US Constitution, and serving as the American ambassador in Paris (Fig. 13.2).

It was natural to make a comparison with Newtonian gravity when analyzing the electric attraction or repulsion. In addition to the two kinds of charge, the electrical force is a stronger version of Newton's force law making studies easier. English theologian and physicist Joseph Priestley (1733–1804) was the first to demonstrate that the force law between charges was indeed an inverse square law like in Newton's law of gravity. The most thorough studies of the electric force law were carried out by Charles Augustin Coulomb (1736–1806) in France; the force law has been named after him as Coulomb's law.

The discovery of the electric battery by the Italian physicist Alessandro Volta (1745–1827) opened up the field for sweeping research which changed the picture completely. Earlier this strong electric currents were generated only temporarily during electric discharges. Now every laboratory could equip itself with a powerful battery (Fig. 13.3). The power of the electric current for research increased by 10 000 fold. New secrets of nature were thus revealed.

Fig. 13.3 The big battery in the vaults of the Royal Institution was used, e.g., by Humphrey Davy in his experiments

Electricity and Magnetism are Combined

The next big discovery happened almost by accident. Hans Christian Ørsted (1777–1851), professor of physics at university of Copenhagen, was preparing a lecture on electricity and magnetism, and for that purpose he had brought a battery to the class to demonstrate the effects of an electric current. Next to it he had placed a compass needle for the demonstration of magnetic forces. He had noticed earlier that there may be some connection between electricity and magnetism, e.g., in the form of a compass needle flipping during a thunder storm. Since he had extra time, he decided to make a little experiment before the beginning of the lecture. Ørsted put an electric current close to the compass needle, and indeed his suspicions were confirmed: the compass needle moved at the introduction of the current. Thus the two separate phenomena, electricity and magnetism, which so far has been considered totally different, had after all some connection to each other. Ørsted continued his studies and published the results in 1820.

News about Ørsted's discovery spread fast. His article was read in the meeting of the French Academy of Sciences later in the same year. Among the members of the audience was Ampère who started immediately to work on explaining Ørsted's finding. The theory was ready in a week, and it provided the foundation for combining electricity and magnetism into the theory of electromagnetism.

André-Marie Ampère (1775–1836) was born near Lyon. His father was a well-to-do merchant who was executed during the French revolution when he held the position of Justice of the Peace in Lyon. One may still visit Ampère's home which is now a museum. Young Ampère did not go to school, but acquired knowledge by reading. He had a rare ability for memorizing and learning as described by the following incident. As a small boy he went to the library of Lyon and asked to read the works of the famous mathematicians Euler and Bernoulli. The librarian explained to the boy that they were difficult mathematical books which he could not possibly understand, and moreover they were written in Latin. The last part about Latin took Ampère by surprise, but he did not let his lack of Latin stop him. After a few weeks he came back to the library knowing Latin and started to read the books.

Ampère married when he was 24 and started to support his family as a school teacher. In 1808, he was appointed as inspector of schools, a position which he held for the rest of his life. In addition, he worked also as a professor in Paris. By 1820, when Ampère became interested in electromagnetism, he was already well known for his work in mathematics and chemistry. This versatile scholar started as professor of mathematics, then moved to the professorship of philosophy and later became professor of astronomy! Since 1824, Ampère was professor of physics at College de France.

Ampère was not satisfied to merely to explain Ørsted's results but started experiments of his own. For example, he demonstrated that by rolling up the electric wire into a coil it was possible to create an artificial magnet, an electromagnet, which corresponded fully to the natural magnets. Ampère concluded boldly, but quite correctly that natural magnets hold inside them small permanent current coils which act together to create natural magnetism.

Ampère realized right away the great significance of electromagnetic phenomena in information transfer. By turning current on and off one can make the compass needle move instantly in a far-away place. Messages can be sent as far as one can make the electric current flow. The development of telegraph machines working on this principle started soon. One of the first telegraph lines was established in Göttingen in 1834 between the physics laboratory of Wilhelm Weber and the astronomical observatory of Carl Friedrich Gauss. In the same year, the first commercial telegraph line between Washington and Baltimore in the USA was started by Samuel Morse (of Morse code fame).

Another scientist who immediately understood the great significance of Ørsted's discovery was the Englishman, Michael Faraday. Faraday was the son of a blacksmith and received only minimal formal education. At 13, he became an apprentice to a book binder. Besides binding books, he also read them. One of his customers gave him a free ticket to public lectures by Humphry Davy (1778–1829). Faraday made careful notes of the lectures, bound them nicely, and sent them to Davy with a note of enquiry whether Davy might have any job for him. Faraday was surprised when Davy invited him for a visit. The notes were neatly written and Davy got a good impression of the boy, so he decided to offer him a position as an assistant at the Royal Institution of London in 1820. Thus began one of the most remarkable

Fig. 13.4 Michael Faraday (1791–1867), in a painting by Thomas Phillips

carriers in science. It was said that Davy's greatest scientific discovery was Faraday (Fig. 13.4).

Faraday learned his science directly from Davy. When Davy made a one-and-half year tour of the continent, he took Faraday along. Here he met, among others, Ampère and Volta. And while Davy worked with Louis Gay-Lussac studying the new element iodine in Paris, Faraday acted as an assistant. The practical chemistry experiments were part of his duties also at home.

With the exception of a short period of interest in electromagnetism inspired by Ørsted's discovery, Faraday was a professional chemist up to year 1830. In 1833, he became professor of chemistry in the Royal Institution. But by then his scientific interests had changed. Faraday was convinced that if an electric current can cause a magnetic force, then a magnet must be able to create electric current. Here he agreed with many others, among them Ampère who however was not able to confirm the intriguing idea.

Faraday carried out different experiments on electromagnetism over 10 years. In 1831, he put two coils together, one inside the other. When electric current was put through one of the coils, it became an electromagnet. Faraday studied whether the magnet would cause electric current in the other coil. A current was indeed generated, but only momentarily when the electromagnet was turned on or off. This led Faraday to an important discovery: a *changing magnet*, for example a magnet with changing power or a rotating magnet, *generates electric current* in the nearby coil. It was crucial that the magnet was changing.

This is how Faraday discovered the electric generator, a simple dynamo which formed the basis of the electric industry of the future. Once when he was explaining a discovery to William Gladstone who was Chancellor at the time he was asked, "But after all what use is it?" Faraday quickly responded, "Why sir, there is every probability you will be able to tax it."

Force Fields

One of Faraday's big achievements was his new interpretation of how a force is transmitted between bodies. He saw lines of force penetrating through space instead of the action at distance. Faraday continued to develop his concept of lines of magnetic or electric force through the 1830s and 1840s. Because this novel concept was not mathematical, however, it was rejected by most scientists. Two important exceptions were William Thomson and James Clerk Maxwell. Thomson demonstrated how lines of force could be interpreted mathematically, and he also showed how Faraday's concepts of magnetic and electric force could be treated analogously to theories of heat and mechanics, thus laying the mathematical foundations of field theory. Faraday recognized the support from these "two very able men and eminent mathematicians" and "it is to me a source of great gratification and much encouragement to find that they affirm the truthfulness and generality of the method of representation."

For Faraday the concept of lines of force came naturally from experiments with magnets. When he sprinkled needle-like iron filings on a piece of paper lying on a bar magnet, he found that the filings lined up in very definite directions, depending their position relative to the magnet (Fig. 13.5). He thought that poles of the magnet are connected by magnetic lines, and that these lines are visualized by the help of the iron needles which line up parallel to the lines. For Faraday these magnetic lines were real, even though invisible.

Faraday generalized the concept of lines of force also to electric forces, and he believed that gravity could be treated similarly. Rather than saying that a planet knows by some strange reason how it has to move around the Sun, Faraday introduced a *gravitational field* which guides the planet in its orbit. The Sun generates a field in its vicinity, and planets and other celestial bodies feel the field and behave accordingly. Similarly, a charged body generates an *electric field* in its surroundings. Another charged body recognizes the field and reacts to it. Also there is a *magnetic field* associated with magnets.

In Newton's view, the basic entities are particles bound together by forces; the space between is empty. Faraday visioned both particles and fields in interaction with each other, our current understanding. One cannot say that particles are more

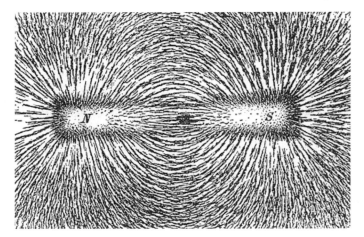

Fig. 13.5 Magnetic lines of force of a bar magnet shown by iron filings on paper

real than fields. It is customary to represent fields by lines which point to the direction of the force at each point in space (Fig. 13.6). The more densely spaced are the lines, the stronger is the force. Let us take the gravity of the Sun as an example. One may say that a whole lot of lines of force end at the Sun, and that they come equally from all directions. We may draw spheres of different radii centered on the Sun, hence the same lines of force cross every sphere. The area of the spheres increases as the square of the radius; thus the density of the lines decreases as the inverse square of the distance. Thus the concept of field lines leads us directly to Newton's law of gravity (and to Coulomb's inverse square law for the electric field of a stationary charge; Fig. 13.7).

A few simple rules must be followed when using the concept of force field (for gravity, as an example):

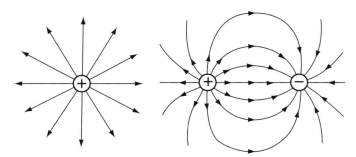

Fig. 13.6 Lines of force of a single positive charge and lines of force between a positive and a negative charge

Fig. 13.7 Gravitational lines of force associated with a spherically symmetric distribution of mass. The number of lines of force crossing a similar area decreases inversely proportional to the square of the distance to the mass center

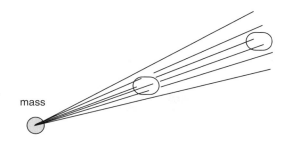

mass

1. The acceleration of gravity takes place along the force field which passes through the body.
2. The magnitude of acceleration is proportional to the density of lines at the point in question.
3. The lines of force can terminate only where there is mass. The number of lines terminating at a given point is proportional to the mass contained at that point.

Now it is easy to prove a result which caused Newton a lot of trouble. While comparing the accelerations at the Earth's surface and at the orbit of the Moon, Newton assumed that the Earth attracts bodies as if all of its mass were concentrated at its center. Why is this so?

Assume for simplicity that the Earth is completely spherically symmetric. Thus, all parts of its surface are equally covered by incoming lines of force. The number of the lines is by the third rule only dependant on the mass of the Earth. If all the Earth's mass were at its center, all the same lines would continue to the center. Thus, the Earth's gravity field is quite independent of how the mass is distributed under its surface, as long as the spherical symmetry is valid. In particular, the Earth mass concentrated at the center causes exactly the same gravity as the real Earth.

Similar deductions apply to the electric field. Since there are two kinds of electric charges, positive and negative, the direction of the line of force changes to the opposite when the sign of the charge is switched. Lines of force start from a positive charge and end at a negative one as seen in Fig. 13.6.

Electromagnetic Waves

The lines of force were seen intuitively by Faraday, but he was not able to put the theory in mathematical form. This task was completed by James Clerk Maxwell, the greatest theoretical physicist of the nineteenth century. Maxwell had a strong educational background; he was enrolled at the University of Edinburgh only at the age of 15. Three years later he moved to the University of Cambridge graduating in 1854. Two years later he became professor of physics at University of Aberdeen in Scotland from whence he moved to London. In 1865, he moved to his country

Fig. 13.8 James Clerk Maxwell (1831–1879) predicted electromagnetic waves and Heinrich Hertz (1857–1894) demonstrated their existence

estate Glenlair, not far from Glasgow, where he wrote his famous work *Treatise on Electricity and Magnetism*, published in 1873 (Fig. 13.8).

In the meantime, the University of Cambridge received a large donation from the heirs of Henry Cavendish (1731–1810) who was renowned for his studies of electricity, for the purpose of setting up a laboratory of physics. Up to then the physicists in the university had carried out their experiments in their own college rooms. There was a new professorship associated with the donation; Maxwell was chosen to the job in 1871. He started the distinguished series of Cavendish professors whom we will discuss later: John Strutt, better known as Lord Rayleigh, Joseph Thomson, and Ernest Rutherford. Over the years about 30 Cavendish Laboratory scientists have been honored with the Nobel prize in the fields of physics, chemistry, and physiology.

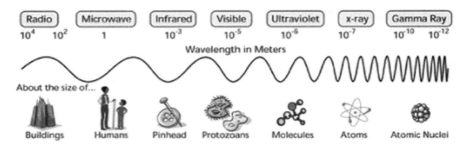

Fig. 13.9 Different kinds of electromagnetic waves and their wavelengths (credit: NASA)

Maxwell combined the separate laws of electromagnetism discovered by Coulomb, Ampère, and Faraday into what is known as Maxwell's equations, treating electricity and magnetism together as a single phenomenon, electromagnetism. From Maxwell's equations one could deduce that vibrating electric and magnetic fields can proceed through space with a high speed which Maxwell calculated. The value was so close to the measured velocity of light that Maxwell wrote in a long letter to Faraday (1861): "I think we now have strong reasons to believe, whether my theory is a fact or not, that the luminiferous and the electromagnetic medium are one...." And in a later paper he wrote: "The agreement of the results seems to show that light and magnetism are affections of the same substance, and that light is an electromagnetic disturbance propagated through the field according to electromagnetic laws."

Thus, light is made of electric and magnetic fields which oscillate perpendicular to the direction of propagation agreeing with the previous discovery of polarization. In a remarkable experiment in 1887, Heinrich Hertz tested Maxwell's hypothesis of *electromagnetic waves*. He was able to produce and detect another form of electromagnetic radiation, radio waves. The only difference between radio waves and light is that in the latter the oscillations of electric and magnetic fields have a much higher frequency than in radio waves. The consequence of rapid oscillations is a short wavelength; in typical light the wave crests are separated by half-a-micrometer (= 0.0005 mm). In radio waves the crest separation is from 1 mm upward, all the way to kilometer-long waves.

Between radio and light, infrared heat radiation has wavelengths between a micrometer and a millimeter. Waves too short to be detected by eye just beyond violet light are called ultraviolet radiation. In 1895, Wilhelm Conrad Röntgen (1845–1923) discovered x-rays by accident. They appered to go through matter like it was nothing. By placing his hand in front of the x-ray tube and a screen he was surprised to see the bones of his hand (the first x-ray examination). X-rays are electromagnetic radiation with wavelengths shorter than the ultraviolet. The very shortest wavelength radiation called gamma radiation was discovered a few years later during studies of radioactive elements (Fig. 13.9).

Chapter 14
Time and Space

Remember the quote by Thomas Huxley in our discussion of the successes of celestial mechanics, "science is nothing but trained and organized common sense." During the 1700s and 1800s, common sense was applied to atoms. Following Newton, we may imagine atoms as small billiard ball-like spheres, interacting by bouncing off each other. In many ways, this application was successful, but, early in the last century it became obvious that to describe nature at the atomic level, it appears that some microlevel phenomena and those involving high speeds do not "make sense." As a sign at the gate of an English physics department warned: "Beware. Physics may expand your mind!"

The Strange Speed of Light

The first "nonsense" result of physics came from the American physicists Albert Michelson and Edward Morley in 1887 who tried to measure the motion of the Earth through space by studying from which direction the light comes at highest speed. Indeed, one would expect that the light should appear to come faster from the direction toward which we are heading as compared with other directions. This is based on everyday experience when moving through air. Michelson and Morley calculated that the travel time back and forth between two parallel mirrors should be at its greatest when the line connecting the center points of the mirrors is parallel to the direction of the motion of the Earth, and at its smallest when the ray of light between the mirrors travels perpendicular to this motion (Figs. 14.1 and 14.2).

Michelson and Morley estimated that the travel time difference in their experiment should be small, but easily measurable. However, no time difference was seen. The conclusion is that light travels always at the same speed independent of the state of motion of the measuring apparatus. An experimenter at rest measures the same speed of light as another going toward the source of light, or away from the source.

P. Teerikorpi et al., *The Evolving Universe and the Origin of Life*
© Springer Science+Business Media, LLC 2009

Fig. 14.1 (**a**) Albert A. Michelson (1852–1931) and (**b**) Edward W. Morley (1838–1923)

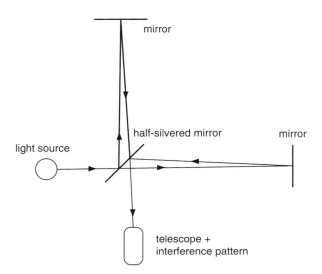

Fig. 14.2 Michelson's interferometer. Light from a source is split into two beams using a semi-transparent mirror. The two beams travel in perpendicular directions, and are reflected back from two mirrors. The reflected light beams are directed back through the same semitransparent mirror with the combined beams going to a telescope. By analyzing the fringes arising from the interference of the two light beams one can determine how the speed of light depends on the motion of the Earth (and the apparatus) through the space

Rowing in a river offers an everybody analogy illustrating this unexpected insensitivity of the motion of light to the "ether stream". In a peculiar competition one boat races back and forth across the river while another one races an equal distance up and down river. Assume that both boats have the same speed relative to the water. The speed of the second boat is enhanced downstream and slowed down upstream. A simple calculation shows that the person rowing crosswise to the current clears the round trip faster than the equally strong competitor who goes parallel to the current. However, light does not behave in this "reasonable" manner.

Michelson and Morley (and other experimenters, too) demonstrated without doubt that light is not an ordinary wave propagating in an ordinary medium. If the principle of their experiment is applied to sound waves or any other waves propagating in a medium (like water), then the difference in speed and the direction of motion can always be determined. Maxwell had proposed that light can be understood as vibrations of the electromagnetic field, believing that vibrations take place in ether. Giving up the ether did not solve the problem of the unchanging speed of light. It was necessary to introduce an entirely new way of thinking of the nature of space and time, as was done by Albert Einstein.

Albert Einstein

Einstein was born in Ulm, Germany, to a Jewish family. He had difficulty in adapting to the school system, which he was forced to leave at 16. His father, hoping that his son would get into business, looked for alternatives to continue his education. The Technical University of Zurich admitted Albert without the German high school diploma (on the second try), and at the age of 21, in 1900, he graduated. Einstein's next problem was finding a job. After 2 years of temporary employment, he finally became a technical officer of the patent office of Bern. It turned out to be a suitable job, and while working, Einstein finished his PhD which was accepted after some initial trouble.

There was nothing in Einstein's early career to anticipate the 1905 miracle: three articles in the esteemed journal *Annalen der Physik* which made Einstein perhaps the most famous scientist of the century leading to the Nobel Prize. The articles dealt with Brownian motion, "light gas," and Special Relativity. The first article gave crucial arguments in favor of matter consisting of atoms, a fact by no means generally accepted at the time. The second article gave a new interpretation of the nature of light and the third, most famous, article discussed in a novel way the concepts of space and time, and, among other things, later led to a prediction of the enormous reserves of energy hidden in matter.

Einstein's research was not unnoticed, but it took a while before it became general knowledge among professionals. Einstein was appointed a docent in the University of Bern in 1908, but his university career started properly a year later when he became associate professor at the University of Zurich. He moved to Prague in

Fig. 14.3 Albert Einstein (1879–1955) and Hendrik Lorentz (1853–1928) in Leiden in 1921

1911. The time in Prague was significant to Einstein's career since he learnt there new mathematical methods with the help of his assistant Georg Pick. These were necessary for his next great step forward in physics.

Only a year later Einstein returned to Switzerland, to his *alma mater* in Zurich where he started developing the General Theory of Relativity together with Marcel Grossmann. This is a new theory of gravity which improves on the Newton's theory. Einstein became so famous that he was invited in 1914 to become the head of the Physics Department at the Kaiser Wilhelm Institute in Berlin and a member of the Prussian Academy. Here he published the foundations of the General Relativity in 1916. During the solar eclipse of 1919, British delegations organized by Arthur Eddington observed the bending of light predicted by Einstein, thus making his theory a serious rival of Newton's theory.

By now the name of Einstein had become widely known among the general public, not entirely a positive thing (Fig. 14.3 shows him at the early 1920s). There was a campaign against him because of his Jewish origins. In 1922, a Jewish friend was murdered and there was a rumor that he would be next. Einstein was also well known for his antimilitarist comments. During Hitler's coup in 1933 Einstein was safely in California. He gave up his German citizenship before it was taken away. In Germany, Einstein was berated as an enemy of the state, and his books were burnt among other literature deemed dangerous. In 1934, Einstein settled at Princeton, New Jersey, living there rest of his life, working to unify electromagnetism and gravity under a single theoretical framework. He did not succeed, and neither have others. In his last years, he fought to ban nuclear weapons. Their development was not only based on his principle of equivalency of matter and energy but the United States nuclear program was started as a result of a 1939 letter Einstein sent to President Roosevelt after hearing of German research on the fission of uranium.

There are many stories about Einstein. One evening there was a phone call to the president of Princeton. The president was out, so the caller asked: "Perhaps you could tell me where Dr. Einstein lives." He was told that the information was confidential in order to protect Einstein's privacy. The voice in phone continued: "Please don't tell anybody but I am Dr. Einstein. I am about to leave for home but I have forgotten where my house is." Einstein had recently moved to his new home and had not yet learnt to remember its coordinates . . .

Four-Dimensional World

In his Special Theory of Relativity, Einstein accepted Michelson and Morley's observation that the speed of light is a constant, c, independent of the state of motion of the observer. He did not ask why, but rather addressed what consequences derive from this odd fact. What are space and time? The constant speed of light does not make sense in everyday life; our idea of what makes sense originates from our experiences where many features of reality are blurred. In fact, the familiar formula "speed = distance/time" suggests that the speed of light can be the same for everybody only if space and time are linked in a way which nobody had anticipated.

The entanglement of space and time coordinates means that we live in a special kind of *four-dimensional* world (see Box 14.1). The nature of time is different from the three spatial dimensions (length, width, height), and not only because we measure time by a clock, and distance by a ruler. Hermann Minkowski (1864–1909), one of Einstein's teachers, explained this view in 1908 as follows: "From henceforth, space by itself, and time by itself, have vanished into the merest shadows and only a kind of blend of the two exists in its own right."

Everyone has one's own four-dimensional space–time which differs from the space–time of others by larger amounts for faster relative motions. Generally, the differences become significant only when the motion is close to the speed of light. Since our own motion never reaches such speeds, we fail to notice the real

Box 14.1 Space, time, and event

In relativity theory, and in everyday life, we use space and time together to describe an event. Let the event be, say, the signing of a document. Then the "coordinates" are written like this: In Turku, Finland on 26 March, 2007. The spatial location is given by the name of the city Turku while the time coordinate is 26.3.2007 (if we are not too particular about the exact place and time of signing).

Space and time are not absolute quantities. They are described by coordinates like the positions in a map. We may say that the city of Pori is 115 km north and 20 km west of Turku or that Pori is 0 km north and 105 km west of Tampere. Both coordinates of Pori, (115,20) and (0,105) are correct, but remember that the former coordinates have been measured from Turku, the latter from Tampere. But the distance between two places is independent of the reference system. The coordinates of the cities, Turku (0,0) and Pori (115,20), tell that their difference in the north–south is 115 km and in the east–west 20 km. Thus the distance (in kilometers) between them is the square root of $115^2 + 20^2$, or 117 km.

After space coordinates and distances, let us now inspect space–time coordinates. As an example, calculate the space–time interval between two events; let the difference in time be 40 s and the position difference 15 light seconds, then the interval is the square root of $40^2 - 15^2$. It is 37 s, and it is independent of what coordinate system has been used. Note that when we calculated the space–time interval, we used a minus sign inside the square root. If we are dealing with ordinary spatial (Cartesian) coordinates, then the square root must have a plus sign inside, as follows from Pythagoras' theorem. The minus sign underlines the different nature of space and time; it is because of this minus sign that the entanglement of space and time appears to go beyond "common sense."

Note that in this formula for calculating the interval, the distance is measured in light travel time units, in light seconds. It is the distance traveled by light in 1 s, a little less than the distance between the Earth and the Moon. One could also use light years which is the distance traveled by light in a year; the nearest star Alpha Centauri is about 4 light years from us. Using these units, the interval between two events also comes out in time units.

The special nature of the interval between two events is highlighted by a simple example. Let event one be the occasion when a beam of light starts its journey from some point in space, and let the event number two be when the same light arrives at another point in space. Then interval between the events is zero. The explosion of a nova in our Galaxy, and the arrival of the information to us about the explosion are two events whose space–time separation is (surprisingly!) zero.

connection between space and time. We assume that our time ticks away with the same rate as our neighbor's time, but this is true only as long as we are in the same state of motion as our neighbor.

One totally unexpected fact in relativity theory is that two observers may measure different spatial distances and time intervals between two events if the observers are in motion relative to each other. The formulae for these differences, so-called *Lorentz transformations*, were derived already in 1887 by Woldemar Voigt (1850–1919), on the basis of Maxwell's equations, and later by Hendrik Lorentz (see Fig. 14.3), who thus laid out the mathematical groundwork for relativity. We remember that the constant c already appears in Maxwell's equations. It is curious that the first relativistic theory was Maxwell's electromagnetism which was constructed before relativity theory itself! When he invented his famous equations, Maxwell did not know that they were hiding a treasure, the theory of relativity.

Time Dilation

The passing of time is measured by intervals between events, such as swings of a pendulum. This time appears to slow down in a fast moving clock in comparison with the time counted by a stationary observer with a clock. The clock carried by the observer measures the "correct" time (called proper time), while moving clocks indicate longer time intervals. This strange effect is called time dilation.

To test the reality of time dilation, in 1971, Americans Joe Hafele and Richard Keating sent four accurate atomic clocks in commercial airlines around the Earth, once eastward and once westward. Even though an airplane is much slower than light, it causes a tiny slowing down of time as compared with time passing on the ground. The difference can be detected by comparing the clocks which made the round trip around the Earth with clocks that stayed in one place all the time. As the ground itself is in rapid motion due to the eastward rotation of the Earth, the time dilation depends on whether we travel eastward or westward. A person traveling westward, opposite to the rotation, actually goes around the Earth more slowly than a person staying on the ground. For that reason, the clock which has flown around the Earth was seen to be fast in relation to the clock on the ground by 0.27 millionths of a second. When traveling eastward, the airplane speed is added to the ground speed. As a result, the traveling clock was found to be slow by 0.06 millionths of a second after the 3-day trip. The measurements agree well with Einstein's theory according to which the four clocks should have lost 40 billionths of a second on the eastbound trip and gained 275 billionths of a second on the westbound. The actual results were only 5% off on the eastbound and no more than 30% on the westbound flight.

Time dilation may become useful in future long space flights. We are accustomed to the acceleration of the Earth's surface gravity. Thus, if the spacecraft is constantly accelerated with that same acceleration, we should feel quite comfortable; the floor facing the back of the spacecraft would appear to pull us just like the Earth pulls us.

If we want to stop at our destination, we must start slowing down the spacecraft half way through the journey by reversing its orientation, and if we again use the same amount of deceleration, we will feel the familiar pull.

If by this method, we visit the Andromeda galaxy 2.5 million light years away from us, the round trip takes roughly five million years, since a good part of the journey happens with nearly the speed of light. But time in the spacecraft is dilated so much that on return, the travelers are only about 60 years older than when they left. Meanwhile, there has been 5 million years of unpredictable evolution on the Earth!

Time dilation stays hidden in our usual low-speed life; however, elementary particles in the laboratory can move with high speeds. The alpha particles emitted in radioactive decay travel at about 10% of the speed of light. In accelerators, particle speeds are only a fraction of a percent below the speed of light. For high-energy physics, time dilation and other relativistic phenomena are everyday events.

Mass and Energy

A celebrated result found by Einstein is the connection between mass and energy. All matter has hidden energy by the amount

$$\text{Energy} = \text{mass} \times (\text{speed of light})^2.$$

Since the speed of light has a big number, this formula implies that even a small bit of matter contains a huge amount of energy. If 1 g of matter could be turned totally into energy, it would provide 10^{14} J – roughly the same amount of energy is liberated

---

Box 14.2 Addition of speeds in Relativity Theory

Let the speed of a rocket be V with respect to the ground, and the speed of a bullet shot from the rocket forward v. Then the Special Theory of Relativity gives the speed of the bullet relative to the ground:

$$v' = (v+V)/(1+vV/c^2),$$

where c is the speed of light. If $v = 0.75c$ and $V = 0.75c$, then $v' = 0.96c$. If we take a beam of light in place of the bullet, then $v = c$ and the formula gives us $v' = c$. This is consistent with Einstein's assumption that the speed of light is independent of the speed of the emitter (or receiver) and explains the result of Michelson and Morley's experiment. Note that if the speed of light were infinite, then we would have the usual velocity addition formula. Note also that if the velocities V and v are both very small relative to the velocity of light, then $vV/c^2 \ll 1$ and the result is practically the same as the familiar velocity addition formula $v' = v+V$.

by burning 10,000 barrels of oil. The enormous power of nuclear energy is based on liberating a small fraction of the mass of the atomic nucleus into energy. The energy of the Sun is produced by nuclear reactions where four protons combine to make a helium nucleus. We will discuss these reactions in Chap. 19.

The mass of a stationary body is called its rest mass; when the body acquires a state of motion, its mass increases, until it grows many times relative to the rest mass at very high speeds, close to the speed of light. The growth of mass helps us to understand why material particles cannot reach the speed of light. In theory the mass (and energy) of a body traveling at the speed of light is infinitely large and that is obviously impossible.

We might think that it is easy to exceed the speed of light by sending out a rocket with the speed of 75% of the speed of light, and by then shooting a bullet from the rocket in forward direction, say, again with 75% of the speed of light. No fundamental problems with increasing mass appear at these speeds. By the usual algebra, the speed of the bullet should be 1.5 times the speed of light relative to the

Fig. 14.4 Henri Poincaré was a pioneer in chaos theory and stated the Principle of Relativity

ground. But it is not, because the algebra of nature gives the puzzling result $0.75 +$ $0.75 = 0.96$ when we are adding speeds relative to the speed of light (Box 14.2).

Principle of Relativity

We conclude this excursion to relativistic phenomena by discussing briefly the Principle of Relativity which lies at the heart of the Special Theory of Relativity. We learned in Chap. 7 about Galileo's principle of relativity – an observer participating in a uniform motion cannot detect this motion by mechanical experiments. Michelson and Morley showed that one cannot detect one's uniform motion relative to absolute space (or ether) even by the means of light rays. Such results prompted the mathematician Henri Poincaré (1854–1912) to formulate in 1904 the principle of relativity "according to which the laws of physical phenomena should be the same, whether for an observer fixed, or for an observer carried along in a uniform movement of translation; so that we have not and could not have any means of discerning whether or not we are carried along with such a motion." Originally, in 1902, Poincaré spoke about "the principle of relative movement." Here we can see the root of the word "relativity" – we study phenomena measured by observers moving at different uniform speeds relative to each other (Fig. 14.4).

In his 1905 paper, Einstein emphasized that "the phenomena of electrodynamics as well as of mechanics possess no properties corresponding to the idea of absolute rest" and "the same laws of electrodynamics and optics will be valid for all frames of reference for which the equations of mechanics hold good." In addition to this Principle of Relativity, Einstein stated that "light is always propagated in empty space with a definite velocity c which is independent of the state of motion of the emitting body." From these two postulates, Einstein derived his Special Theory of Relativity where a "lumiferous ether" proved to be superfluous and no absolute space was needed.

Chapter 15
Curved Space and Gravity

Our ordinary view of space is such that it resembles Euclidean geometry. In fact, in the Special Theory of Relativity the spatial part of the four-dimensional space-time is flat (Euclidean). Euclid, who worked in Alexandria around 300 BC (practically nothing else is known about his life), developed a system of geometry which is still part of our mathematics curricula. It was based on five "obviously true" axioms, out of which a rich collection of 465 theorems were derived (the essential knowledge of geometry). Among the five axioms, the most widely discussed is the last axiom which states that

- Through a given point in a plane one can draw one and only one line parallel to a given line in this same plane.

Remember that lines are parallel if they lie in a common plane and never cross each other. Euclid and many of his followers had misgivings about this *Parallel Postulate*. Though it seems intuitively true, there was no way of confirming it experimentally. Let there be a straight line through point *P* which is parallel to another line *S*. If we now rotate our line ever so slightly, how do we know that it does cross the line *S* after this rotation? In practice, we are always dealing with limited segments of straight lines, and cannot observe the whole of the straight line. But perhaps it could be inferred from the other four axioms? In fact, for two millennia mathematicians tried to demonstrate that the fifth postulate is implied by the others. All these attempts failed.

Discovery of Non-Euclidean Geometries

Not until the nineteenth century did it become clear that the fifth axiom can be replaced, ending up with other systems where geometric relations are different from what we are used to. Among the many possibilities, there are two interesting cases: *hyperbolic geometry* which was discovered independently by Carl Friedrich Gauss, Nikolai Lobachevski, and Janos Bolyai (Fig. 15.1), and *spherical geometry*, invented by Georg Riemann. Besides the Euclidean *flat geometry*, these two are the

Fig. 15.1 The inventors of hyperbolic geometry Karl Friedrich Gauss (1777–1855) (*center*), Nikolai Lobachevski (1792–1856) (*right*), and Janos Bolyai (1802–1860) (*left*)

only possible descriptions of a universe which is homogeneous and isotropic, that is, where all places and directions are equivalent. They are all important for modern cosmology.

The Russian Lobachevski, professor and rector of University of Kazan constructed a logically consistent geometrical system in which Euclid's parallel postulate was replaced by another axiom:

- Through a given point in a plane, there can be drawn an infinite number of lines, which do not intersect a given line in the plane.

He called this system "imaginary" geometry (also "pangeometry") and entertained the philosophy that there is no field in mathematics, however abstract, which cannot one day be applied to the real world. Gauss, Bolyai, and Lobachevski were unaware of each other's work. However, Lobachevski was the first to publish an article on the new geometry. It appeared in *Kazan Messenger* in Russian in 1829 and passed unnoticed. Trying to reach a broader audience, he published it in French in 1837, then in German in 1840, and then again in French in 1855. Lobachevski was successful in his job as the head of Kazan University and even received an award from Tsar Nicholas I. However, in 1846, he retired (some say he was dismissed by the University) and not until years after his death was his name associated with the discovery of the non-Euclidean geometry. The last government citation received by Lobachevski a few months before his death was for the discovery of a new way of processing wool!

At the same time, unaware of Lobachevski's work, Hungarian Bolyai "created a strange new world out of nothing." Both Bolyai and Lobachevski first tried to prove the fifth postulate but, in time, felt the task impossible: Bolyai in 1823, Lobachevski in 1826. Janos Bolyai's father, Farkas – a friend of Gauss and a mathematician himself – had studied the same problem. When he read his son's work, Farkas urged him to publish it and included it as a 26-page Appendix to his book that appeared in 1832.

Gauss, in a letter to Farkas Bolyai, approved of the work but claimed to have developed the same ideas some 30 years earlier. Janos was crushed by Gauss' letter.

He had lost his priority, and he never wrote anything on the subject afterward. Gauss invented the term "Non-Euclidean Geometry" but did not publish anything on the subject since "he was most unwilling to get involved in something that would put him under criticism," as he explained in his letter of 1829. In a private letter of 1824, Gauss wrote: "The assumption that (in a triangle) the sum of the three angles is less than 180° leads to a curious geometry, quite different from ours, but thoroughly consistent, which I have developed to my entire satisfaction."

Riemann developed the mathematical methods which are required for calculations in the non-Euclidean geometry. This field of mathematics, which took a while even for Einstein to learn, is now known as tensor calculus. Tensors are complicated quantities compared with vectors used to describe the electric fields. As an example of a tensor, we may mention the curvature tensor which is used to describe the way space curves, i.e., how it differs from Euclidean space. For four dimensions, the curvature tensor is described by 20 components. In contrast, the electric field can be described by only three components.

Already as a child, Georg Riemann (1826–1866) exhibited exceptional mathematical skills. He also studied the Bible intensively and enrolled at the University of Göttingen in 1846 to study theology, following his father's wishes. After attending some mathematics lectures, he asked his father if he could transfer to study mathematics. His father agreed, and Riemann then took courses from professor Gauss, among others. His Ph.D. thesis was also supervised by Gauss. Riemann was then employed by Göttingen University while he worked for his Habilitation. To complete the process, in 1854, Riemann had to give a Habilitation lecture. The lecture *On the hypotheses that lie at the foundations of geometry*, a classic of mathematics,

Fig. 15.2 Georg Riemann developed mathematics which paved the way for General Relativity

discussed the definition of the curvature tensor and posed deep questions about the relationship of geometry to the world we live in. What is the dimension of real space and what geometry describes our space? Riemann proposed that space itself could have measurable properties (Fig. 15.2).

The lecture was too far ahead of its time to be appreciated by most scientists. The general line of thought up to then, supported by Newton, was that space is some kind of rigid background against which the measurements are carried out. Among Riemann's audience, only Gauss was able to appreciate the depth of the young mathematician's thoughts. Returning to the faculty meeting, he spoke with the greatest praise to the professor of physics, Wilhelm Weber, about the originality of Riemann's work.

Properties of Non-Euclidean Geometries

Whether the universe is finite or infinite, there are difficulties in "seeing" it. Euclidean geometry splendidly describes our usual measurements. But infinity is hard to visualize even in everyday geometry. On the other hand, it is also painful to try to imagine a finite world with a spherical geometry, even though its finiteness is easy to understand mathematically.

The common way to visualize non-Euclidean geometry is to use surfaces as an example. Our three-dimensional universe, when we ignore the time, is flat for all practical purposes, and we can easily see the curvature of ordinary surfaces in it. However, it is difficult to imagine a four-dimensional space, not to mention what its curvature might mean. Our brains are not used to tackle such problems; thus it is best to limit ourselves in looking at two-dimensional surfaces.

The spherical universe has the odd property that the space has a finite volume even though one cannot find an edge in any direction. This is easier to understand by thinking of the surface of a sphere, which helps us to grasp also another interesting property of the spherical geometry: a traveler who is going straight ahead will come back to his/her starting point after having traveled right around the world. On the Earth, if you go straight ahead along a great circle you will return to your original position, a strange experience if you believed that the Earth is flat!

As one might guess, the two-dimensional counterpart of a spherical space is the surface of a sphere. It is not necessary to be aware of the third dimension or to go around the sphere to infer the curvature of the spherical surface. A creature living on the spherical surface which does not possess any third dimension off the surface, and who does not even understand what the third dimension would mean, can carry out geometrical measurements on the surface to find out the overall geometry. One may draw a triangle, and measure the sum of the internal angles. If the result is more than 180°, that determines right away that the creature lives on a spherical surface (Fig. 15.3). Alternatively, he may draw a circle and measure it. If the ratio of the

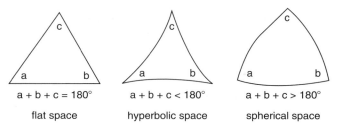

Fig. 15.3 Triangles in flat, hyperbolic, and spherical space. The sum of the angles differs in different spaces

circumference to the diameter of a drawn circle is less than π (pi $= 3.141592\ldots$), the creature would know that he lives in a spherical geometry.

In the contrary case, when the sum of the internal angles of a triangle is less than 180 degrees, when the ratio of the circumference of a circle to its diameter is greater than π, and when one can draw any number of lines through a given point which are parallel to another line, the creature knows that he lives in hyperbolic space. Hyperbolic space continues to infinite distance and does not have a counterpart in everyday experience. A saddle, more exactly its central part, curves more or less like a limited area of the hyperbolic surface.

The important borderline case between a spherical surface and a hyperbolic surface is a flat surface, or a two-dimensional Euclidean space. The laws of Euclidian geometry which we are familiar with, are valid in this and only in this geometry: the sum of the internal angles of a triangle is exactly 180 degrees, the ratio of the circumference of a circle to its radius if exactly π, and one can draw one and only one straight line through a point, parallel to another straight line (Fig. 15.4).

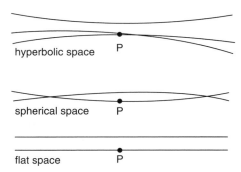

Fig. 15.4 Parallel lines in different spaces. In flat space one can draw only one straight line through a given point P which is parallel to another straight line. In hyperbolic space, one can draw any number of such straight lines. In spherical space, all straight lines cross one another, i.e., one cannot draw a parallel line at all

The Significance of the Curvature of Space

The mathematician, William Clifford (1845–1879) translated Riemann's works to English, and during this work became impressed by Riemann's ideas about the link between physical phenomena and geometry. He developed the ideas further; in a lecture to the Cambridge Philosophical Society on "the science of space" he discussed our ability to deduce the geometry of space at astronomical distances and in space too small (i.e., particles) to be observed, stating that "small portions of space are in fact analogous to little hills on a surface which is on the average flat, namely that the ordinary laws of geometry are not valid in them." Further he thought that "this property of being curved or distorted is continually being passed on from one portion of space to another after the manner of a wave" and that "this variation of the curvature of space is what really happens in that phenomenon which we call the motion of matter."

Clifford concluded that the entire physical world (*motion of all matter)* was a result of this property of space. His ideas were revolutionary at the time because *space* was not yet a concept that many scientists recognized. Clifford died young, in the same year that Einstein was born, and he did not have time to develop the theory further. His vision of space preceded General Relativity Theory by 40 years.

The starting point of Einstein's General Relativity was Galileo's law that all bodies of different masses accelerate equally fast (if the friction of air can be neglected). This empirical observation can be understood as following from Newton's second law of motion (force equals mass times acceleration) and his law of gravity (gravitational force is proportional to the mass of the body). Both these laws have the same coefficient of proportionality, the mass of the body, so the acceleration of a falling body is independent of its mass. Since we are dealing with two independent laws of nature, we have to wonder how both of them happen to have the same coefficient.

According to Einstein, this is no accident. Galileo's law has a deep meaning: it shows that gravity is not a true force but only an apparent one. Apparent forces are already familiar to us, like the Coriolis force explained by the French physicist Gaspard de Coriolis (1792–1843). In the northern hemisphere, southward flowing winds tend to turn toward the east and northward flowing winds toward the west; it causes a counterclockwise rotation of winds around low pressure areas. The Coriolis force is simply a reflection of Earth's rotation around its axis and not a true force. It is typical of apparent forces that they accelerate all bodies equally, independent of their properties such as mass, size, electric charge, etc.

In the same way, the acceleration by gravity is independent of the properties of bodies. An apparent force is easy to eliminate (in principle); if we stop the rotation of the Earth the Coriolis force disappears. Gravity may be eliminated by going to free fall. In a freely falling capsule, we experience weightlessness, say, in an elevator whose cable snaps and whose brakes fail. Far from Earth, a force like gravity at the Earth's surface may be generated artificially in a spacecraft which accelerates by $9.8\,\mathrm{m/s^2}$ which is equal to the acceleration of gravity that we normally experience (see Fig. 15.5).

Fig. 15.5 Newton and Einstein thinking about the falling of an apple. Both are in an enclosed room, Newton at the Earth and Einstein in a spacecraft which is accelerated by $9.8\,\mathrm{m/s^2}$. In both cases, the fall of the apple happens in the same way (credit: Ursa Astronomical Association and J. Nykänen)

Einstein concluded that if the acceleration of gravity can be created and removed so easily, it must be a reflection of some deeper phenomenon. According to Einstein, this phenomenon is curvature of space. Matter makes the surrounding space curve, and bodies react to this curvature in such a way that there appears to be a gravitational attraction.

Consequences of the General Theory of Relativity

From the known geometry of space, it is possible to calculate the orbit of a body that is not influenced by anything else besides gravity. Now we do not view gravity as a force – we are describing force free motion. In flat space such motion happens on a straight line, but in a curved space the free motion can create practically closed orbits. Take a planet circling around the Sun. It moves forward as straight as possible, but because the Sun has curved the space, the orbit becomes an ellipse. Figure 15.6 illustrates this with a stretched horizontal sheet of rubber ("flat space"). An iron ball placed in the middle of the sheet causes a dip in the surface. Now roll a child's marble along the sheet. With a push in the right direction, you may get the light marble to roll around the big one, perhaps in an elliptic orbit. It appears as if there is a central force pulling the marble, when in fact the orbit arises from the

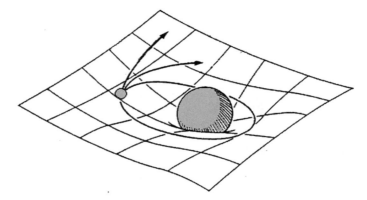

Fig. 15.6 A heavy ball makes dip in a stretched sheet of rubber. The curvature of the sheet makes it possible to roll a small ball around the heavy one much in the way if there were a gravitational force between the balls. Three different orbits of the small ball have been illustrated

form of the surface. (The analogy is not quite perfect as it involves an extra force, the gravity of the Earth.)

In case of the motion of planets around the Sun, both Newton's theory and Einstein's theory give practically the same result. The most important difference arises with Mercury, orbiting close to the massive Sun. As we said earlier, the major axis of Mercury's orbit precesses slowly due to influences of other planets. But Einstein's theory gives an extra precession by 43 arcsec per century as compared with Newton's theory. In fact, this little bit extra had already been observed and was a knotty problem at the time (Fig. 15.7)!

Fig. 15.7 The precession of the orbit of Mercury. Since the central force affecting Mercury is not exactly inverse square force, the orbital ellipse does not close. The point in the orbit furthest away from the Sun (the aphelion) precesses slowly, actually much more slowly than in the figure

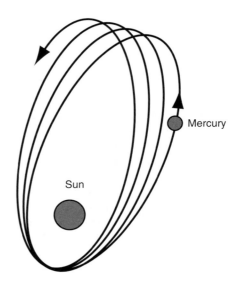

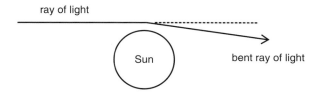

Fig. 15.8 While passing by the surface of the Sun, the light ray deviates from its original direction by 1.75 arc seconds (the figure exaggerates the bending)

The explanation of the motion of Mercury was the first success of Einstein's new gravitation theory. Another consequence is the bending of light rays when they pass close to the surface of the Sun. Because of it, stars appear to shift away from their usual places in the sky when the Sun is in the foreground close to them. Normally we don't see the Sun and the stars at the same time, but during a solar eclipse it is possible. When the shift of stars by the expected amount was detected during the solar eclipse of 1919, it was hailed as victory for Einstein (Fig. 15.8).[1] Nowadays, using radio wave emitting sources, the most accurate results agree with Einstein's theory with 1% accuracy.

The third prediction of General Relativity was verified much later. According to the theory, time slows down in curved space, or in other words, in a strong gravitational field of attraction. Thus time goes a little more slowly in the basement than in the attic of a house since the attic is more distant from the center of the Earth and the gravity is slightly less than in the basement. In 1960, the Americans Robert Pound and Glen Rebka measured the difference in a vertical 22.5 m distance. The result agreed with Einstein's theory within 10%; nowadays the prediction has been verified to the accuracy of 0.01%.

Strange Properties of Black Holes

The world as described by General Relativity has many oddities; one of the most peculiar among them is a *black hole*. When a body is compressed to a smaller and smaller volume, the gravity at its surface becomes stronger and stronger. Let us consider the Earth. Its average diameter is 12 742 km. The escape velocity from its surface, the starting speed which we need to give to a spacecraft, for example, on the way to Moon, is about 11 km/s. If a huge giant would come and squeeze the Earth until it is only the size of a tennis ball, the escape speed would increase to 70,000 km/s.

If the giant decides to keep on squeezing, then the escape speed increases until at one point it equals the speed of light (300,000 km/s). Then the Earth is only less

---

[1] At that time only two competitors to General Relativity were known. The theory of Finnish physicist Gunnar Nordström (Chap. 18) did not predict any bending of light. Newton's theory did predict a bending, but only half of the General Relativity value.

than 2 cm across. Now the giant will be surprised: after this light is not able to escape from the Earth which becomes invisible. The Earth will continue to collapse on its own, until it is totally crushed at its center point. Some estimates say that the density at the center point would become 10^{94} g/cm^3, a number completely beyond common sense. But there would be another surprise in store: the Earth by now is an invisible ball, a black hole, which would start to suck matter from the giant's fingers close to the hole. At this point, the giant would probably want to get rid of his new creation!

Many aspects of the deductions above could be carried out on the basis of Newton's theory. John Michell (1724–1793), Rector of The church of St. Michael and All Angels at Thornhill, near Dewsbury in England, noted the possibility of a black hole in 1784. Such an object would not be directly visible, but could be identified by the motions of a companion star if it was part of a binary system. William Herschel became interested in Michell's black holes. He even thought that he had found one, but it turned out to be a misinterpretation of observations. Laplace suggested the same idea of high-gravity objects trapping light in his *Exposition du Système du Monde* in 1796.

Karl Schwarzschild (1873–1916) was the first to apply General Relativity to the black hole problem. He was the director of the Potsdam observatory and the leading astronomer in Germany at the outbreak of World War I. He joined the army where he served first on the Belgian and later on the Russian front. While in the latter service post, he wrote in 1916 two studies on Einstein's new theory where he defined the so-called Schwarzschild radius. This quantity is proportional to the mass of a body and it tells the minimum radius of a body before it collapses into the black hole. For the Sun, this critical radius is about 3 km and for a star ten times more massive, it is 30 km. Later in the same year Schwarzschild became ill and died at the front.

Some properties of a black hole can be understood only via the use of General Relativity. Now space is so strongly curved that space–time closes on itself around the black hole. In a way, it is a universe of its own, connected to the outside world only through its gravity. The black hole pulls surrounding matter into itself. As a result, its mass increases, as does the "throat" of the black hole which is measured by its Schwarzschild radius. Gulping surrounding matter just teases the appetite of the black hole!

We may go back to the 2D example of a rubber sheet to illustrate a black hole (Fig. 15.6). Let the heavy ball placed on the sheet gradually shrink in size. Then the pressure against the surface increases, and a deeper and deeper dent develops in the surface. In the end, the rubber surface curves all around the ball, and the ball hangs on to the rest of the sheet via a narrow neck. The sheet of rubber has not been much affected far from the ball, but the local curvature has increased a lot in the process of shrinking the ball. The part of the sheet with the extreme curvature imitates the space around a black hole.

The conditions inside the Schwarzschild radius of a black hole are quite incomprehensible. The roles of the space and time coordinates switch places there. For example, normally time flows only toward the future. Inside the black hole, time can go back and forth while in space we can move only in one direction, toward the center of the black hole. Our brains have not developed to fathom a world like this, even though we can treat the situation mathematically.

Because of the strong curvature of space, time slows down near a black hole. If we were able to follow a falling clock into a black hole, say, through a telescope, and if the clock would survive the infall intact, we would see the clock advancing more and more slowly when it approaches the black hole. At the distance of the Schwarzschild radius, the clock appears to stop altogether. Thus, observing from far away, time seems to freeze at the border of a black hole; however a person traveling into the black hole with the clock does not notice anything peculiar about the progress of time. This is yet another example of the absence of a rigid absolute time; every observer experiences the run of time in his/her own way.

The rays of light near a black hole also behave oddly. The rays may be bent by a large angle, or even start circling around the black hole. Some light rays disappear inside the black hole forever. We would find it very difficult to comprehend what we see near a black hole since the "data processing" of our vision assumes that rays of light travel in straight lines. Even small everyday deviations from a straight line, as in the case of mirages, baffle us.

Black holes in nature have one more property not been mentioned yet. They may rotate. The bodies which collapsed were in all likelihood rotating. The *conservation of angular momentum* dictates that a black hole arising from such a body must also rotate, even much more rapidly. The curvature of space around a rotating black hole was first calculated by the New Zealand mathematician, Roy Kerr, in 1963.

The rotation of the black hole shows up as a rotation of the nearby space: the black hole drags the space along like a whirlpool centered on it. In the plane of rotation, the speed of the whirlpool can be as high as the speed of light at the Schwarzschild radius. Consequently, a body at rest in space will be observed (from a distance) to rotate around the black hole with the speed of light. Well beyond the black hole's Schwarzschild radius or near an ordinary rotating object, the motion of an orbiting body will be perturbed. Close to the black hole the whirl is overpowering; even traveling backward at the speed of light cannot prevent a body from being dragged around in the direction of rotation of the black hole.[2]

The rotation of a body around another body in space is easy to understand. But how can one understand that space itself is dragged around the central body? This falls outside common sense. We commonly think of space as a rigid background against which we measure the motion. Instead, real space, as revealed by General Relativity is elastic, a characteristic with observable consequences.

The dragging of space around rotating bodies was proposed by Austrian physicists Joseph Lense and Hans Thirring in 1918. Not until 2004 was it possible to measure this effect in space surrounding the rotating Earth. By following the motions of two Earth orbiting satellites LAGEOS I and II, a team led by Ignazio Cuifolini of

---

[2] For every black hole, there is a maximal speed at which it can rotate. The limiting surface for a maximally rotating black hole is only half a Schwarzschild radius from the center. Outside the limiting surface there is a region called *ergosphere* where the space whirl exceeds the speed of light. Under suitable conditions particles may extract a little bit of the rotational energy of the black hole in this region and fly off, carrying the energy with them.

University of Lecce, Italy, and Erricos Pavlis (University of Maryland) found that the planes of the orbits of the satellites have shifted by about two meters per year in the direction of the Earth's rotation. The result is in agreement with the prediction of Lense and Thirring within the 10% accuracy of the experiment. The satellite Gravity Probe B, specially designed for the measurement of space dragging by Stanford University and NASA, is currently attempting to confirm these results.

Gravitational Waves

One of the phenomena related to the elasticity of space is gravitational waves, small changes in the curvature of space which propagate in space with speed of light. There has been no confirmed direct detection of gravitational waves even though in 1967 American physicist Joseph Weber (1919–2000) claimed to have detected them. Weber was for long years a lonely pioneer in this field.

His detector was a 1.5 ton aluminum cylinder, which was hanging in a vacuum container, isolated from outside as well as possible. When a gravitational wave hits the cylinder, it starts to oscillate with its own specific frequency. The oscillation amplitude is expected to be very small, only about 10^{-15} cm, or 1% of the diameter of a proton. It is understandably difficult to measure such a short distance. Moreover, any kind of vibrations in the surroundings, from passing traffic to earthquakes, can also

Fig. 15.9 The gravitational wave observatory LIGO in the USA: the aerial view of the Hanford antenna consisting of two long vacuum tubes each extending over 4 km from the station. Its twin antenna is operating in Livingston (credit: LIGO Laboratory)

make the cylinder oscillate. Since nobody else has been able to detect gravitational waves, it is usually thought that Weber's oscillations were from the surroundings. However, the effects of such moving ripples of space are so tiny that the lack of detection so far does not mean that the waves are not there.

In a new type of detector, a laser measures the distance between suspended test masses (mirrors). The LIGO antenna (Laser Interferometric Gravitational Wave Observatory) in the USA consists of two detectors of this kind, separated by 1,000 km. In contrast to local "noise" at either detector, a genuine gravity wave passing through the Earth should be seen at both sites (Fig. 15.9). A similar gravitational wave observatory VIRGO is operating in Italy.

At the moment, evidence for gravitational waves is indirect. The binary neutron star system PSR 1913 + 16 appears to emit gravitational waves. Observations show that the binary system does lose energy which cannot be explained in other ways beside gravitational wave emission. The loss rate of energy matches rather well what is expected in the General Theory of Relativity. This coincidence is usually taken as proof that gravitational waves do exist, even though the radiation from PSR 1913 + 16 is not directly measurable by gravitational wave antennas.

A promising case for direct detection is the binary black hole system of quasar OJ287 to be discussed later. It is a distant extragalactic object and here one of the members is more massive than stars by a factor of 10^{10}. Thus waves from this source should be much more powerful than PSR 1913 + 16. The rate of loss of energy in this binary system was recently confirmed by an international team lead by astronomers from Tuorla Observatory. The confirmation happened dramatically when OJ287 suddenly increased its brightness on 13 September 2007 as if 10,000 billion Suns would have lit up in this quasar, just as was predicted on the basis of Einstein's theory. The next generation of gravitational wave antennas should be able to confirm the emission of gravitational waves from OJ287. A new important window to the universe is ready to be opened.

Chapter 16
Atoms and Nuclei

We now understand light as vibrations of electric and magnetic fields which somehow propagate through space. Obviously, we need to discuss the nature of light further, but before we do so, we should first ask, "What is matter?" The Greek philosopher Empedocles (Chap. 2) had many interesting ideas on the workings of nature. For example, he envisioned that light travels at a finite (very large) speed, which gained acceptance only much later. He also proposed the idea that matter is made of four elements – earth, water, air, and fire. It was the mainstream view through the Middle Ages up to the seventeenth century.

Robert Boyle was critical of the four-element theory. He thought that matter consists of different kinds of particles and that gross matter was composed of clusters of particles and that chemical changes resulted from the rearrangement of the clusters. In the *Sceptical Chymist* (1661), the Irishman criticized alchemists who tried to make gold from other elements. He defined an element as a substance which could not be further broken down by any means, thereby originating chemistry as a scientific subject.

Conservation of Energy

Boyle also realized that heat is an expression of internal motions of particles in matter. Let us hit a nail into a plank of wood. As long as the nail advances forward, it does not heat up noticeably. But if the nail is hammered even after it is in the wood up to its head, the nail starts to heat up. The hammering does not drive the nail forward, but causes swift motions inside the nail, observed as heat or thermal energy (Fig. 16.1).

Much later the German medical doctor Julius Robert von Mayer (1814–1878) interpreted heat as a form of energy. His tryst with physics was unusual. As a ship's surgeon on a voyage to Java, he noticed that the venous blood of the sailors was redder than at home. He knew the theory proposed by Lavoisier that body heat is generated by a burning process for which the blood gives oxygen. Perhaps the blood

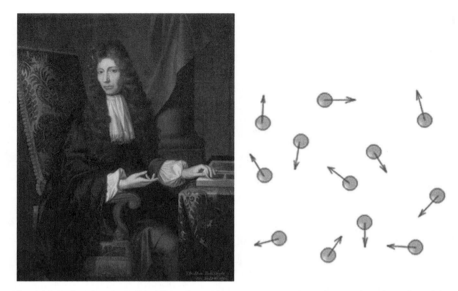

Fig. 16.1 Robert Boyle (1627–1691) viewed heat as an expression of state of motion of particles. In hot gas, the speeds of the molecules moving in different directions are greater on average than in cool gas

was redder because less burning was needed in the tropics? This made Mayer think about the relation between heat and mechanical work performed by muscles. He reasoned that heat and work are two forms of energy. There are different kinds of energies and their total sum is conserved in a physical process, and ultimately, in the whole universe. Thus he was the first to state the conservation of energy in all generality. However, Mayer's ideas were published in private pamphlets and were overlooked. Later it was painful for him when similar ideas were credited to Joule.

Independently, James Joule (1818–1889) came to the same conclusions. His skillful experiments on heat, electricity, and mechanical work were needed for the scientific community to accept energy conservation. This wealthy English brewer could dedicate much of his time to his hobby, comparing different forms of energy.

Developments in Chemistry

In the eighteenth century, it was thought that a burning substance loses a fire element, phlogiston, which explains why a candle becomes shorter when it burns. The credit for discovering the true nature of burning goes to Antoine-Laurent Lavoisier, a generalist scientist who specialized in mathematics, meteorology, and geology (see Fig. 16.2). He was elected to *Académie des sciences* at only 25, and around the same time he obtained a comfortable job as tax collector. He studied chemistry in his later position in the Royal Gunpowder Administration and experimented, among other

Fig. 16.2 Antoine-Laurent Lavoisier (1743–1794) and his wife Marie-Anne Pierrette Paulze (1758–1836). As a result of her close work with her husband, it is difficult to separate her individual contributions from his (a painting by Jacques-Louis David)

things, with the burning of phosphorus and sulfur. He found that results of burning weigh more than the original substance and that the difference can be accounted for by the loss of air of the same weight. Lavoisier recognized and named the active substance in air as *oxygen*. Phlogiston was no more needed.

Lavoisier published his results in *Traité Élémentaire de Chimi*, in 1789. This classic presented a unified view of new theories of chemistry and clarified the concept of an element as a simple substance that could not be broken down by known methods of chemistry. He also theorized on how elements form chemical compounds and stated that matter is neither created nor destroyed (i.e., mass is conserved).

Lavoisier continued as tax collector even after the beginning of the French Revolution. During the Reign of Terror he was condemned to death in the guillotine together with the other 27 French tax collectors and was executed on 8 May 1794 in Paris. It did not help that some years earlier he had criticized the revolutionary leader Jean-Paul Marat and his views about burning.

After Lavoisier there was a systematic attempt to search for and catalog new elements. Joseph Gay-Lussac (1778–1850) in France and Humphry Davy in England were especially known for this work. Much attention was paid to the relative amounts of elements needed to form a compound. They concluded that compounds are always made of elements in constant proportions. For example, to make 9 g of water (H_2O) one needs 8 g of oxygen (O) and 1 g of hydrogen (H); the necessary proportions not to have any oxygen or hydrogen left over.

This is different from, say, baking a cake, where it is not so critical to have all ingredients in exactly the right proportions. The cake may taste a little different, but it is a cake anyway. The discovery of this *law of constant proportions* is credited to the Swedish chemist Jöns Jakob Berzelius (1779–1848). He showed that inorganic substances are composed of different elements in constant proportions by weight. Based on this, in 1828 he compiled a table of the relative atomic weights, which included all elements known at the time. This work provided evidence in favor of the *atomic hypothesis*: that chemical compounds are composed of atoms combined in whole number amounts. To aid his experiments, he developed a system of chemical notation in which the elements were given simple labels – such as O for oxygen, H for hydrogen, etc. – the same basic system used today.

The law of constant proportions helped John Dalton (1766–1844) to develop his theory of atoms. He had studied many fields from meteorology to physics, but it was the atomic theory that aroused his interest in chemistry. In his *A New System Chemical Philosophy* (1808) Dalton stated: "These observations have tacitly led to the conclusion which seems universally adopted, that all bodies of sensible magnitude, whether liquid or solid, are constituted of a vast number of extremely small particles, or atoms of matter bound together by a force of attraction…" He continued: "Therefore we may conclude that the ultimate particles of all homogeneous bodies are perfectly alike in weight, figure, &c. In other words, every particle of water is like every other particle of water; every particle of hydrogen is like every other particle of hydrogen, &c." However, he recognized that atoms of different elements were not the same and had different weights.

Dalton supported himself as a school teacher in Manchester. In 1800 he became secretary of the Manchester Literary and Philosophical Society and served as a public and private teacher. Later he became president of the Philosophical Society, an honorary office that he held until his death.

According to Dalton, atoms join each other always in the same way in chemical compounds. This creates new identical combinations of atoms, now called molecules. The law of constant proportions derives from here; the same proportions prevail already in the molecules. We know two atoms of H combine with one of O, to create water, H_2O. However, Dalton's conclusions as to how exactly molecules are made of atoms were often in error.

Correct chemical formulae and atomic weights were discovered after Gay-Lussac found in 1808 that not only elements combine in given weight ratios, but they also combine in given volume ratios when the elements are in gaseous form. For example, 2 liters of hydrogen and 1 liter of oxygen always produces 2 liters of water vapor (not three!). Gay-Lussac's rule was explained by Amedeo Avogadro (1776–1856), professor of physics at the University of Turin. In 1811, he published an article that drew the distinction between the molecule and the atom, pointing out that Dalton had confused the concepts of atoms and molecules. Dalton's "atoms" of hydrogen and oxygen are in reality "molecules" containing two atoms each, H_2 and O_2. Thus two molecules of hydrogen can combine with one molecule of oxygen to produce two molecules of water.

Avogadro suggested that equal volumes of all gases at the same temperature and pressure contain the same number of molecules, now known as Avogadro's principle. Using this rule we may right away deduce that a water molecule has two hydrogen atoms for every oxygen atom, i.e., its chemical formula is H_2O. When we add that the weight ratio of oxygen to hydrogen in a water molecule is 8:1 = mass of one O atoms divided by two H atoms, we find that O/H mass = 16 or the oxygen atom weighs 16 times more than the hydrogen atom.

The Periodic Table of Elements

New elements and their measured atomic weights started to reveal intriguing regularities in the properties of elements. The British chemist John Newlands (1837–1898) noticed that if you order the elements according to their atomic weights, the chemical properties start to repeat themselves after seven steps forward. He called it "The Law of Octaves"; the musical analogy did not enhance the idea's credibility. Since noble gases were not known at the time, every sequence of seven elements was missing one.

The *periodic table of chemical elements* was introduced by Dmitri Mendeleev in a two-volume textbook *Principles of Chemistry* (1868–1870). Mendeleev grew up in Tobolsk, Siberia, as the 14th child of the Master of the Tobolsk Gymnasium. He studied in St. Petersburg and Paris, and became professor of chemistry in St. Petersburg University in 1863 (see Fig. 16.3). With the help of his periodic table, Mendeleev was able to predict new elements, filling the gaps in the system (the first one to be discovered was gallium in 1875). A similar system was developed independently by German Lothar Meyer (1830–1895).

Fig. 16.3 Dmitri Ivanovich Mendeleev (1834–1907)

Box 16.1 The periodic table and atomic weights for the first 56 elements

I	II	III	IV	V	VI	VII	0	VIII
H 1.0							**He 4.0**	
Li 6.9	Be 9.0	B 10.8	C 12.0	N 14.0	O 16.0	F 19.0	**Ne 20.2**	
Na 23.0	Mg 24.3	Al 27.0	Si 28.1	P 31.0	S 32.1	Cl 35.5	**Ar 40.0**	
K 39.1	Ca 40.1	**Sc 45.0**	Ti 47.9	V 50.9	Cr 52.0	Mn 54.9		Fe 55.9
								Co 58.9
								Ni 58.7
Cu 63.6	Zn 65.4	**Ga 69.7**	**Ge 72.6**	As 74.9	Se 79.0	Br 79.9	**Kr 83.8**	
Rb 85.5	Sr 87.6	Y 88.9	Zr 91.2	Nb 92.9	Mo 95.9	**Tc 98.9**		Ru 101.1
								Rh 102.9
								Pd 106.4
Ag 107.9	Cd 112.4	In 114.8	Sn 118.7	Sb 121.8	Te 127.6	I 126.9	**Xe 131.3**	
Cs 132.9	Ba 137.3							

Those discovered after Mendeleev in boldface. You may inspect the modern, complete version of the periodic table, e.g., in http://www. chemicool.com/

The English chemist William Prout (1785–1850) suggested as early as 1815 that atoms are composed of smaller units. The unit appeared to be the hydrogen atom. However, some atomic weights are *not* multiples of the atomic weight of hydrogen; for example, chlorine atom weighed 35.5 in hydrogen atom units (Box 16.1). This problem was solved by Frederick Soddy (1877–1956) in 1913. He discovered atomic *isotopes*, i.e., chemically equal atoms but with different atomic weights. For example, chlorine was found to consist of two kinds of atoms: 77.5% of them weigh 35.0 units and 22.5% weigh 37.0 units, almost exact multiples of the hydrogen atom. The average value is 35.5 units.

Even though the gaps in Mendeleev's system were gradually filled in, it remained unclear whether all elements necessarily fit into the system at all. One could imagine, for example, that there are one or two elements between hydrogen (atomic weight one) and helium (atomic weight 4). Not until 1913 when Henry Moseley (1887–1915) invented the *atomic number* could one do the final inventory of elements. Moseley specialized in the University of Manchester in the measurement of x-rays. Looking at x-rays from different elements, he found the integral atomic numbers associated with each element. Soon after this discovery the young scientist joined the army and died in the battle of Gallipoli.

According to Moseley's measurement, the atomic number of calcium is 20. Since it is also the element number 20 in the periodic table, it is obvious that there are no more unknown elements lighter than calcium. At Moseley's time there were four atomic numbers missing; the elements corresponding to these numbers have since been discovered. The periodic table now contains 117 known elements, out of which 94 occur naturally on Earth. Among the abundant natural elements, uranium (92) is the heaviest; neptunium (93) and plutonium (94) have been found in trace amounts on Earth. Heavier elements are unstable and must be made artificially in particle accelerators (lighter ones technetium (43) and promethium (61), rare on Earth, also belong to such synthetic elements).

Discovery of the Electron

The atomic number is related to the electric properties of an atom, as was first revealed in electrolysis. The method was discovered in 1800. William Nicholson and Anthony Carlisle put two electric wires in water such that each wire was connected to the opposite electrodes of an electric battery. They found that the surroundings of the negative electrode started to generate hydrogen gas, while the vicinity of the positive electrode created oxygen gas. Apparently, water molecules were broken into their elements by this process (Fig. 16.4).

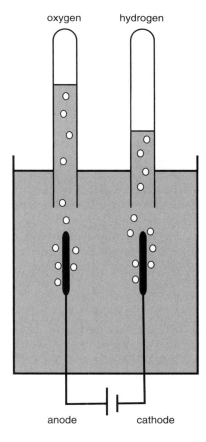

Fig. 16.4 Electrolysis. Two wires are connected to a direct current source (battery) and placed into water. At the end of the positive electrode (anode) oxygen gas is generated and collected in a tube. At the same time, hydrogen gas is generated and collected at the cathode. The amount of gas collected is directly proportional to amount of electric charge passing through the wire, as Faraday demonstrated. Also the hydrogen volume will be twice the oxygen according to Avogadro's principle

Fig. 16.5 Joseph J. Thomson
(1856–1940), the discoverer
of the electron

Humphry Davy experimented with electrolysis, but his colleague at the Royal Institution, Michael Faraday, explained the phenomenon as follows: a small fraction of water molecules is always dissociated into two electrically charged groups – hydrogen atoms of positive charge, and molecules composed of hydrogen and oxygen with a negative electric charge. Faraday called these charged particles "ions." The wire connected to the negative electrode attracts positively charged hydrogen ions. When they touch the wire, they pick up enough negative charge from the wire to convert the ions into neutral hydrogen atoms. Then the hydrogen gas bubbles up from the water. A slightly more complicated process produces oxygen at the wire connected to the positive electrode.

The charge passed on to the atoms of the liquid is replaced by new charge via a current flowing from the electric battery. Collecting and measuring the gases generated in the process gives the amount of mass per unit charge of the current. Since we are dealing with hydrogen, we can calculate the ratio of the mass and the charge of the hydrogen ion. Using the units of mass and charge, kilogram and coulomb, the ratio is about 10^{-8}. The mass and charge cannot be found separately here, only their ratio.

We could get the mass of the hydrogen ion, our unit of atomic weight if we knew its charge, though it was not possible in Faraday's times. It was not clear what kind of particle carries the charge. In 1874 George Stoney suggested the name *electron* for the unit charge in electrolysis. The particle associated with this charge (and known by the same name) was discovered in 1897 by Joseph J. Thomson (Fig. 16.5).

Thomson studied at the University of Cambridge where his second rank at mathematics tripos was good enough to win him a Fellowship at Trinity College, Newton's old college. Thomson worked for the rest of his life in the college and became finally the Master of Trinity. He started out in mathematics; so his appointment in

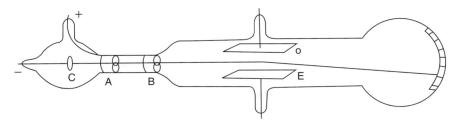

Fig. 16.6 Thomson's experimental setup. Particles are emitted from the cathode C. Their stream is deviated from its straight line path by the electric field created by plates D and E (From Thomson's publication in 1897.)

1884 to the professorship of experimental physics at Cavendish Laboratory came as a big surprise. He never mastered experiments; one of his assistants told he was "very clumsy with his hands, so I try to keep him far from experiments." Nevertheless, Thomson became one of the leading experimental physicists since he had an intuition about suitable problems to work on.

Since becoming a Cavendish professor, Thomson started to study electric discharges in vacuum tubes. The most familiar discharge is lightning, difficult to use in research! Instead it was noted in the eighteenth century that one could create huge discharges in glass tubes cleared of air. Different colors of light are generated, depending on what gaseous element is put inside the tube which is used in neon lights and other applications.

Heinrich Geissler (1814–1879) invented a pump which could reduce the air pressure inside the tube only to one-thousandth part of the atmosphere outside. Using Geissler's pump Julius Plücker (1801–1868) manufactured a discharge tube and connected it to a strong voltage source. The glow in the tube disappeared, except for around the negative electrode, the cathode, as if some particles got loosened from the cathode where they caused the glow. Afterward they passed through the tube and were collected in the positively charged electrode. Eugen Goldstein (1850–1930) showed that it did not matter what material the cathode is made of; thus, *cathode rays* are not atoms breaking loose from the cathode.

Plücker showed that one can bend the cathode rays by using a magnet; thus they must be charged particles. Then Thomson made the particles go through magnetic or electric fields which made the particle stream change the direction. Then he let the particles stream freely until they hit the opposite end of the tube (Fig. 16.6). By measuring the distance of the target point from the central axis of the tube he was able to calculate both the speed and the mass-to-charge ratio of the particles. A similar particle stream arises in a television tube where a cathode ray beam sweeps quickly over the screen to create pictures. In the television tube, electric fields are used to direct the cathode ray beam.

In his *Philosophical Magazine* article in 1897, Thomson calculated that the speed of the cathode rays was about 10% of the speed of light and that the electron mass-to-charge ratio was about 10^{-11} kg/C. Now assume that both the hydrogen ion and the cathode ray have a charge of the same magnitude. As Thomson measured the hydrogen ion mass-to-charge ratio to be 10^{-8} kg/C, the mass of the cathode ray charge cannot be more than 1/1,000 of the mass of a hydrogen ion (modern value: 1/1,840). He concluded:

> We have in the cathode rays matter in a new state, a state in which the subdivision of matter is carried very much further than in the ordinary gaseous state: a state in which all matter...
> is of one and the same kind; this matter being the substance from which all the chemical elements are built up.

His result was first doubted, but later research confirmed the existence of the electron. Thomson and his colleagues measured the charge of the electron: 10^{-19} C. Now we know the more exact value 1.602×10^{-19} C. With the modern mass-to-charge ratio 0.57×10^{-11} kg/C, we may deduce that an electron weighs only 9×10^{-28} g. The hydrogen atom is about 1,840 times heavier.

Toward the Atomic Nucleus: Radioactivity

Electrons are negatively charged, but atoms which seem to include the electrons are electrically neutral. There must be a positive charge somewhere in the atom to neutralize the negative electrons. The next task was to find out where the positive charge lies inside the atom. Thomson was in favor of the "raisin bun model" where the positive charge fills the whole atom, and electrons are in it like raisins in a bun. The Japanese scientist Hantaro Nagaoka (1865–1950) suggested that there is a positively charged particle in the middle of the atom, around which lighter electrons circulate like planets orbit the Sun. In both cases, the attraction between the positive and negative charges binds the electrons to the atom.

Discovering which of the two models is correct was left to Ernest Rutherford. He grew up in New Zealand coming to the Cavendish laboratory to study in 1895. Three years later he was appointed professor at McGill University in Canada where he worked until 1906. Then he moved to Manchester which was one of the leading research centers of physics. Here he concentrated on the study of the structure of atoms. In 1919 Rutherford returned to the Cavendish laboratory to become its director (Fig. 16.7).

After arriving at Cambridge, Rutherford began to study *radioactivity*, discovered a few years earlier (1896) by Henri Becquerel in Paris. Becquerel had tried to generate x-rays from various materials by exposing them to sunlight. One of his materials was a uranium compound. Even though the sample was kept away from sunlight, it "exposed" photographic plates which had also been kept in the dark. Some rays were emanating from uranium! A few years later Marie Sklodowska-Curie and Pierre Curie, patiently handling a ton of uranium ore (pitchblende) in their modest Paris laboratory, discovered the element radium. It radiates millions of times stronger than uranium (Fig. 16.8). Rutherford found three kinds of rays in radioactivity: alpha-rays, beta-rays, and gamma-rays, as he called them. These offered a key to the atomic nucleus. As mentioned earlier, gamma-rays are very short wavelength electromagnetic radiation, but what are alpha-rays and beta-rays?

Fig. 16.7 (a) Henri Becquerel (1852–1908) and (b) Ernst Rutherford (1871–1937)

Becquerel measured the mass-to-charge ratio of beta-rays and found that these negative charges have the same ratio as the electrons. Beta-rays thus must be electrons thrown off from the radioactive material. Rutherford was able to measure the mass-to-charge ratio of the positively charged alpha-rays; it turned out to be twice as big as for the positive hydrogen ions. If the charge of alpha-rays is one unit, then their mass must be twice the hydrogen atom mass. However, Rutherford made the correct conclusion that alpha-particles have charges of two units which makes

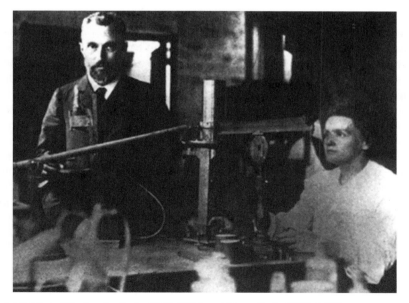

Fig. 16.8 Marie Sklodowska-Curie (1867–1934) and Pierre Curie (1859–1906)

their mass four atomic mass units. That is, an alpha-particle is an ionized helium atom. This was confirmed by Rutherford's colleagues in McGill University, William Ramsay (1852–1916) and Frederick Soddy, who observed helium being generated by a radium compound.

In their 1903 study, Rutherford and Soddy explained radioactivity: it is a process where one element becomes another. When an atom expels an alpha-particle, its atomic number decreases by two in the table of elements; if it expels an electron, its atomic number increases by one. This was a radical idea: ever since the death of alchemy the permanency of elements was never questioned. It was a basic axiom that an element cannot be created or destroyed. However, Rutherford's and Soddy's proposal was based on accurate measurements showing that a radioactive element always turns into another element in a standard way in any environment. For example, radioactive thorium transmutes into radon gas which itself is radioactive. The activity of this radon gas goes down fast: after 1 min the activity is halved, after 2 min it is one quarter, after three minutes one-eighth of the original, etc. Rutherford and Soddy showed that this is connected to disappearance of the radon gas: one half of the sample disappears after the first minute, of the remaining one half disappears in the next minute, and so on. We say that radon gas has *half-life* of one minute (to be exact, 54.5 s). The half-life varies a lot from one radioactive substance to the next. It is 1,600 years for radium, 1.4×10^{10} years for thorium, and 4.5×10^9 years for uranium. The disintegration of radioactive elements is used in the age determination. We return to this topic when we discuss the age of the Earth (Chap. 29).

Rutherford Discovers the Nucleus of the Atom

After returning from Canada Rutherford started new kinds of experiments: bombarding atoms by alpha-particles. The collision itself happens in such a small scale that we cannot observe it directly, but we may infer a lot about it by looking at the consequences. As a result of the collision, the speed and direction of the alpha-particle is changed; the target atom suffers a similar fate. The process is called *scattering*. The speed and direction of motion of the alpha-particle before and after the collision can be measured using suitable equipment. Thus we can also calculate what has happened to the target atom.

In Rutherford's experiment the alpha-particles came from a radioactive sample; a narrow beam of emitted particles was selected by using shades made of thick sheets of lead with a hole in them. A sheet of gold was placed beyond the hole; thus the alpha particles were hitting gold atoms. Since the alpha-particles came with a high speed, it was expected that they would travel through the gold sheet with only a slight change of orbit. The alpha-particles were detected using a sheet of zinc sulfide which produces a small flash of light when a particle hits it (Fig. 16.9).

Rutherford had an assistant from Germany – Hans Geiger. In 1909 the group was joined by Ernest Marsden as a student. What happened next in 1910 was told by Rutherford:

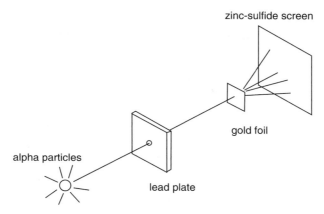

Fig. 16.9 The scattering experiment of Hans Geiger (1882–1945) and Ernest Marsden (1889–1970). Alpha particles scatter from the sheet of gold and cause a flash of light in the zinc sulfide detector when they hit it

One day, Geiger suggested that a research project should be given to Marsden. I responded, 'Why not let him see whether any particles can be scattered through a large angle?' I may tell you in confidence that I did not believe they would be since we knew that the alpha particle was a very massive particle with a great deal of energy. Then I remember two or three days later Geiger coming to me in great excitement and saying, "We have been able to get some alpha particles coming backwards." It was quite the most incredible event that has ever happened to me in my life. It was almost as incredible as if you fired a 15-inch shell at a piece of tissue paper and it came back and hit you.

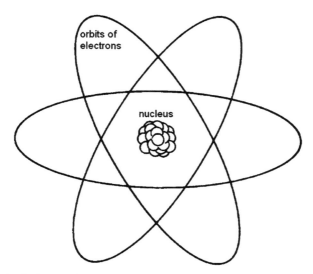

Fig. 16.10 Rutherford's model of an atom. The heavy nucleus is made of many nuclear particles and it is orbited by electrons

After a few weeks of brainwork Rutherford declared: "Now I know what happened in the experiment, and I also know the structure of the atom." He said that almost all the mass, and with it the positive electric charge, is concentrated in the atomic nucleus which is no more than 1/10,000 of the size of the atom. The rest of the atom is empty except for the electrons with their negative charges (Fig. 16.10).

Nagaoka's view of the atom turned out to be basically correct. In the Solar System, the Sun contains most of the mass. Similarly, the mass of an atom is mostly in its nucleus. Just as the Solar System is mainly "empty" space between the Sun and the planets, so also the atom is mostly "empty" between the nucleus and the electrons. In atoms the concentration to the center is even more extreme; in the scale of the Solar System the size of the nucleus would be no bigger than a planet. There is no firm knowledge about the size of an electron, but in this model it could not be bigger than a smallish asteroid.

Chapter 17
Strange Microworld

Having delved rather deeply into the secrets of matter, we can now return to light. We have described the ascendancy of the wave theory in the nineteenth century over Newton's early particle ideas. However, a wave requires a medium through which it can move. Sound waves require air, and there is no air and sound in outer space. Ether, which presumably filled space, was proposed as a medium for light waves, but this idea encountered problems. An important step forward was Einstein's first 1905 article where he demonstrated that in some circumstances light strangely behaves like a particle, now known as the *photon*.

Particles and Waves Unite

Maxwell's theory explained light as electromagnetic vibrations. But there were problems using this theory to understand the spectrum of black body radiation. It was known that the black body radiation is strongest at a definite wavelength and that it becomes weaker on either side of the spectrum away from the maximum. The dip toward high frequencies was not expected from classical theory. The German physicist Max Planck found a reason why the black body spectrum has this shape: one had to assume that atoms radiate energy only in packages of certain size. The energy related to radiation was like particles: you can only emit one, two, or three particles, etc., but not a fraction of a particle.

The minimum package of energy proposed by Planck is proportional to the frequency of the wave: the higher the frequency, the greater the energy in the package. The constant of proportionality is called *Planck's constant*. Thus

$$\text{Energy} = \text{Planck's constant} \times \text{frequency}$$

Since frequency and wavelength are inversely proportional to each other, the energy package is inversely proportional to the wavelength. Planck's constant is small, so we do not notice the packages of light during everyday life just as we do not notice that matter is actually made of tiny atoms when it appears continuous.

P. Teerikorpi et al., *The Evolving Universe and the Origin of Life*
© Springer Science+Business Media, LLC 2009

a   b

Fig. 17.1 (**a**) Max Planck (1858–1947) and (**b**) Niels Bohr (1885–1962)

Max Planck came from the city of Kiel, but he did most of his studies in Munich where he obtained his doctorate (see Fig. 17.1). Before that, Planck had followed the lectures of Kirchhoff and Helmholtz in Berlin. Rather surprisingly, he was elected as successor of Kirchhoff in Berlin. Planck specialized in the study of black body radiation which led to his great discovery in 1900. Apparently Planck did not quite appreciate the significance of his discovery that energy can only appear in packages of certain size called *quanta*. He thought that this was a property of atoms, thinking there to be no reason why the electromagnetic wave itself could not carry any amount of energy whatsoever.

The next step was left to Einstein, who described how the quantization of energy into packages is not only associated with vibrations in atoms, but also with electromagnetic radiation itself. As evidence for light quanta or photons, Einstein explained the *photoelectric effect* in which light can cause electrons to be emitted from a metal. This phenomenon was discovered by Heinrich Hertz as an unexpected byproduct of his experiments with radiowaves in the 1880s. Ultraviolet, high-energy photons can knock electrons out of a metal even if the light is of very low intensity. The few high-energy quanta of the high-frequency light are each sufficient to do the job for one electron. However, the individual low energies of red or infrared low-frequency light quanta (even when numerous or intense) are each insufficient to knock electrons loose. A rather rough analogy is the relative consequences of having a bucket of sand grains thrown in one's face versus getting hit by a large boulder.

Light is made of a kind of particles, as Newton said, but we cannot ignore the evidence that light consists of waves. There must be a wave-particle duality of light and electromagnetic radiation, difficult to understand using everyday experience. We are used to connecting waves and particles to separate kinds of phenomena.

Somehow, on the scale of atoms both descriptions fit the same phenomenon. It is no use trying to visualize a creature which is a wave and a particle at the same time.

To make matters even more peculiar, the French prince and physicist Louis de Broglie (1892–1987) proposed in 1924 that the electron is not only a particle but also a wave. His 1922 doctoral thesis *Recherches sur la théorie des quanta* introduced his theory of electron waves. This was soon confirmed experimentally; electrons behave in many ways like light waves. For example, the previously described interference in which waves in the same phase of oscillation strengthen each other, and those in opposite phases cancel each other, is seen in experiments using electron beams impinging on crystals. De Broglie's waves are routinely used in electron microscopes to get sharper images than optical ones, as the wavelengths of electrons are shorter in comparison with light.

The Bohr Atom

The Danish physicist Niels Bohr applied the new quantum concepts to atoms. Bohr was born in a wealthy Copenhagen family. In his youth he was outstanding in football; together with his brother he played at the top national level. Bohr studied in Copenhagen University receiving his Ph.D. in 1911. A turning point in his career was working in England after completing his thesis. First, Bohr went to Cambridge, but after meeting Rutherford he decided to go to Manchester. It was just at this time that Rutherford had confirmed the "solar system model" of an atom in his alpha-particle experiments.

The atoms of the same element are identical to each other, but there is no rule in the simple solar system model as to where the electrons should be placed. In the Solar System itself, there is no definite physical constraint on how far from the Sun the planets should be. The Earth's orbit could be a bit larger or smaller than it actually is. Moreover, the orbiting electron is like an oscillating charge in an antenna, and thus it should radiate outward energy with its orbital frequency. But there is no external source of energy for the electron like the antenna of a radio station has. The resulting loss of energy should make the electron plunge into the atomic nucleus.

These are the problems that worried Bohr in Manchester. He arrived at his solution two years later. One of his friends urged him to take a look at the formula for the spectral lines of hydrogen, which Balmer had discovered a few decades earlier. "When I saw the formula, the whole matter became clear immediately," Bohr said years later. He assumed that in a hydrogen atom an electron is in orbit around a proton with the electrical attraction binding the two. In contrast to the planets in our Solar System, Bohr said that *only certain orbital radii are allowed* for all the atoms of an element. Otherwise the electron may follow the rules of mechanics.

The second deviation from standard physics was Bohr's requirement that *an electron following the allowed orbit does not radiate*. This was contrary to the theory of electromagnetic radiation. Instead, Bohr associated radiation with a different phenomenon, a change of orbit of an electron. Every circular orbit has an associated energy which is greater the further from the proton the electron orbits. An electron may jump from an upper (i.e., more distant) orbit to a lower orbit by radiating a photon whose energy corresponds to the energy difference of the two orbits. In the same way, an electron may "steal" a photon of suitable energy from a passing beam of light in order to climb to a higher orbit.

Since only orbits of given energies are allowed only very definite energy differences and photons corresponding to them are possible. Think of a set of stairs. You cannot stand or jump half a step, you can step only integral numbers of steps. Since the value of the photon energy is related to its wavelength, only certain wavelengths can appear in the light emitted by the hydrogen atom. In the Balmer formula, the wavelengths are codified using whole numbers; Bohr recognized that these are the numbers of the orbits in order of increasing distance from the nucleus. For example, Balmer lines are born when an electron in a hydrogen atom jumps to orbit number 2 from some higher orbit (Fig. 17.2).

After returning to Denmark, Bohr wrote an article about his discoveries and sent it to Rutherford. Rutherford had doubts about Bohr's theory, but he sent it anyway to *Philosophical Magazine* for publishing. The article's acceptance varied from the esteemed Lord Rayleigh's remark, "I don't see anything useful in it," to Einstein's enthusiasm. Einstein said that he had had similar thoughts, but did not have the courage to push the matter further.

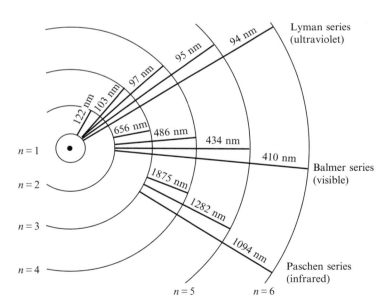

Fig. 17.2 Bohr's electron orbits in a hydrogen atom, and the transitions of electrons from one orbit to another. The spectral lines associated with the transitions have been grouped in line series according to the inner orbit. For example, the Balmer lines of hydrogen are associated with transitions from the second level upward (absorption lines) or from an upper level to the second level (emission lines). (Credit: NASA)

Box 17.1 Bohr's model and the spectroscopic laws of Kirchhoff

The Bohr model of an atom nicely explained the experimental laws of spectroscopy discovered by Kirchhoff. In thin hot gas atoms collide with each other, raising electrons to higher-level orbits. Soon they drop back downward to lower-level orbits. As a result, the atom radiates photons whose energy corresponds to the energy difference of the orbits. Thus the spectrum of the gas shows bright emission lines (Kirchhoff's II law). When radiation passes through thin gas, those photons which have the right energy to raise an electron from a low orbit to a higher orbit are absorbed. Thus absorption lines are created in exactly the same places in the spectrum where the bright emission lines appear (Kirchhoff's III law). In dense gas and in solids, the atoms are so close together that they perturb each other's electron orbits; the orbits are shifted from their usual orbital radii. As a result, energy jumps of all kinds can appear, and photons of all wavelengths are emitted. Thus a continuous spectrum is observed (Kirchhoff's I law).

Bohr was appointed professor of theoretical physics at Copenhagen in 1919. A special institute was founded to further his research; it became one of the leading centers of the study of *atomic physics*, a place where researchers from different parts of the world could meet, not always easy in the post-World War I atmosphere.[1]

Bohr's model explained the radiation of atoms so well that gradually it was accepted as fact (see Box 17.1). But the assumptions made by Bohr had no real basis in physics. Many of the physical laws in the microworld are quite different from the laws found in our usual environment. Neither Newton's mechanics nor Maxwell's electromagnetic theory could be directly applied to the phenomena at the atomic level.

Mechanics of Atoms

The new theory of mechanics for the atomic level became known as quantum mechanics. The first breakthrough in its discovery was made by the German physicist Werner Heisenberg. A little later quantum electrodynamics was developed to describe electromagnetic phenomena in the world of atoms. These new theories are connected to older, so called classical physics theories in such a way that when one

---

[1] Arnold Sommerfeld (1868–1951) started using elliptical orbits of electrons to model the atom. He assumed that the electron can have, besides the circular orbit, also elliptical orbits of the same diameter. Later the description of an electron's motion along an orbit was given up; what remained from these early models was the concept of the orbits as energy levels. An atom can go to a higher energy level or become *excited*. When the excitation is released, the atom radiates a photon.

Fig. 17.3 (**a**) Werner Heisenberg (1901–1976) and (**b**) Erwin Schrödinger (1887–1961)

moves from the atomic scale to large scale, the results of classical physics are obtained at the limit. In this sense quantum physics offers a deeper view of reality than
classical physics.

Werner Heisenberg (Fig. 17.3) worked at University of Göttingen in a group
led by Max Born (1882–1970) whose goal was to clarify the strange behavior of
electrons inside atoms. In June 1925 there was optimism in the air: the breakthrough
must be near. But just then Heisenberg had such a bad case of hay fever that he had
to leave Göttingen and travel to the austere surroundings on the island of Helgoland
in the North Sea where his hay fever passed. There the 23-year-old Heisenberg
continued thinking about the problems. At last everything fit together and an exact
mathematical description of the behavior of electrons was born. Heisenberg said
later that one morning at three o'clock

> ...I could no longer doubt the logic and unity of the branch of quantum mechanics at
> which my calculations pointed. At first I was very restless; I felt that I was looking through
> surface of atomic phenomena into their strangely beautiful interior, and I had a dizzying
> feeling when I was allowed to study this bounty of mathematical structures which nature
> had generously spread in front of me.

After returning to Göttingen, Heisenberg was too timid to advertise his discovery.
He wrote his result in a scientific article, and gave a copy to Born and another one
to his friend Wolfgang Pauli in Hamburg. Born send the article to the magazine
Zeitschrift für Physik for publication. Heisenberg had to take a trip and left Born to
ponder about the significance of the tables in the article.

Born noticed that Heisenberg's tables were matrices, the basic quantities of
a class of mathematics known as matrix algebra. Together with his colleague

Pascual Jordan, Born started to translate Heisenberg's theory to the matrix language. Heisenberg, now temporarily in Copenhagen, took part in the finalizing of the theory. Around the same time, Paul Dirac at Cambridge gave the theory another mathematical form, and a year later Erwin Schrödinger developed one more representation (to be described later). Those were hectic times for quantum physics!

Nebulous Particle: Heisenberg's Uncertainty Principle

The essential feature of quantum mechanics is its probabilistic nature formulated by Max Born in 1926. Instead of talking about exact values of physical quantities, one can only describe their probability distributions. This is related to the *uncertainty principle*, given by Heisenberg in his 1927 publication. It had become clear to Heisenberg that the existence of a particle at the same time as a material body and as a wave makes a fundamental restriction on the position of the particle. One cannot say that an electron is located at a specific distance from the atomic nucleus at a given time. Both things cannot be known at once. The electron is "spread out" to the surroundings of the nucleus. One can only say that there is more probability of the electron in some distance and direction than in another distance and direction – the "planetary orbits" in the simple Bohr model represent only the most likely regions where one might find the electron. This does not apply only to electrons bound to an atom, but to all electrons and to all particles. As a general rule one can say that the more "spread out" a particle is, the lighter it is. The "spreading" of an everyday object like a tennis ball is unnoticeable.

The "spreading" of a particle may sound abstract, but as a matter of fact it has very concrete consequences. For example, in the emission of alpha radiation, the particle *tunnels* its way out of the radioactive nucleus. The alpha particle is bound to the nucleus by the strong nuclear force which should keep it absolutely tied to the nucleus. However, from time to time we see an alpha particle leaving the nucleus. George Gamow (who also studied cosmology and the genetic code; Chaps. 24 & 28) explained this by using the quantum theory such that the alpha particle is not only "spread out" in the area of the nucleus but also to a small extent outside it. "Spreading" means that the particle has a small probability of being found anywhere within the area where it has "spread." Thus the alpha particle is inside the nucleus with slightly less than 100% probability, but at the same time, it is also outside the nucleus with a small probability. Therefore occasionally the positively charged alpha particle materializes outside the nucleus, outside the range of the strong nuclear force, whereby the electric repulsion of the positive nucleus pushes it away.

Also the fusion of helium inside the Sun, and hence the sunshine we enjoy, is based on tunneling. Since protons repel each other by electric force, one must drive them together with very high speed before they come close enough to be affected and bound together by the strong nuclear force. But the protons in the Sun are too slow. How do we solve the dilemma? Since also protons have "spread out" around their mean position, it happens occasionally that protons materialize closer to each other than their mean position would suggest. Then to their great surprise, the protons may suddenly find themselves within the reach of the strong nuclear force, even though it supposedly was impossible.

Consider now the following event. We bounce a tennis ball off a brick wall of a house. Then suddenly the ball goes through the wall and appears inside the building. There was no hole in the wall, and no hole is created either; the ball has tunneled through the wall. The fact that this never happens is due to the large mass of the tennis ball in comparison with a proton! Now it is also clear why the electron cannot be a constituent particle of the nucleus. As a light particle, the electron has spread out to such a wide area that it is quite impossible to keep it trapped inside the nucleus.

The Structure of Atoms

Quantum theory finally made it possible to understand the structure of atoms, why each atom has its chemical properties, why atoms form chemical compounds, and much else. The calculations in quantum mechanics are based on the Schrödinger equation, discovered in 1926 by the Austrian Erwin Schrödinger while working in Zurich (Fig. 17.3). Since it had already become obvious that electrons have to be regarded as waves, Schrödinger treated the electrons in atoms also as a phenomenon of vibration. He showed that only certain kinds of vibrational states are long lasting, like a musical instrument with long lasting, discrete notes. The "notes" of an atom correspond to Bohr's electronic orbits (Fig. 17.4).

Bohr's hydrogen atom evolved further into a *shell model* for atoms which explains the periodic table of the elements. Instead of Bohr's orbits we now talk about shells of an atom. Elements heavier than hydrogen have several electrons which are placed in different shells. However, electrons cannot freely choose the shells where they go, but in its most bound state the shells of an atom are filled up with electrons going from below to the top (from the nucleus outward), until all the electrons of the atom in question have been placed. The chemical properties are determined by the level of occupation in the outmost shell. Atoms try to complete the level of

Fig. 17.4 Electron waves circling around the atomic nucleus. If there is a whole number of waves per revolution, the wave enforces itself, and an allowed electron orbit is created. If the phases of the wave do not match after one revolution, interference between waves destroys them. Then this orbital radius is not possible in Bohr's model of an atom

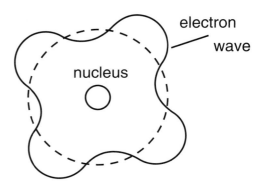

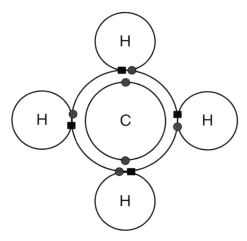

Fig. 17.5 A sketch of the covalent bonds in the methane molecule. The nuclei are represented by the element symbols C and H and the shells of electrons by circles. The electrons are shown by filled circles (carbon) and filled squares (hydrogen). Four hydrogen atoms each share their one electron with carbon to form four covalent bonds and make a methane molecule. Notice that all the shells are full in the result: two electrons in the inner shell and eight electrons in the outer shell

occupation of their outer shell by borrowing electrons from their neighbors or by sharing electrons with them. This creates chemical bonds. In the chemically inert noble gases the outer shells are full so that they feel less need for association with other atoms.

For example, the bond between two hydrogen atoms which allows the hydrogen molecule to exist is based on sharing the two electrons of both atoms. It is the *covalent chemical bond*, discovered by German physicists Walter Heitler and Fritz London in 1927. The covalent bond is important in complex molecules, such as the molecules that life is based on (we will discuss the structural elements of life in Part IV). This is because there can be several bonds per atom with definite orientations relative to each other. Moreover, the covalent bond is strong. Especially important is the carbon atom which is missing four electrons in its outer shell. The carbon atoms complete their outer shells in many different ways which (with the bond strength) can lead to very complicated chains of atoms (Fig. 17.5).

The reason why each shell can only have a limited number of electrons, and the maximum number per shell, was discovered in 1925 by the Swiss physicist Wolfgang Pauli (1900–1958). Arnold Sommerfeld and Niels Bohr had studied the same problem earlier.[2] Pauli concluded that the number of electrons in different shells is based on what came to be called the *Pauli Exclusion Principle*: No two electrons in an atom can have identical quantum states.

---

[2] Although the reason for the shell structure was not yet known to Bohr, he was able to predict that the unknown element number 72 (hafnium) should be chemically like zirconium (40). Inspired by the prediction, the new element was soon discovered at the Niels Bohr Institute by the Dutch physicist Dirk Coster and the Hungarian chemist Georg von Hevesy.

The states of an electron are described by whole numbers which correspond to the electron orbits of Bohr and Sommerfeld. In addition, an electron possesses a *spin* or rotational state. Every orbit can have at most two electrons, one spinning around its axis in the same sense as the electron orbits the nucleus (as most planets do in the Solar System), and the other spinning in the opposite sense. The spinning of an electron around its axis cannot be taken too literally; rather it is just a way to describe the two spin states. Atomic-level phenomena do not have clear-cut everyday parallels.

The Pauli Exclusion Principle is responsible for the structure of the cloud of electrons around the atomic nucleus as well as for the differences in chemical properties of the elements. It also makes atoms hard spheres which cannot easily penetrate each other, in spite of the fact that the atom can be described as mostly empty space in the Bohr model.

Common Sense and Reality

Quantum physics has turned out to be extremely accurate in accounting for the properties of matter, and in that sense it is "correct." However, the conceptual foundations of the quantum theory are still the subject of discussion and research. The phenomena are so different from what we are used to in the macroscopic world and in "common sense" that it makes us wonder that a deeper layer of reality may be reflected in quantum physics. One of the most influential thinkers of the philosophical aspects of quantum mechanics has been Niels Bohr.

A free particle which moves with a constant, exactly known velocity was a basic entity for the old physics. But then Heisenberg's Uncertainty Principle tells that we do not know anything about its position; it is anywhere and nowhere in the universe! A classical particle simply cannot live in the quantum realm. Similarly, the familiar concept of an orbit becomes dubious.

Consider an electron which has left point A and is later observed in point B (Fig. 17.6). Laplace, the advocate of Newtonian mechanics, would calculate an orbit between the two points. He could tell you exactly where in the orbit the electron has been at every instant of the journey, and what the speed of the electron was. The Uncertainty Principle prevents this kind of continuous description of the journey. The electron has been observed in points A and B, but we really do not know where it has been at intermediate times. The best we can do is to make probability calculations of any electron orbit between the two points.

If the electron has no definite orbit, how does it know where it is going? We may say that the electron tries all possible routes all at once. Every route is represented by an electron wave. When the waves from all routes are combined, at most points the waves cancel each other. Only in some points the waves interfere constructively and a high probability of finding the electron remains; point B is just this kind of a point. So what was the real route from A to B then? All routes, or no route at all, as you like it. The concept of an orbit has lost its significance. When we discuss

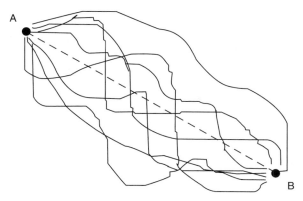

Fig. 17.6 The travel of a particle from point A to point B. To find the shortest path the particle tests all possible tracks. The waves related to the particle destructively interfere with each other everywhere except on the straight (dashed) line connecting A and B. In quantum theory, the particle may be found with greater probability, but not necessarily, on this line

heavier bodies, then we approach the classical orbit. Then the interference pattern from all orbits produces a high probability narrow line connecting points A and B. In everyday phenomena we may safely apply Laplace's concepts.

What happens to Laplace's clockwork universe which progresses in a fully predictable way once set in motion? The Uncertainty Principle destroys the clockwork even before you can set it in motion. Laplace's assumption, "if the positions and speeds of all bodies were known at an initial moment of time," cannot be realized since there is fuzziness both in the position and the speed of the body; and even if one of them were measured momentarily, the other would remain undetermined. The accidental materialization of a particle even beyond an impenetrable "wall," as in tunneling, makes the prediction of the future impossible.

This may seem difficult to accept, and for many "old guard" physicists it was impossible. Even though they used the mathematical methods of quantum physics, these physicists could not accept their philosophical consequences. This is somewhat similar to the initial period after Copernicus' time when his methods of calculation were widely used while his Sun-centered system was not accepted.

Perhaps the foremost doubter of the interpretations of quantum mechanics was Albert Einstein who said: "God does not play dice." To disprove the philosophical foundations of quantum physics, he developed thought experiments on how one would get around the Uncertainty Principle. For most such arguments, Bohr and other developers of quantum philosophy had a ready answer. However, there was one experiment which had to be carried out before we knew who was right and who wrong. This experiment was proposed by Einstein together with his colleagues Boris Podolsky and Nathan Rosen.

The idea of Einstein, Podolsky, and Rosen was substantially as follows (presented somewhat differently by them): Let two particles collide and then separate from each other. Because of the collision, both the positions and the speeds of the two particles become interdependent. If we measure the speed of particle 1, then the speed of particle 2 is easily calculable without a measurement. On the other hand, by measuring at some other (later) time the position of particle 1, the position of particle 2 is determined through calculations. This would indicate that particle 2 has a well-defined speed and well-defined position at every moment of time since the collision. This is an apparent conflict with the Uncertainty Principle. Einstein, Podolsky, and Rosen used this example to claim that the system of quantum mechanics is incomplete. However, Niels Bohr argued that even though the position of particle 1 can be measured, the simultaneous measurement of its speed was not possible due to the inherent disturbance of the measuring process on the speed of the particle. Neither could one then calculate the speed of particle 2 with certainty; the Uncertainty Principle would apply also to particle 2.

In 1964 the Irish physicist John Bell (1928–1990) transformed the described thought experiment to a form where it could be tested in reality. In 1982 Alain Aspect carried out the experiment in Paris. It showed that Einstein and his associates had been wrong. You cannot fool particle 2: it "knows" about the measurement carried out on particle 1, even when there is no chance of transmitting information between them even with the speed of light. The two particles are really part of the same system.

Thus it has been shown that the Uncertainty Principle is a fundamental principle of Nature, and you cannot get around it. It is also exciting that one may apply it to situations which would be hard to understand without this principle. An example is the *vacuum*.

What is a vacuum? Take away all matter, radiation, and force fields from space; what is left over could be called a vacuum. Boring? On the contrary, vacuum is full of happenings. Heisenberg tells you that the energy involved in some "happening" is the more uncertain, the shorter time the phenomenon lasts. Even though the average energy in vacuum may be zero, over short intervals of time the Uncertainty Principle

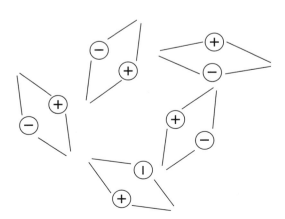

Fig. 17.7 Particle–antiparticle pairs are born and annihilated even in the vacuum of space

allows particles to be born out of nothing and to disappear into nothing. Particles like that are said to live on "Heisenberg loan."

In this way a vacuum is automatically filled with particles. Although every particle lives only a minuscule length of time, new ones are constantly born to replace the ones which have disappeared. The more permanent ordinary particles swim through this "sea" of particles (Fig. 17.7). Later we will find out that a vacuum can have even stranger properties that rule the evolution of the whole universe.

Chapter 18
Elementary Particles

By the year 1932, the view of the structure of matter had reached a simple form. The atomic nucleus was thought to contain protons and some electrons. Electrons served to neutralize the electric charges of some of the protons. (This is because excepting ordinary hydrogen, the atomic weight of an element always exceeds its atomic number which tells the electric charge of the nucleus). Beta radiation, where an electron is fired out of the nucleus, supported the view of the electron as a part of the nucleus. All matter was made of two units, *elementary particles*: the light negatively charged electron and the heavy positive proton. By combining these, one may assemble the nuclei of all elements. By adding suitable number of electrons to circle around the nucleus, one obtains all elements. By chemical bonding one may make all matter out of the elements, in all its different forms.

This simple picture collapsed in the "crazy" year of physics, 1932. The first major event of the year was the discovery of a new elementary particle, the neutron. The main honor for this goes to James Chadwick (1891–1974). He was a student of Rutherford in Manchester and was vice director of the Cavendish Laboratory during that year.

Nuclear Force

When a beryllium target is bombarded by fast alpha rays, beryllium starts to radiate unknown, very penetrating radiation. Chadwick found out first that it does not consist of electromagnetic radiation, but of particles. Then, he reasoned that the particle cannot have electric charge since it penetrates matter much better than protons. Finally, his collision experiments produced the mass of the particle, approximately equal to the proton mass. Chadwick called the particle the *neutron* because of its electric neutrality. His notes are on display at the Trinity College Library, Cambridge. In them he writes "Eureka I have found it!" He didn't believe that it was an elementary particle but thought that it consisted of a proton and an electron (Rutherford had suggested this in 1920).

P. Teerikorpi et al., *The Evolving Universe and the Origin of Life*
© Springer Science+Business Media, LLC 2009

In 1932, in the magazine *Nature*, Chadwick already hinted that the neutron could be a new elementary particle. This view was supported by the value of the mass of the neutron which he and Maurice Goldhaber measured two years later. It is slightly heavier than proton and electron together, in contrast to the model of a combination particle. Moreover, it was soon found that there is a previously unknown force between nuclear particles, the nuclear force that does not discriminate between protons and neutrons. Therefore the neutron is an elementary particle in the same way as a proton.

This nuclear force had to be attractive so that the charged protons do not expel each other from the nucleus. Inside the nucleus the force must be stronger than the electric repulsion between protons; according to current views the nuclear force between two protons is about 100 times the electric repulsion. On the other hand, the grip of this force cannot extend much outside the atomic nucleus where the electric force must dominate to hold the electrons bound to the atom. Thus the nuclear force must weaken with distance much faster than in the inverse square law of the electric force.

This strange force was explained by Hideki Yukawa (1907–1981), the first Japanese to receive the Nobel Prize in physics in 1949. He suggested a new concept of how the influence of the nuclear force is transmitted between particles. Namely, a particle makes its presence known by emitting messenger particles to its surroundings. When the messenger meets another particle, it relays the information concerning its host, and the receiving particle knows how to react appropriately. When the particles keep in touch with each other by emitting messengers, they know to keep together and not wander off.

This idea was not just a fanciful plot for describing the nuclear force: the electromagnetic force can be understood in the same way. Pieces of the electromagnetic field, energy packages, fly between charges carrying their messages. Viewed in this way, the electromagnetic force field consists of photons.

According to Yukawa, the essential difference between the electromagnetic field and the field of the nuclear force lay in the mass of the messenger. The photon of the electromagnetic field is massless, while the messenger of the nuclear force field is a particle with nonzero mass at rest. Yukawa predicted that the particle in question is 200–300 times heavier than electron; thus, the messenger of the nuclear force should be intermediate between nuclear particles and electrons (a proton is 1,836 times and a neutron 1,839 times heavier than an electron). Particles in this category are called *mesons*, from the Greek "meso" (middle). The range of influence of the messenger particle depends on its mass. The heavier is the messenger, the shorter is the range. Only massless messengers like photons can extend their influence to any distance.

In today's physics, the bouncing of messengers back and forth has replaced the whirls of Descartes, action at distance by Newton, and Faraday's lines of force and waves in ether. Of course, the descriptions of Newton and Faraday are still useful for the gravity and the electromagnetic force. However, the new forces discovered in the twentieth century are better described by the method of Yukawa, the *strong nuclear force* being the first example.

Yukawa made his prediction about the messenger particle in 1935.[1] Twelve years passed before Cecil Powell's (1903–1969) group at Bristol University got the first sight of Yukawa's particle. It is called the *pion*, and it comes both in charged (mass: 273 electrons) and neutral (mass: 264 electrons) varieties.

Pions are short lived according to everyday standards. But we should really compare its lifetime to the "nuclear year," the time it takes the nuclear particles to revolve around the nucleus, which is only 10^{-22} s. Then the pion mean lifetime of 2.6×10^{-8} s, huge 10^{14} times the "year," looks like eternity in the nuclear time scale. Even the neutral pion, living 10^{-16} s on average, is long lived from this point of view. If we consider the "purpose" of pions in nature as messengers of the nuclear force, there is no need for them to live any longer.

Phenomena of Atomic Nuclei and the Weak Force

We have arrived at a picture where the atomic nucleus consists of one or more nuclear particles, and they move around each other in the small volume of the nucleus because of the attraction of the nuclear force. There are two kinds of nuclear particles: protons and neutrons. We may visualize a cloud of pions crossing between nuclear particles resulting in the nuclear force. The nucleus may also possess tight agglomerations of two protons and two neutrons, which may escape from radioactive nuclei as alpha particles. In analogy with the electrons jumping between energy levels, the nuclear particles may reorganize their orbits in such a way that energy is liberated as high-energy gamma radiation. The energies in the nuclear phenomena are much greater than in atomic phenomena, typically by a factor of a million. This explains the benefit of nuclear fuel per gram in nuclear power stations with respect to power stations using chemical reactions. The same fact explains the huge power of nuclear explosives.

If there are no electrons in the nucleus, what about the beta radiation where electrons are emitted from nuclei? This was explained by the brilliant Italian physicist Enrico Fermi only a year after the discovery of the neutron (see Fig. 18.1). According to Fermi, there is also another nuclear force operating inside the atomic nucleus called the *weak force*. It causes an electron first to be *born* and then to be emitted from the nucleus; at the same time, a neutron turns into a proton. We will understand this process better after we discuss the internal structure of a neutron and a proton.

Fermi's theory is remarkable also because it predicts the existence of a new elementary particle called the *neutrino*. This "little neutron" is unaffected by the

---

[1] Only two years later a new particle was found among the *cosmic rays* (particles arriving from space to Earth). About 200 times heavier than the electron, this particle was a good candidate for the messenger. The particle is short lived; it disintegrates into other particles in two microseconds on average. However, further research has shown that this particle, called the muon, is a heavy form of an electron (mass: 207 times the electron mass) rather than the wanted messenger.

Fig. 18.1 Enrico Fermi (1901–1954) made important contributions to nuclear physics (Credit: NARA)

electromagnetic force or the strong nuclear force; its main link to the outside world is the weak force. The range of the weak force is very short, only one percent of the diameter of a proton, and its strength is only one part in 100,000 of the strength of the strong nuclear force. Thus a neutrino must come very close to its neighbor before they affect each other. For this reason the existence of neutrinos was only deduced indirectly at first, as a particle connected with a strange loss of energy during the beta decay. Wolfgang Pauli realized that the missing energy escapes away in the form of penetrating particles. The mass of the common variety of neutrino is less than 10^{-4} electron mass; as we will learn later, there are other varieties of neutrinos whose mass is even less certain.

A neutrino is so unlikely to collide with another particle that it can travel through a sheet of lead light-years thick without a collision. Only when huge amounts of neutrinos are created, a few of them may be captured by instruments. In 1955, the first neutrinos were detected near the Savannah River nuclear fission reactor in the United States. In recent decades there have been detections of neutrinos from the "fusion reactor" of the center of the Sun and also from other astronomical sources. It is believed that neutrinos are among the most common particles in the universe, but extremely difficult to observe.

Just one day after Chadwick had sent his article about the discovery of neutron to *Nature* magazine, *Physical Review* magazine received the news of the second big discovery of 1932 by a team led by Harold Urey (1893–1981), a chemist, physicist, and astronomer. The work mentioned here was carried out in Columbia University, New York.

Remember the explanation for the strange atomic weight of chlorine, 35.46, almost half way between two whole numbers. In nature there are two kinds of chlorine, two "isotopes" with atomic weights of 35 and 37. Most elements have several isotopes. There are just over a hundred known chemical elements, but the number of known isotopes is well over 2,000; only about 280 of them are stable. In the chlorine nucleus, the 17 protons may associate themselves either with 18 or 20 neutrons. Thus the atomic number which determines the chemical properties is 17 in both cases, but the atomic weights $17 + 18$ and $17 + 20$ differ (in addition, there are rare chlorine isotopes $17 + 19 = 36$ and $17 + 23 = 40$).

Before 1932 there had been suggestions that hydrogen might have several isotopes since the atomic weight of hydrogen in nature exceeds the weight of a proton. The difference is so small (excess of about 10^{-4}) that there was no certainty about the question. One would have to isolate the heavy form of hydrogen to be sure, which is difficult because the chemical properties of the isotopes are identical. However, Urey and his colleagues succeeded in doing it. It was then easy to show that the isolated heavy hydrogen had atomic weight of 2, which means that its nucleus consists of one proton and one neutron. This form of matter was given the name *deuterium* even though it is really hydrogen, the heavy variety of it. In fact, deuterium deserves its own name. It has had a key role in the study of the nuclear force since the two-body motion is much easier to figure out than, say, the three-body motion (remember the complicated three-body problem under the force of gravity, Chap. 11).

Not all combinations of proton and neutron number are possible; for massive nuclei the suitable number of neutrons is a little bigger than the number of protons. If we try to reduce the neutron number artificially so that it goes outside the suitable range, we get unstable nuclei which transform through radioactivity until they become stable nuclei (Fig. 18.2).

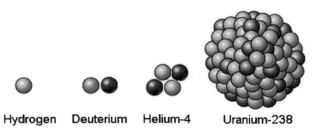

Fig. 18.2 Nuclei are made of protons (*gray balls*) and neutrons (*dark balls*). The uranium nucleus has 92 protons and 146 neutrons. It is among the heaviest known atomic nuclei

Particles and Accelerators

Even though the atomic nucleus had been known since 1911, one may assign the actual beginning of nuclear physics to the year 1932. In addition to the findings we discussed, the most important tools of this research, *particle accelerators*, started operation during this "crazy" year. Up to this time atomic nuclei had been bombarded by using particles emitted by radioactive materials. The cost of radium, for example, $100,000 g^{-1}$, made the generation of a strong stream of particles very expensive. Moreover, the splitting of heavy nuclei required particle streams which move much faster than the particles emitted by natural sources.

Charged particles may be accelerated by letting them fall through a strong voltage drop. If an electron falls through voltage drop of 1 V, it accelerates so much that its energy becomes 1 electron volt (eV, a convenient unit of energy). In chemical reactions, one usually deals with energies per atom of about 1 eV. In nuclear reactions, energies are typically millions of electron volts (MeV) per atomic nucleus.

John Cockcroft (1897–1967) and Ernest Walton (1903–1995) built in the Cavendish Laboratory an accelerator where the voltage drop was 700,000 Volts. In 1932, they let protons accelerate through this drop and then hit a lithium target. Behind the target they had placed zinc sulfide plates which registered the flashes caused by alpha particles. The impacting protons managed to break the lithium nuclei into alpha particles (helium nuclei), the first artificial transformation of an atomic nucleus to another kind (Fig. 18.3)!

At the same time, the American Ernest Lawrence (1901–1958) developed an even stronger accelerator called the cyclotron (Fig. 18.4). Lawrence graduated at Yale University and then moved to the University of California. There he happened to spot an article by the Norwegian Rolf Wideroe who suggested that particles could be easily accelerated in stages if the particles travel in closed circles due to magnetic deflection. During every cycle the particle travels through a voltage drop which increases its speed. Lawrence and his student constructed an accelerator based on this principle and used it to accelerate protons to energies above 1 MeV in 1932. They could confirm the results of Cockcroft and Walton. The cyclotron soon became common with about twenty machines in operation five years later.

In early 1950s, the cyclotron was further developed into the *synchrotron* which could reach collision energies above 1,000 MeV (that is 1 GeV, Giga electron volt). At present, the largest accelerator is at the European Organization for Nuclear Research (CERN) in Geneva. Most of the activities at CERN are currently directed toward a new collider, the Large Hadron Collider (LHC) and the experiments for it. This occupies a 27 km circumference circular tunnel. The tunnel is located about 100 m underground, in the region between the Geneva airport and the nearby Jura mountains. The large ring protons, travelling in opposite directions in two pipes, will reach speeds of about 0.999,999,99 (!) times the velocity of light.[2] The LHC

---

[2] Actually the particles are accelerated through a series of steps that successively increase their speeds and energies, before they are finally injected into the main circular accelerator. These preparing stages include a linear accelerator, a proton synchrotron booster, a proton synchrotron, and a super proton synchrotron.

Fig. 18.3 In early particle accelerators a Cockcroft-Walton voltage multiplier made the required strong voltage drop. This picture shows such a device built in 1937 by the firm Philips and currently residing in the National Science Museum in London (PD/Wikipedia)

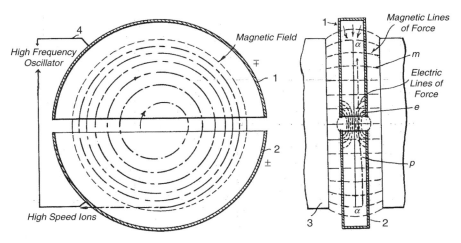

Fig. 18.4 Diagram showing cyclotron operation from Lawrence's 1934 patent

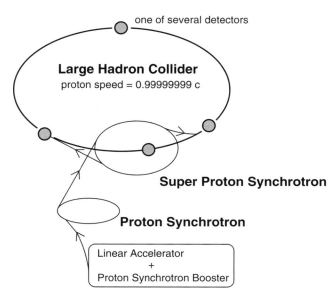

one of several detectors

Large Hadron Collider
proton speed = 0.99999999 c

Super Proton Synchrotron

Proton Synchrotron

Linear Accelerator
+
Proton Synchrotron Booster

Fig. 18.5 Sketch of the Large Hadron Collider of CERN. Protons are accelerated gradually to their high speeds through several systems. The collider tunnel contains large detectors to register the interactions of proton beams travelling in opposite directions around the ring (based on information at CERN home pages: http://public.web.cern.ch/Public)

will collide protons at an energy of 7 TeV (TeV = 1,000 GeV) each, with the total collision energy of 14 TeV. Each proton will have the kinetic energy of a flying mosquito – for a proton this is a huge amount! At these energies, millions of times greater than what Lawrence was able to reach, new kinds of particles may appear (Fig. 18.5).

In 1932 the particles were detected using a "cloud chamber" which has water vapor in supercritical stage so that water droplets condense along the path of a charged particle. By taking a photograph one finds the tracks of the charged particles that have just then passed through the chamber. A magnetic field in the chamber will make the tracks curve: the amount of curvature and finding which way the track bends helps to identify the particle. The bubble chamber came into use as an improved detector in the 1950s. There the particle tracks show up as sharp lines of liquid bubbles. They can be photographed from different directions and analyzed. Today there are many new more automatic detection methods in use.

The fourth big discovery of the year 1932 was done by Carl Anderson (1905–1991) who studied the tracks of cosmic rays in a cloud chamber. Among them the American physicist found one that was just like a track of an electron except it curved in the wrong direction in a magnetic field, i.e., it had a positive charge (Fig. 18.6). Anderson checked his surprising result in many ways and published it. He had detected the *positron*.

Anderson was not aware that the English physicist Paul Dirac (1902–1984) had many years ago predicted the existence of the positron. Not only the electron, but

Fig. 18.6 A gamma ray enters a bubble chamber from above and creates an electron–positron pair. Due to the magnetic field the orbit of the positron curves to the left and the orbit of the electron to the right. Another particle pair was born at the tip of V. Also the orbit of another electron is seen. Based on a Lawrence Berkeley Laboratory bubble chamber photo

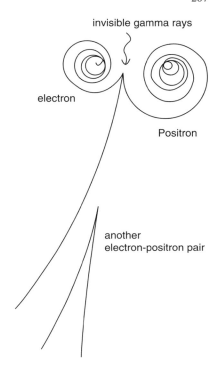

invisible gamma rays

electron

Positron

another
electron-positron pair

also all elementary particles should have similar counterparts except for the opposite charge. These are called *antiparticles*. Besides the electron, the proton should have its antiparticle also. In principle, there could be a whole "antiworld" where the atomic nucleus has negatively charged antiprotons and the nucleus is circled by a cloud of positively charged positrons. All chemical phenomena would happen just like in our world.

Antimatter, consisting of antiparticles, does not exist in significant amounts. This is easy to understand since matter and antimatter cannot coexist peacefully. When a positron and an electron meet, they destroy each other with only gamma radiation left over. Similarly protons and antiprotons annihilate each other. The antiproton was found in 1955. As every particle must have its antiparticle, the list of known particles doubled in length. Thanks to the discoveries of the exceptional year 1932, Chadwick, Anderson, Urey, Lawrence, Cockcroft, and Walton joined the ranks of Nobelists between 1934 and 1951.

Quark: At Last the Fundamental Building Block?

For some time protons and electrons were viewed as indivisible true "atoms." But Nature turned out to be not that simple. The number of known elementary particles grew all the time with the increase of the power of particle accelerators. As it had happened with the chemical elements a century earlier, systematic trends were noticed among elementary particles. The particles may be divided into three main categories: *leptons*, *hadrons*, and *photons*. Leptons are not affected by the nuclear force, and as far as we know, they are so small that they behave like mass points in collision experiments ("lepto" means "small" in Greek). The leptons include electrons, muons, and neutrinos.[3]

Hadrons feel the nuclear force: they are the nuclear particles (protons and neutrons) and related particles, called *baryons*; and messengers of the nuclear force, pions, and their relatives, called *mesons* as a group. In 1960s it became evident that hadrons are not really elementary but consist of smaller parts, *quarks*. When protons and neutrons are bombarded by electrons and muons, they behave as if they were mostly empty, except for a few point-like centers (a parallel with Rutherford's experiments!). The diameter of a proton is about 10^{-12} mm; it is actually the region where quarks move about. Quarks themselves are much smaller, possibly point-like.

Since 1950s, Murray Gell-Mann had been searching for order among the elementary particles, and, like Mendeleev before him, he discovered rules and predicted new particles with success. Gell-Mann and Georg Zweig, both at the California Institute of Technology, proposed independently in 1964 that both proton and neutron are made of three quarks. In fact, the quarks were first viewed as convenient mathematical tools to make calculations in complex elementary particle physics. The quark idea did not catch on since no isolated quarks were found.

The detection of quarks should have been rather easy since they have fractional electric charges. The most important quarks are the *up quark* with the electric charge $+ 2/3$, and *down quark*, with charge $-1/3$ (as usual, the charge of an electron is -1 in these units). However, in the bubble chamber pictures no such fractional charges can be identified; all particles have the electronic charge or its multiples. Nevertheless, the hard cores inside a proton and a neutron match so well with the quark theory that apparently quarks exist at least there. The idea today is that the quarks in the nuclear particles are in a forced union; unlike other particles, they cannot exist alone, but always require a partner or two.

In the quark model, baryons are made of three quarks. The proton is made of two up quarks and one down quark, while the neutron has two down quarks and one up quark. The mesons are made of two quarks, one of which is an ordinary particle and the other its antiparticle. For example, the neutral pion is a combination of an up quark and its antiquark, while the positively charged pion has an up quark and antidown quark (Fig. 18.7). In the original quark model there was also a third quark called *strange quark*. It was needed to explain the so called strange particles.

Three kinds of quarks were enough to explain all known hadrons until the early 1970s. Then Burton Richter's group at Stanford University and Samuel Ting's group at Brookhaven National Laboratory discovered a new particle which could not fit in the Gell-Mann and Zweig system. Richter called it psi; Ting, J. Even though J/psi

---

[3] The *tauon* was discovered in 1977; even though it is 3,510 times heavier than the electron, it is still classified as a lepton because of its other properties. There are three kinds of neutrinos which raises the number of known leptons to six, and if the antiparticles are counted, then there are altogether 12 leptons.

Fig. 18.7 The proton (*left*) is composed of two up quarks (u) and one down quark (d). On the right, we have a pion composed of an up quark and an antidown quark

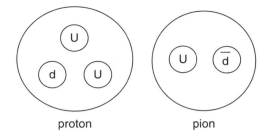

proton   pion

is a meson, its mass is about three times the mass of a proton. To understand it, one has to introduce a new quark called the *charm quark*: J/psi is then a combination of a charm quark and anticharm quark. The reality of the charm quark was soon confirmed when other particles were discovered where the charm quark was a partner.

Only two years passed and a new quark, the *bottom quark*, had to be introduced again. The team led by Leon Lederman at Fermilab near Chicago found a particle named upsilon whose mass is about ten times bigger than the proton mass. It is a big mass meson, a combination of two quarks, a bottom quark and an antibottom quark. The last of the quarks, it is believed, is the *top quark*, discovered at Fermilab in 1995. This makes the total number of quarks six, like the number of leptons (or 12, if we count the antiparticles). The up, charm, and top quarks have the electric charge $+2/3$, while the other three (down, strange, and bottom) possess the charge $-1/3$.

Remarkably, out of all basic particles, only four are needed as the building blocks of ordinary matter: the electron and the electron neutrino among leptons and the up and down quarks. The rest of the elementary particles appear superfluous. The four important particles are said to be of the first generation, the remaining eight are classified as second- or third-generation particles. We do not know why nature repeats itself with two generations of greater mass particles (Box 18.1).

Box 18.1 Particle generations

Generation	Leptons	Quarks
I	Electron (1)	Up ($\sim$5)
	Electron neutrino ($\sim$10^{-6})	Down ($\sim$10)
II	Muon (207)	Charm ($\sim$3,000)
	Muon neutrino ($\sim$10^{-6})	Strange ($\sim$200)
III	Tauon (3,536)	Top ($\sim$350,000)
	Tauon neutrino ($\sim$10^{-6})	Bottom ($\sim$8,000)

Ordinary matter is made up by the Generation I particles: the electron, the electron neutrino, and the up and down quarks. The mass is in parentheses, in electron units. The masses of quarks are uncertain; the neutrino masses practically unknown. Note that up and down quarks are much lighter than proton and neutron which they make. Much of the mass of a nuclear particle is associated with the binding of the quarks together.

Messengers of the Weak Force

What is the previously mentioned weak force affecting neutrinos? In the 1960s Steven Weinberg at Harvard University and Abdus Salam (1926–1996) at Imperial College in London proposed a theory that the weak force and the electromagnetic force are two aspects of one single force called the electroweak force. Just as Maxwell had put together earlier electric and magnetic phenomena as two sides of the electromagnetic force, continuing the same process the weak force was now added under the same umbrella.

The Weinberg-Salam theory made an important prediction: the weak force should be carried by superheavy particles ("W" and "Z"). The groups led by Carlo Rubbia and Simon van der Meer started a new experiment at CERN in the late 1970s to raise the collision energy high enough to create such particles. In January 1983 the first evidence of W was found. A few months later the Z-particle was also found.

W is either positively or negatively charged, and weighs as much as 88 protons, while Z is neutral and a little heavier, with about 99 proton masses. When the messenger of the weak force has such a high mass, it is no wonder that the resulting force is weak. The effect of the force is based on bouncing the messenger particles back and forth. A heavy particle does not fly far, and such a particle cannot be thrown about too often. Thus the likelihood that a passing particle is hit by one of these messenger particles W or Z is very low, and the force is weak.

At first sight it would appear that things get more complex when we add three more heavy particles to our list. But the good news was that this proved the unification of electromagnetic force and weak force which simplifies the overall physics picture a lot. We see that the electroweak force is carried by four different particles: photons, positive W, negative W, and Z. Since the photon is massless, its influence goes to a great distance; the other three "photons" carry their force only over a very short range. How these "photons" get their mass is speculative. A theory by Peter Higgs involves "Higgs particles" which are yet to be found (one of the subjects of the Large Hadron Collider of CERN). They would "lend" their mass to the weak interaction photons.

Weinberg and Salam shared the Nobel Prize with Sheldon Glashow (Harvard University) for the idea of unifying the electromagnetic and weak forces. Glashow had proposed the idea of the four kinds of photons in 1961. Rubbia and van der Meer were also similarly rewarded, just one year after they discovered the W and Z particles!

With quarks and the weak force, the radioactive beta radiation can be explained. During this process, one down quark inside a zero charge neutron is transformed into an up quark by the influence of the weak force. As a result the neutron is changed to a positively charged proton. A negatively charged electron and a zero charge neutrino escape so that the electric charge and total energy is conserved in the process. The reaction obeys one of the basic laws of physics, the *conservation of electric charge*. The total charge of all particles before and after the reaction must be the same. The neutrino emitted in beta decay is called the electron neutrino, since it is associated with the electron. It also must have an antiparticle, the antineutrino (Fig. 18.8).

Fig. 18.8 Radioactive beta decay. A neutron is made of an up quark and two down quarks. One of them shoots off a negative W particle (*left*). It makes the down quark turn into an up quark, and the neutron becomes a proton. The W particle decays into an electron and an antineutrino

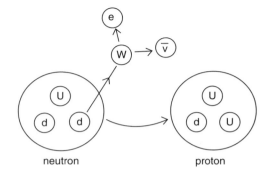

neutron   proton

An Excursion Still Deeper: Does Gravity Live in Many Dimensions?

The force of gravity is intimately related to curvature and dimensions of space. It turns out that all forces of nature may have a link to higher dimensions. How do we count the number of space dimensions? Just draw straight lines so that they are perpendicular to each other. On a sheet of paper you may draw two perpendicular lines; thus the plane is two dimensional. You may imagine a third line coming straight up from the plane, perpendicular to the lines in the plane, defining a third dimension (Fig. 18.9). However much we try, we cannot draw a fourth straight line that is perpendicular to all the three others. Thus our space has three dimensions. If a fourth spatial dimension exists, it must somehow be unobservable.

Einstein's way of expressing gravity as curvature of space was so elegant that it made physicists wonder if other forces could not also be manifestations of space curvature. Upon the completion of the General Relativity Theory, the only other known force was the electromagnetic force, well understood in Maxwell's theory. Einstein had a feeling that somehow gravity and electromagnetism must be associated with each other. He spent much of his later years in search of a unified theory.

Fig. 18.9 The edges of a rectangular box form three lines perpendicular to each other. In a three-dimensional universe one cannot identify a fourth straight line that is perpendicular to all three of them

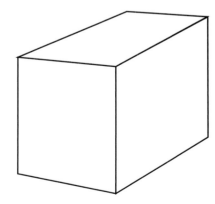

This view was also shared by the Finnish physicist Gunnar Nordström (1881–1923) who published in *Physikalische Zeitschrift* in 1914 a unified theory of gravity and electromagnetism where space has four dimensions (not three), and time is the fifth one. He was the first to introduce an extra dimension to our space-time so that gravitation would be just an aspect of electromagnetic interactions in five dimensions. When projected to the known four dimensions, gravity and electromagnetism appear as separate forces. Unfortunately, the theory of gravity that Nordström used did not turn out to be the correct one. But the basic concept of unification by using additional space dimensions was born.

Gunnar Nordström was a contemporary of Albert Einstein. An engineer by training, he turned to chemistry leading him to Göttingen in 1906 to study under Walther Nernst. In Göttingen, the young Nordström became a wholehearted believer in relativity. After only one paper in chemistry, Nordström's remaining published work was focused almost exclusively on relativity, electrodynamics, and gravitation.[4]

After returning to Helsinki, Nordström became a "Docent" of theoretical physics at the university, at the same time teaching elementary physics at high school level. Between 1916 and 1918, Nordström worked in Leiden, Holland. In 1918, he became a professor of physics at Helsinki University of Technology. No tradition of theoretical physics existed in Helsinki prior to Nordström. The level of understanding of his work is reflected by a negative reply to his request for travel funds: "One can study the fourth dimension at home, without any trips abroad."

In 1921 the German physicist Theodor Kaluza (1885–1954) independently came to the same idea of a unified theory via a fifth dimension. In Kaluza's work electromagnetism is also a consequence of the curvature of space-time, but again it has to be a five-dimensional space that is curved. Thus electromagnetism would also be a sort of gravity.

How is it possible to have five dimensions, four space dimensions plus time as opposed to the four-dimensional gravity (three space dimensions plus one time dimension) we see and describe using Einstein's theory? Saying that everything would be nice if we add one more *ordinary* space dimension would lead to problems. In 1747 Immanuel Kant showed that the law of gravity is related to the number of space dimensions. If gravity weakens with distance like some inverse power of distance, say, n, then the number of space dimensions is $n + 1$. In Newton's law of gravity this power is $n = 2$; thus, the number of space dimensions is $2 + 1 = 3$. If we calculate how bodies move in different force fields, with different values of n, one can show that orbits in which n is greater than 2 are unstable. For example, if the force of gravity around the Sun were to decrease such that $n = 3$, then any tiny perturbation would cause the Earth to dive into the Sun or fly away. Similarly, if the

---

[4] We mention that Nordström constructed the first relativistic theory of gravitation, a precursor to General Relativity. He presented the theory in 1912 and modified it in 1913 while working with Einstein in Zurich. It was given a new formulation in 1914 by Einstein and A. D. Fokker. The ultimate failure of Nordström's theory was that it did not predict the bending of light discovered in 1919. By then Nordström himself had given up his theory and worked on Einstein's General Relativity.

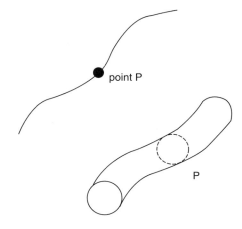

Fig. 18.10 A curled dimension. The upper line appears fully one dimensional, but when we magnify point P in it, we see that in fact the line is a two-dimensional tube. The second curled dimension is hidden. In Klein's theory the dimensions higher than three are hidden in the same manner

electric force had $n = 3$, the shells of electrons could not exist around atomic nuclei. Complex chemistry and life would be impossible.

After Nordström and Kaluza, the Swedish physicist Oscar Klein (1894–1977) wrote down a theory of five-dimensional gravity. To remedy the aforementioned problems, Klein suggested that the extra spatial dimension was "compactified." Specifically, he curled up the fifth dimension on a circle, a circle so microscopically small that it is not directly observed even inside atoms. The remarkable result of Nordström, Kaluza, and Klein was that their theory of five-dimensional gravity unified gravity with electromagnetism.

How does the curling of dimensions happen in Klein's theory? Consider a piece of wire as an example. Seen from far away, it looks one dimensional, its only dimension is its length. Only when we take a closer look we realize that the wire has a certain circumference; another dimension is required to describe the length of the circle around the wire. This second dimension is curled (Fig. 18.10).

In Klein's view there is a fourth dimension associated with every point in our three-dimensional space. It is a small circle around the curved fourth dimension. We do not see these circles all around us because they are minute in size, as much smaller than a proton as the proton is smaller than a planet. If such a dimension exists, no wonder we do not observe it directly.

The theory of Nordström, Klein, and Kaluza was forgotten for years. But when new forces were discovered, physicists started to ask why all forces could not be described as phenomena of space curvature in higher dimensions. This is what is done in the theory of *supergravity*.[5] In this concept, there are actually no forces, only space curvature which appears in different forms or influences ("forces"). There is

---

[5] Supergravity is related to the highly theoretical and much studied *string theory*. It postulates that all matter and energy are composed of excruciatingly minute filaments called strings (instead of point particles as usually thought) and membranous entities called branes. By replacing the point-like particles with strings it may be possible to unify the known forces (gravity, electromagnetic, weak, and strong nuclear forces).

no final supergravity theory yet, but current models employ as many as ten spatial dimensions (plus time). All but three of those dimensions would be somehow "compacted" in a tiny volume, e.g., curled in a seven-dimensional ball of size around 10^{-32} cm. One should not try to imagine this tangle of dimensions as existing in our space; the other dimensions are completely outside our three-dimensional reality.

Some years ago, Savas Dimopoulos of Stanford University and his colleagues Nima Arkani-Hamed and Georgi Dvali came up with a bold proposal. Perhaps some of those extra dimensions were not so tightly confined. Given that no experimental evidence precluded the possibility, an extra dimension might be even as relatively big as a millimeter in radius, roughly the size of a poppy seed, they argued.

In this new hypothesis of large extra dimensions resides a possible solution to a long-standing puzzle: Why is gravity so much weaker than the other forces? Although electromagnetism and the weak and strong forces are comparable in strength to each other, they are as much more powerful than gravity as a mountain is larger than one of those fantastically teeny extra dimensions of string theory. To bridge that vast gap, Dimopoulos and his colleagues hypothesized that not only are there large extra dimensions, but also that gravity is the only force that permeates all the dimensions (for example, the photons which carry the electromagnetic force cannot "leak" out of our three-dimensional space). Consequently, gravity is not really so weak. Rather, we feel it so weakly because gravity actually lives in many dimensions. Gravity is diluted by this enormous extra space that we do not feel.

We have shown you some glimpses of difficult new territories in physics, in order to give a feeling of what kinds of ideas inspire modern physicists. Many dimensional spaces may sound quite fantastic, but it is good to remember that the roots of modern supergravity and string theories go to the 1910s, when the Theory of General Relativity was born.

The microcosmos deals with very short sizes. The diameter of a proton is about 10^{-12} mm, but this is an unbelievable giant in comparison with the spatial scales of mere 10^{-31} mm boldly considered in supergravity theories. When we now turn our eyes to the heavens, we do well to change the "$-$" sign in the exponents of ten to "$+$." Namely, the diameter of our Sun is about 10^{+12} mm and the diameter of the observable universe is about 10^{+30} mm. In this sense the world of human beings lies midway in scale between the subatomic scale and the realm of stars and galaxies.

Part III
The Universe

.

Chapter 19
Stars: Cosmic Fusion Reactors

Now that we have the tiny world of elementary particles in hand, we return to the big universe. At this point, we will examine the most common cosmic objects, stars. Our Sun is a typical star, and we have learned many facts about stars by studying the Sun. But there are many different kinds of stars, some very different from the Sun. In fact, those very differences have helped us to understand the structure of the stars and the physics determining their life cycle. We begin with the spectra of their light.

Spectral Classification of Stars

Hydrogen lines are frequently seen in the spectra of stars. Their strength is a useful way of classifying stars. In 1863, Jesuit Father Angelo Secchi of the Vatican observatory, a pioneer in astronomical spectroscopy, put stars in four spectral classes. In the United States, Edward Pickering (1846–1919) of Harvard College observatory started a classification project in 1886 that lasted for decades. In this work, a prism was placed in front of the telescope, and the sky was photographed. This produced spectra of all the stars that were in the field of view of the telescope at once. Thousands of spectra were collected, most quite different from the spectrum of the Sun.

Based on the unique observations, workers at Harvard, most notably Annie Jump Cannon (1863–1941), developed a system of spectral classification that is still in use. Cannon alone studied and classified more than 250,000 spectra! In the original alphabetical system, a star was put in A-class if the Balmer lines were particularly strong in the spectrum. Slightly weaker Balmer lines made a star of class B, etc. When the Balmer lines were hardly noticeable, the spectral class was registered as M or even O.

It is easy to notice that stars are of different colors. Betelgeuse in Orion is clearly red, while Sirius near to it in the sky shines blue. It was soon realized that the spectral class and color are connected with each other. This led to a modification of the system. If one orders stars according to their color, then O, B, and A stars are bluest, while stars of types K and M are red. The yellow Sun is of spectral type G.

P. Teerikorpi et al., *The Evolving Universe and the Origin of Life*
© Springer Science+Business Media, LLC 2009

Some letters were dropped. In all, the Harvard scheme is O, B, A, F, G, K, M, which generations of astronomy students have learnt by heart by repeating O, Be A Fine Girl, Kiss Me (Box 19.1 and Fig. 19.1).

Box 19.1 Spectral classes of stars

Class	Color[a]	Surface temperature	Visual Spectrum[b]	Examples
O	Blue-white	over 25,000	Balmer lines weak, lines of ionized He	lambda Orionis
B	Blue	11,000–25,000	Balmer lines stronger, lines of neutral He	Rigel, Spica
A	Blue	7,500–11,000	Balmer lines strong ionized Fe, Mg, Si	Sirius, Vega
F	Blue-yellow	6,000–7,500	Balmer lines weaker, ionized Ca, Fe, Cr	Canopus, Procyon
G	Yellow	5,000–6,000	Balmer lines weak, ionized Ca strong	Sun, Capella
K	Orange-red	3,500–5,000	Lines of heavy elements	Arcturus, Aldebaran
M	Red	below 3,500	Lines of titanium oxides	Betelgeuse, Antares

[a]The eye can discern the color of a star only if the star is sufficiently bright
[b]Some characteristic spectral lines

The color is important – it tells about the temperature of the star. As we know, a hot solid object or a thick gas emits light at all wavelengths or colors, violet to red, but the color of peak emission changes with hotness. If we heat a piece of iron it first becomes red hot (with the peak in long waves). As the temperature rises, its peak color becomes yellowish. Then, for very hot objects, the peak is at short wavelengths resulting in a blue-white appearance to the eye.

Among stars, the hottest are the O type stars where the surface temperature may exceed 25,000°C; at the other end, M stars may be cooler than 3,200°C. The light from the O type star is predominantly blue, but not entirely. The stellar light has all colors, but in different proportions: in O stars the blue end of the spectrum dominates, while in the M stars the weight is in the red end. In main features this is the same type of behavior as we learned in the case of black bodies. Thus we may use a single parameter, the surface temperature, to classify the stars. But this is not yet enough to characterize all stars.

We know that stars are made mostly of hydrogen. This fact was not always obvious – a hundred years ago it was still thought that iron is the predominant element in the Sun. We owe the breakthrough in understanding star composition to Cecilia Payne-Gaposchkin (1900–1979). She was born Cecilia Payne in England, and in 1934 married Sergei Gaposchkin. Her Harvard/Radcliffe 1925 Ph.D. dissertation

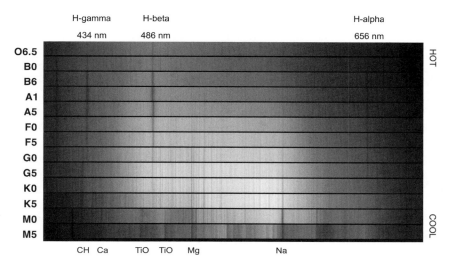

Fig. 19.1 Stellar spectra arranged in order of surface temperature. Spectral lines of some elements and compounds are indicated. Stars are divided into subclasses denoted by numbers after the main spectral type letter. Note the systematic change in the Balmer lines (H-alpha etc.) of hydrogen along the spectral sequence from hot to cool stars. For Sun-like stars (around G) the Balmer lines are relatively weak (credit for the spectra: NOAO/AURA/NSF)

was said to be the best one in twentieth-century astronomy. Not being discouraged by a series of low-paying, low-status positions, she ultimately became the first woman to become a full professor at Harvard. In her dissertation, she showed that most of the wide variation in spectral line intensities of stars was not due to widely different elemental abundances, but different temperatures.

Once all effects of temperature were included, one could finally infer element abundances in stars, obtaining that hydrogen is far and away the most abundant, with helium a distant second ending with small amounts of the other elements. This "cosmic composition," typical of the stars, is quite different from the abundances here on Earth. This was a great discovery.

Dwarfs and Giants

At the end of the nineteenth century there were two main ideas about evolution of stars. One theory regarded that stars are born hot and blue, and they gradually cool during the evolution to become red. In its rival theory, stars are big and red initially, and then they gradually shrink and become hot and blue.

One could not decide from observations which theory was correct. However, one could try to solve the problem by calculations. Among the first to try it was American physicist Jonathan Lane (1819–1880) of United States Patent Office who asked what would happen to a cloud of gas as big as the Sun if it is kept together by its own gravity. He found that the ball of gas would not resemble the Sun. Nevertheless, this was the first *star model*: it gave the description of pressure, temperature, and density of gas inside the Sun at different distances from its center. Notwithstanding this early disappointment, research on spheres of gas continued. Robert Emden of Technical University of Munich published a book entitled *Gas spheres* in 1907 which summarized the current knowledge of the subject. At that time, atomic theory had not yet advanced to the stage where stars could be understood as balls of gas. Moreover, it was not known what makes stars shine.

Ejnar Hertzsprung (1873–1967) from Denmark showed that while some stars are medium-sized stars like our Sun, others are much bigger red giant stars. This was deduced indirectly since stars were too small in the sky to be seen as a disk. A cool star gives off much less energy per second from a square meter of its surface than a hot star. However, some red stars emit hundreds of times the total energy per second as our Sun. To do this they must have much larger surface area than the Sun. In 1906 Hertzsprung estimated that Arcturus is as big as the orbit of Mars around the Sun! With special techniques and then directly with the Hubble Space Telescope, the huge disk of the red giant Betelgeuse was finally seen, verifying the indirect calculation. Henry Norris Russell (1877–1957) of Princeton University compared the properties of giants and other stars. He found that their masses are very similar in spite of their different sizes. This means that giant stars are made of gas that is much thinner than the gas in the Sun, on average much thinner than even the Earth's atmosphere. The cores of giants can, however, be dense.

These studies led Hertzsprung and Russell to the conclusion that there are two kinds of stars: *main sequence stars* and *red giant stars*. One may build the so-called

Fig. 19.2 Arthur Eddington (1882–1944)

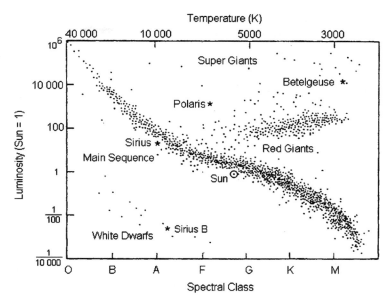

Fig. 19.3 Hertzsprung–Russell diagram separates different kinds of stars: main sequence stars, red giants, and white dwarf stars are in their own regions of the diagram. The *horizontal axis* gives the surface temperature (and the spectral class) and the *vertical axis* indicates the luminosity in terms of the brightness of the Sun. Decoding the message of this diagram was one of the success stories of twentieth-century astronomy

Hertzsprung–Russell diagram (HR diagram for short) so that the temperature or spectral class is the horizontal axis and the luminosity the vertical axis. In the main sequence (MS) the bluer (hotter) stars are more powerful radiators. The main sequence stars are clearly separated from the giants in the HR diagram. In Fig. 19.3 some well-known stars have been plotted. We find that Betelgeuse is situated among red gaint stars in the diagram, while Sirius, on the other hand, is a main sequence star hotter than the Sun. Also shown in the lower part of the HR diagram are white dwarf stars which will be discussed later.

What is the reason for the narrow sequence of stars? Could it be that stars evolve along the main sequence, e.g., cooling and shifting from left to right? But this would require huge losses of mass during the process, as cool MS stars are much less massive than the hot ones. Therefore it appears impossible for the same star to go through the whole main sequence during its evolution. Arthur Eddington, professor of astronomy at Cambridge University since 1913, was one of the pioneers of stellar studies in the era of quantum mechanics (Fig 19.2). He calculated that the brightness of a star depends on its *mass* in the first place: the more massive is a sphere of gas, the brighter it shines.[1] The main sequence is indeed a sequence of different masses.

---

[1] At the time this was not so clear and the topic caused interesting debates in the meetings of the Royal Astronomical Society between Eddington and James Jeans (1877–1946), the leading English theoretical astronomer of the time. In the end, Eddington was right, even though many details of stellar evolution remained unclear to him.

The mass, the luminosity, and the surface temperature all increase when one moves from right to left, that is, from light main sequence stars to heavy ones.

Internal Structure of a Typical Main Sequence Star, the Sun

About 4.6 billion years ago the Sun was born out of gases which had perhaps 73% hydrogen, 25% helium, and tiny amounts of heavy elements. The radius of the Sun is now 694,000 km and its energy output is 3.90×10^{26} W. This "light bulb" has apparently kept pretty much the same luminosity and size through historical time and from fossil evidence, for most of the geological history of the Earth.

We cannot look at its deep interior, but the conditions there can be deduced from the fact that the Sun is neither expanding nor contracting. The high density and the temperature at its center are necessary to keep the Sun from collapsing. Its internal properties are described by the gas sphere model in Table 19.1. Inspection of the table reveals that the temperature and the density change very steeply from the center to the surface, while the hydrogen fraction is about the same for the outer two-third of the Sun's radius and smaller only in the innermost core of the Sun (the result of the "burning" of hydrogen).

The Sun does not have a solid surface. Rather its light comes from different depths of a layer called the *photosphere* which is about 300 km in thickness. The temperature usually quoted as $5,500°C$ is a kind of average over different depths of the photosphere.

The coolest part of the Sun is at the top of the photosphere, about $4,300°C$. Outside the photosphere lies the *chromosphere*, a layer about 2,000 km thick. Here the gas is rare and the temperature rises to $100,000°C$ at the top of the chromosphere. Outside the chromosphere starts the *corona* where the temperature is millions of degrees. The gas forming the extensive corona is very rarefied. It emits little visible light that is best seen when the Moon covers up the photosphere during a solar eclipse (see Fig 19.10).

The Sun is losing 3.90×10^{26} W (J/s) into space. If this energy is not replaced the Sun will not be at equilibrium. Now we know that the energy of a main sequence star

Table 19.1 Current internal properties of the Sun

Distance from the center (10^6 km)	Mass within this distance	Temperature (10^6 K)	Density (g/cm^3)	Hydrogen (%)
0.00	0	15.7	158	36
0.10	20	11.3	59	65
0.20	60	7.1	15.2	72
0.32	90	3.9	1.84	73
0.48	99	1.73	0.117	73
0.62	99.955	0.66	0.0063	73
0.694	100	0.0045	3×10^{-8}	73

comes from nuclear reactions that fuse hydrogen nuclei into helium nuclei. In lighter MS stars the basic reaction is the proton-proton chain, also happening in the Sun, while in stars considerably heavier the chain of reactions is more complicated. Such various routes from hydrogen to helium were found by German-American physicist Hans Bethe (1906–2005) in his theoretical studies in the late 1930s; briefly the processes are called *burning of hydrogen* (here "burning" is an energy-generating nuclear process). Bethe was among those scientists with Jewish family roots who had to leave his native country. He received a Nobel prize in physics for his work on stellar nuclear synthesis in 1967.

Life After the Main Sequence

Most of the life of a star is spent at the main sequence stage where the star turns more and more of its hydrogen into helium. Examining Table 19.1 for the interior of the Sun, we see that the photosphere of the Sun has its original 73% mass abundance of hydrogen. However, to construct an equilibrium model the core must have only 36%. This is consistent with the theoretical picture that hydrogen in the core converts to helium during the life of the Sun, but conditions near the photosphere are too cool for fusion to happen.

The rather peaceful life of a star as a stable MS member ends when the hydrogen fuel is exhausted close to the very hot center of the star. Heavy stars use up their fuel much faster than light stars, in spite of having a larger fuel stock to start with. This means that heavy stars stay in the main sequence a much shorter time than, say, the Sun that spends about 10 billion years in this phase of its life.[2] A main sequence star as heavy as 30 suns is 140,000 times brighter than the Sun and stays in the MS phase only about 5 million years. A small star with half the mass of the Sun radiates at a rate of 4% in comparison with the Sun, and its life expectation in the main sequence is pretty long, 30 billion years.

When the available fuel starts to be exhausted at the center of the star, its central layers will contract in order to make the temperature rise. In this way the star gathers new hydrogen fuel from the shell surrounding the central helium core. At the inner edge of this reserve hydrogen burning into helium takes place, while the center of the star is composed of the ashes, i.e., helium. The burning shell is growing in radius. In general the energy production rate of the star increases with time, and in order to be able to radiate at a correspondingly higher rate, the star has to inflate its surface area. The outer layers expand so much that the star becomes a red giant star. This is also the fate of our Sun (Fig. 19.4.).

After the main sequence, the temperature inside the core of a star increases. The highest temperatures that are reached depend on the mass of the star. In Table 19.2,

---

[2] With the rate a star uses its fuel (energy output/sec) being proportional to a star's mass to the fourth power and the amount of fuel proportional to its mass, the MS lifetime for a star ten times the solar mass is only 1/1,000 as long as our Sun's life.

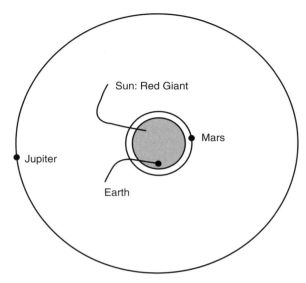

Fig. 19.4 Five billion years from now, the Sun grows and becomes a red giant star. Finally it swallows the Earth inside it and will fill the inner parts of the Solar System

we list the dominating energy-generating nuclear reaction at different temperatures. The first line corresponds to the main sequence stage.

In order for a star to successively go through all stages of nuclear fusion in the table, its initial mass must be at least 15 times the mass of the Sun. In less massive stars, the temperature never rises high enough for silicon fusion. Carbon burning and reactions possibly following from there require a star of at least three times more massive than the Sun. A star which is a quarter of the mass of the Sun or lighter will never reach beyond hydrogen burning, and it ends up as a helium star. Stars between one-fourth and 3 solar masses start helium burning at later stages of their evolution and end up as carbon-oxygen stars. They never get hot enough to go beyond this nuclear reaction.

Table 19.2 Energy-generating nuclear reactions in stars

Temperature (10^6 K)	Process	Burning product
10–20	Hydrogen fusion	Helium
100–200	Helium fusion	Carbon, oxygen
500	Carbon fusion	Neon, sodium, magnesium
1,000	Oxygen fusion	Silicon, sulfur, phosphorus
2,000–4,000	Silicon fusion	Iron, nickel

Little Green Men or White Dwarfs?

A small number of "radio stars" (actually quasars, Chap. 26) were known in the early 1960s. Then at the Cavendish laboratory (University of Cambridge) Anthony Hewish developed a new method to find radio stars by using the scintillation. The ordinary stars twinkle because their light is passing through restless air layers. In the same manner, radio stars twinkle because radio waves have to traverse through the variable solar wind on the way to the Earth. Hewish filled a two hectare field with radio antennas and started to search the sky systematically for twinkling radio stars that could turn out to be quasars. The instrument produced 30 meters of paper tape every day. It was inspected by Hewish's student Jocelyn Bell who had the responsibility for operating the telescope and analyzing the data. She noticed that one of the radio sources twinkled in a special way. The peculiar thing was that there were pulses of radiation arriving at a constant interval of 1.3 s. First Hewish thought that the source is man made, but soon it became clear that the source was in the sky, not on Earth. The next, more exciting idea was that the pulses are generated by other intelligent beings living on a planet circling around their own sun.

However, Bell soon found another pulsing signal from a quite different part of the sky. Now, she reasoned, "it was very unlikely that two lots of little green men would both choose the same, improbable frequency, and at the same time, to try signaling the same planet Earth"! New similar sources were found all over the Milky Way and one had to consider a natural phenomenon (Fig. 19.5).

Before the results were published in *Nature* in early 1968, Hewish gave a seminar in Cambridge and suggested that the pulses come from white dwarf stars. Fred Hoyle, head of the Institute of Theoretical Astronomy, was in the audience, and he replied: "I don't believe they are white dwarfs, I think they are supernova remnants." Nobody could have been more correct after only a few minutes' brain work.

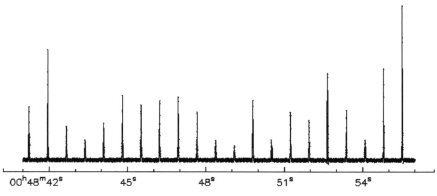

Fig. 19.5 Pulses from the pulsar PSR B0329 as observed with the Nançay radio telescope in France (see color supplement). The interval between the pulses is exactly 0.714 s (courtesy of I. Cognard & G. Theureau)

Table 19.3 Comparison of properties of the Sun and white dwarfs

Quantity	Sirius B	The Sun
Mass	1.05	1.00
Radius	0.008	1.00
Luminosity	0.03	1.00
Surface temperature (K)	27,000	5,700
Average density (g/cm^3)	2.8×10^6	1.41
Central density (g/cm^3)	3.3×10^7	1.6×10^2
Central temperature (K)	2.2×10^7	1.6×10^7
Gravitational redshift (km/s)	89	0.6

What exactly are white dwarfs, supernovae, and their remnants, to which Hoyle referred? At the early part of the twentieth-century observations started to suggest that there are fantastically dense stars, with sizes like the Earth and masses like the Sun. An example is the companion to Sirius, called Sirius B (see Table 19.3). The density of such stars is about a million times greater than the density of ordinary rock! Arthur Eddington remembered how the scientific community reacted: "When the message from Sirius was decoded, it read: I am made of matter which is 3000 times denser than any matter that you know of; a ton of my matter is such a small piece that you can put it in a matchbox. What can you answer to this message? Most of us answered in 1914: Shut up. Don't be silly."

It was not until 1926 that it was realized that Sirius's message was not nonsense. American Ralph H. Fowler applied the newly found Pauli's exclusion principle to an electron gas in white dwarf stars. In the high density prevailing in white dwarfs, the electrons have no room to circle around the atomic nuclei, but they form a gas of their own. A white dwarf star is like a huge atom covered by a cloud of innumerable electrons. Pauli's principle applies to electrons in this cloud just the same as it applies to them in ordinary atoms. Electrons cannot settle in a state which is identical to the state of any other electron in the cloud. When the star cools, all electrons cannot slow down since there are not enough states corresponding to slow motion. Some electrons are bound to have high speeds, and the resulting pressure prevents further shrinkage in the star, even if the temperature were to approach absolute zero.

Referring back to the HR diagram given earlier (Fig. 19.3), we see the white dwarf stars in the lower left part of the diagram, hot and of low energy output compared to the Sun.

Routes to White Dwarfs and Neutron Stars

Nuclear reactions maintain the high pressure and temperature which prevents the gravity to crush the star. However, sooner or later the fuel runs out, the pressure inside a star starts to decrease and the star starts to shrink. What happens next depends on the mass of the star. For stars of three solar masses or less, a carbon/oxygen core forms inside the red giant star. The carbon-oxygen core is extremely hot with a mass

Fig. 19.6 The familiar Polaris in the Little Dipper is actually a triple star. The main star A is a giant star (cf. Fig. 19.3) over 2,000 times brighter than the Sun. It is also a variable Cepheid star. Its faint companion B can be seen through small telescopes, but the third companion, Ab, is so close to the glare of the main star that it could be photographed only in 2006 by the Hubble Space Telescope. The small components B and Ab are main sequence stars. *Credit:* NASA, ESA, N. Evans (Harvard-Smithsonian CfA), and H. Bond (STScI)

comparable to that of the Sun and a size comparable to that of the Earth. This core is surrounded by the diffuse red giant envelope. Via a complicated process, the envelope is gently cast off leaving the core behind. The white dwarf stars are later cooled off stages of this core.[3]

A massive star becomes a red giant at the end of its main sequence evolution just like a lower mass star. In heavy stars the core collapses and becomes hot enough ($>500 - 1,000$ million$°$C) so that nuclear fusion reactions of carbon and oxygen, etc., can proceed. At this stage the star may become a Cepheid variable (see Fig. 19.6), a useful tool to estimate distances of stellar systems as we discuss later. The nuclear reactions proceed until the center of the star is made of iron and nickel. The fusion of still heavier nuclei from iron and nickel does not create energy but rather *consumes* it, which does not help to fight the collapse. The iron–nickel core grows as more silicon fuses at the outer edge of the core. In the end the core becomes so heavy that it collapses under its own gravity, and a supernova explosion is initiated. In the explosion, nearly all the matter of the star is blown into space. The collapsed center becomes either a neutron star or (if the star is massive enough) a black hole. Here we take a closer look at neutron stars.

---

[3] Astronomers have identified clouds of gas flying away from hot cores that are the initial stages of the formation of a white dwarf star. These "planetary nebulae" looked a bit like a planet's disk when seen through early not-so-good telescopes.

Still Denser: Neutron Stars

In 1930 Subrahmanyan Chandrasekhar (1910–1995) calculated that even the pressure of the electron gas is not enough to stop the collapse of a star if its mass is greater than 1.44 times the mass of the Sun. What happens to a star when it collapses to even higher density than the white dwarf? Russian physicist Lev Landau (1908–1968) suggested that the star collapses until it becomes as dense as an atomic nucleus and is primarily composed of neutrons. The Swiss astronomer Fritz Zwicky further speculated that neutron stars are born in supernova explosions that take place at the end of the stellar evolution. He turned out to be right. Then in 1939 Robert Oppenheimer (1904–1967) and his student George Volkoff, a Russian emigré, found out that the star is able to hold back against further collapse if its mass is low enough. Modern calculations put the limit at about 3.2 solar masses. If the mass is greater, nothing can stop the collapse, and the star becomes a black hole.

A typical neutron star is about 30 km in diameter. From this it is easy to calculate that the density in the neutron star exceeds the density of water by 100,000 billion. The whole star is in some ways like a huge atomic nucleus, and it is covered by an unbelievably strong iron shell where the density is 10,000 times that of water. Pulsars and probably also other neutron stars are strongly magnetic with surface field strengths 10,000 billion times stronger than the magnetic field at Earth's surface. The properties of neutron stars are completely outside our range of experience, but it is good to remember that these horrible things were once ordinary stars. When a star collapses, its original magnetic field increases tremendously, by the same amount as the number of magnetic field lines per unit surface area increases. Also the rotation speed grows due to the familiar conservation of angular momentum, in inverse proportion to the contracting radius of the star.

The pulsating star discovered by Bell and Hewish was a neutron star. Neutron stars are so small that they are able to turn around their axis in just a second, and they can emit one or two radiation pulses per revolution. For some reason the pulse is emitted in a bright beam. When the beam sweeps past the Earth, we observe a pulse arriving from the star like from a huge lighthouse; these beacons are called pulsars. The first pulsar was named CP 1919 (CP is short for Cambridge Pulsar and 1919 is derived from its coordinates in the sky). Within a few months three more pulsars were found in Cambridge and today the number of known pulsars is way beyond one thousand. The interval between pulses (likely the rotation period of the neutron star) varies from 0.001 to 4 s. Pulsars were born in fast rotation, perhaps just with the period 0.001 s. Through their strong magnetic fields the pulsars are connected with the surrounding space where electrons are accelerated to high energies and which then radiate into the radiation beam of the pulsar (Fig. 19.7). This process makes the rotation of the neutron star slow down. The faster the rotation, the stronger the radiation. When the rotation has slowed down to about one revolution per 4 s, the beam starts to be too weak to be observed at Earth.

Pulsars can be used as very accurate clocks since their pulses are very regular. However, we have to remember that these clocks slow down, slowly but steadily. Moreover, there are occasional jumps in the clocks which may be related

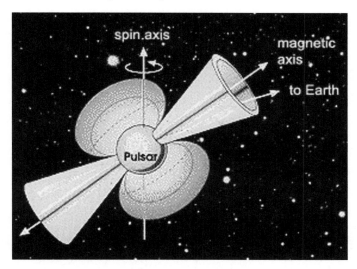

Fig. 19.7 A neutron star rotates fast about its axis (vertical in the figure). The magnetic axis is not generally the axis of rotation. Therefore the radiation beams starting from the magnetic poles sweep around the sky while the star rotates around its axis (credit: NASA)

to "starquakes" in the neutron star surface (corresponding to about 23 in the Richter scale!). Because of the enormous density on the surface of a neutron star, the collapse of even a centimeter-sized "mountain" can cause a noticeable change in its rotation.

Hewish obtained a Nobel prize for the discovery of pulsars. The codiscoverer, Jocelyn Bell (Burnell), was later honored by several organizations. In 2007 she was awarded a DBE (equivalent to a male knighthood) by Queen Elizabeth II.

The Crab Nebula: A Result of Supernova Explosion

Among the most significant pulsars are PSR 0833–45 in the constellation of Vela and NP 0532 in Taurus. Close to both pulsars we observe a nebulous cloud of gas which flew away from the star in an explosion. The latter nebula is known as Crab nebula; it looked crab-like to its discoverer Earl of Rosse William Parsons (Fig. 19.8). These pulsars confirm the connection between supernova remnants and pulsars that Fritz Zwicky first suggested and that occurred to Fred Hoyle in the Cambridge seminar (we will encounter both Zwicky and Hole later on in this book).

But what is a supernova explosion? In fact, there are different types of such explosions in the late phase of a star's life. A star more massive than 15 solar masses becomes a red giant star at the end of its evolution, and it eventually burns silicon into iron and nickel. At the same time other nuclear reactions, requiring lower temperatures, are going on in their own shells in the outer layers of the star. In the end the iron–nickel core becomes so heavy that it collapses under its own gravity and a

Fig. 19.8 Crab nebula, the remnant of a supernova explosion in 1054. Its diameter is about ten light years, and it is expanding at the speed of over 1,000 km/s. Photographed by Jyri Näränen with the NOT telescope at La Palma

supernova explosion is initiated. Nearly all the matter is blown out, spreading heavy elements into space. Many of the elements have been fused already inside the star, but especially elements heavier than iron and nickel are born during the explosion. The collapsed center becomes either a neutron star or a black hole; the heaviest stars are thought to form a black hole.

An equally mighty explosion is initiated in a white dwarf star when matter falls on it from a companion red giant star to the extent that the white dwarf collapses and throws away its outer layers. This overflow of gas can occur at a certain late stage of the evolution of a binary star system. Such a supernova is called type Ia (those described earlier are either type II or Ib supernovae). In recent decades type Ia supernovae have become very important for cosmological studies, because they can be used as accurate "standard candles," having about the same luminosity at maximum. The two famous supernovae in our Milky Way, observed in 1572 and 1604, were probably of type Ia.

The explosion which gave birth to the Crab nebula was seen as a new star in China. Toktaga tells in the history of the Sung dynasty that in 4 July 1054 "a guest star appeared in the southeast corner of the constellation Thien Kuon, and it was several centimeters across. After more than a year it faded away out of sight." The guest star was so bright that it was visible even in the daytime for 23 days. In 1921

Fig. 19.9 The Earth, a solar mass white dwarf, and a neutron star compared with each other. The dot on the right representing the neutron star is ten times too big in order that it can be seen

Knut Lundmark proposed that this event caused the nebula that is seen in the sky at the same place.

There is the interesting possibility that the supernova was also seen by Anasazi Indians who lived in what is now Arizona and New Mexico and who were attentive to the happenings in the sky. One has found petrographs, e.g., in the Chaco Canyon National Park showing a big "star" close to a crescent. In fact, calculations indicate that on the morning of 5 July (1054) the crescent moon came close to the supernova, as seen from Western North America.

The observed supernova explosion, the remaining pulsar, and the gaseous complex-structured nebula around it tell a detailed story of the birth of a neutron star (Fig. 19.9 illustrates its smallness). The star has collapsed in the middle, but at the same time it has thrown a large part of its mass into interstellar space, where it is used in forming new stars. Due to its youth, the Crab nebula pulsar is a very fast rotator, with the period of only 0.033 s. Its pulses are seen in optical and x-ray observations, in addition to radio waves.

X-Rays and Black Holes

As mentioned earlier, if the mass of a neutron star is greater than about 3.2 times the mass of the Sun nothing can stop a further collapse, and the supernova's core becomes a black hole. We have already discussed the theoretical concept of a black hole which was proposed much earlier than one might think. In science, proving that something could exist does not necessarily mean that it did actually form in nature. In a parallel to the proof that neutron stars existed via radio astronomy, it turned out that x-ray astronomy was crucial in giving evidence for the reality of black holes.

X-ray astronomers have to make measurements above Earth's atmosphere. The air absorbs the ultraviolet light and x-rays coming from space; luckily for us, since

Fig. 19.10 Image of solar corona taken during an eclipse in 1999 (photo: © Luc Viatour http://www. lucnix.be)

we could not withstand those doses. On the other hand, the work of UV and x-ray observers is difficult and expensive since the measuring instruments have to be placed in satellites. Another difficult region in the spectrum is infrared. Even though some limited ranges of the infrared spectrum are available from the Earth, at high mountains in dry climates, on the whole also infrared astronomy is an area of space astronomy.

The first source of celestial x-rays was discovered in 1948 during a rocket flight, and it was the Sun. Its x-rays were expected. The Sun's outer layer (corona) extends millions of kilometers above the surface of the Sun (Fig. 19.10). The faint corona is observed during solar eclipses when the Moon covers the bright surface of the Sun. Even before x-ray observations it was known that the gas in corona is very hot, millions of degrees, and such gas radiates mostly in x-rays (see Box 12.2). The gas is even too hot for the gravity of the Sun to hold it. Therefore the corona expands to the surrounding space, and even the Earth is inside this outer corona.

Even though the Sun is a bright x-ray object for us, at the distance of other stars it would be hard to detect. If there were no better sources, the whole x-ray astronomy would have been a science of the Sun. However, all stars are not like the Sun, and neither are all x-ray sources stars. In 1963 Herbert Friedman's team at the US Naval Laboratory found two new sources, Scorpius X-1 and the Crab nebula. The Crab is 1,000 times brighter than the Sun. Its x-rays come from high-speed electrons constantly accelerated by the pulsar in the middle of the nebula (its radio emission has the same origin).

It was much harder to identify Scorpius X-1 in the constellation of Scorpius. Only after the position of the x-ray source had been pinpointed with the accuracy of

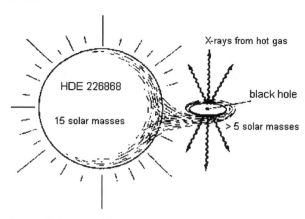

Fig. 19.11 The Cygnus X-1 system. The gas that is pulled out of HDE 226868 ends up in a disk surrounding a black hole. The gas which is close to the black hole is heated up to the extent that it radiates in x-rays (adapted from William J. Kaufmann III, Universe, W.H. Freeman & Co., 1991)

2 arcmin, a weak blue star was discovered which could have something to do with the x-rays. The star is so far away that its x-ray radiation must exceed the Sun's x-rays by 10 billion, if it is really the source! It turned out that the star, or more specifically two stars in orbit around each other, is the correct identification, and it is the fainter (in practice invisible) one of the two stars which emits x-rays. It pulls gas from the brighter star and heats it up to the temperature of millions of degrees in its strong gravitational field. The hot gas circles around the invisible star and radiates x-rays. In 1966 when Scorpius X-1 was first recognized, the only suggestion for the invisible companion was a white dwarf. But since the discovery of pulsars, a neutron star became a better candidate. The matter which falls to the surface of a neutron star arrives there with up to 80% of speed of light. This is an exceedingly efficient machine for producing x-rays.

Up to this point in time all information on x-ray sources came from rocket flights which lasted only a few minutes. Since the results were interesting, Riccardo Giacconi proposed to NASA that it should build a permanent x-ray observatory in a satellite orbiting the Earth. In 1970 the satellite was sent from Kenya to an orbit above Earth's equator; it was called by a Swahili name Uhuru (= freedom).

In two years Uhuru discovered more than 150 x-ray sources. One of the most interesting is Cygnus X-1. It circles around a massive star 15 times heavier than the Sun. However, the x-ray star shows no signs of pulsing which could be related to a rotating neutron star. Studies of the binary motion have shown that the x-ray star consists of at least 5 solar masses of matter; it can be even ten times heavier than the Sun. A neutron star cannot have such a high mass; the only option seems to be that the x-ray star in Cygnus X-1 is a black hole (Fig. 19.11). Other similar black hole candidates have since been discovered. And this is not all. There are much more massive black holes in the centers of galaxies (Chap. 26). Now we turn from stars to galaxies, starting with our own Milky Way.

Chapter 20
The Riddle of the Milky Way

In a dark clear moonless night, far away from city lights, one may see a starlight-full hazy belt which circles the sky. It divides the sky into two halves, passing through the constellations of Cassiopeia and Perseus and between Orion and the Twins. On the other side of Cassiopeia, the Swan and Eagle are in this celestial path with Sagittarius in the most spectacular southern portion. In many languages it is called a "way," for example Milky Way in English which agrees with the Greek name *galaktos*, milk. In Finnish it is the Bird's Way, in Swedish Winter Street. The Chinese call it the Silvery River; the Cherokee, the Way the Dog Ran Away. Unlike the wandering planets, the Milky Way stays fixed relative to the stars, as if it were part of the constellations.

Ideas in Antiquity

Astronomers in antiquity were mostly interested in explaining the motions of the Sun, Moon, and planets in the sky. The steadily rotating sphere of stars, including the Milky Way, did not arouse the same level of curiosity. There were no telescopes, and if a philosopher proposed a new explanation, there was no way of confirming it.

Aristotle discussed the Milky Way in his book *Meteorologica* that some Pythagoreans thought that the Milky Way was the circle along which the Sun had previously traveled and burned its path. He criticized this view saying that the present orbit of the Sun, the ecliptic, would thus be even more badly scorched than the Milky Way, especially since planets move there also. But nothing like the Milky Way is seen in the ecliptic. So what did Aristotle think about the Milky Way? In his world view, stars were fixed to the outermost sphere of crystal, unchanging and perfect above the sphere of the Moon. Aristotle knew that the Milky Way rotates in the sky exactly as the stars. Nevertheless, he placed this irregularly shaped structure below the Moon's sphere, in the lower, imperfect changing world.

P. Teerikorpi et al., *The Evolving Universe and the Origin of Life*
© Springer Science+Business Media, LLC 2009

Aristotle considered the Milky Way as a natural phenomenon like comets whose sudden and somewhat frightening appearance was so puzzling to the ancients (in ancient Greece these celestial signs of bad luck were viewed as souls of dead people). In Aristotle's view, comets could not be situated in the unchanging world above Moon's sphere. He thought that they originated from vapors rising from bogs which glowed via heat that derived from the Sun and stars. Depending on the shapes and rates of burning of the vapors, they could appear as different kinds of comets or even as shooting stars. In the Milky Way zone stars are denser than elsewhere. Thus they could heat the vapors below effectively. Thus Aristotle viewed the Milky Way as a huge permanent comet. However, this idea was never popular in spite of the fact that the rest of his world system formed the basis of science for a long time.

Aristotle's successor as director of the school he founded, the Lyceum, Theophrastus (ca. 370–286 BC), proposed that the Milky Way is a joint where the two halves of the celestial sphere are glued together. This idea could have carried more support if the circle of joining was at the celestial equator (above the Earth's equator), but there is a large tilt between them. The great circle of the Milky Way is also tilted relative to another important great circle, the yearly path of the Sun (the ecliptic). This was a step to the right direction: Theophrastus realized that the Milky Way follows a great circle in the sky far from the Earth. Now astronomers call this circle the Galactic Equator. But the question, "Why does the Milky Way divide the sky into two halves?" was not answered for two thousand years.

Some in antiquity interpreted the milky glow in a way now known as correct. Democritos, the developer of the concept of atoms, believed that a huge number of small stars were responsible for the phenomenon. They were so close to each other that their light united in a uniform glow. This is a fine example of scientific deduction. Even though we cannot see the small stars, we may assume that they exist and then explain quite a different phenomenon.

Belt of Stars

The view that the Milky Way arises from small densely spaced stars appeared from time to time in medieval writings, but the pioneers of new astronomy, Nicolaus Copernicus and Johannes Kepler, hardly mentioned the Milky Way; the motion of planets took center stage in their search for celestial harmony. However, Tycho Brahe's nova, which he observed in 1572, led him to conclude that the nova signified the birth of a new star out of cosmic matter of which the Milky Way was made.

The turning point occurred when telescopes came in use. In the fall of 1609 Galileo Galilei started to survey the sky with his telescope discovering stars which were not visible to the naked eye (see Fig. 20.1). In his book of 1610 *Starry Messenger* Galileo described the Milky Way as follows:

> Third, I have observed the nature and the material of the Milky Way. With the aid of the telescope this has been scrutinized so directly and with such ocular certainty that all the disputes which have vexed philosophers through so many ages have been resolved, and we

PLEIADUM CONSTELLATIO

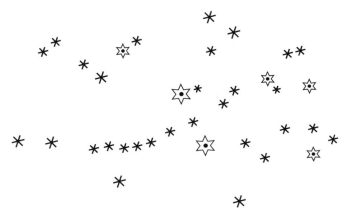

Fig. 20.1 Galileo saw through his telescope many more faint stars in the Pleiades than the six ones usually visible by plain eye and shown as big stars in this drawing from his *Starry Messenger*

are at last freed from wordy debates about it. The galaxy is, in fact, nothing but a congeries of innumerable stars grouped together in clusters. Upon whatever part of it the telescope is directed, a vast crowd is immediately presented to view. Many of them are rather large and quite bright, while the number of smaller ones is quite beyond calculation.

Galileo was happy to simply understand the origin of the milky light, to see with his own eyes what Democritos had envisioned 2,000 years earlier. The same was true of most scientists immediately after him. It took another century and half before the significance of that belt of stars as a cosmic structure was realized.

Toward the Three-Dimensional Milky Way

In 1751 Immanuel Kant, at the time still a student and home teacher (this was three decades before his famous philosophical work *Critique of Pure Reason*), read a newspaper account of the book *An Original Theory or New Hypothesis of the Universe* written by an Englishman, Thomas Wright. The story gave a somewhat misleading impression that Wright said the Milky Way is a flat sheet.

Kant wondered how this shape could be consistent with Newtonian gravity which he had studied at the University of Königsberg in his hometown. He noted the similarity between Saturn's rings and the disk of the Milky Way. Just as the flatness of the rings is the result of rotation around the planet under the force of gravity, similarly the flatness of the Milky Way could arise from rotation. Kant also suggested that other small nebular objects in the sky are really like the Milky Way seen from a large distance. He published these ideas in a book in 1755. Unfortunately, the publisher went bankrupt and the books were seized. Therefore Kant's ideas about the universe took a long time to reach other scientists.

Fig. 20.2 (**a**) Thomas Wright (1711–1786) and (**b**) Immanuel Kant (1724–1804). In addition to these thinkers, Johann Heinrich Lambert also viewed the Milky Way as a projection of a three-dimensional stellar system

But what were the real thoughts of Thomas Wright? Wright was a self-taught astronomer and mathematician who earned his living partly by giving popular science talks. His interest in the Milky Way was connected with his life-long pursuit of a model for the universe. He wanted to see order and harmony in the world created by God and the model should explain the distribution of stars in the sky (Fig. 20.2).

In his *Original Theory* Wright takes it for granted that the universe is infinitely large. He was also convinced that stars are distant suns: if the Sun were seen from a large distance, its 1/2 degree diameter would shrink to a point of light. And this would be true, even if the distant Sun is viewed through a big telescope. He also believed that stars have planets circling around them, just like the Sun does.

Wright estimated how many stars are in the belt of Milky Way: "When the width of Via Lactea (the Milky Way) is on average only 9 degrees, and when we assume that every square degree has only 1,200 stars, then the whole ring-like surface must have almost 3,888,000 stars." His count of stars is far behind the modern figure of 200 billion, but it gave the first good idea of the "astronomical" number of stars. Moreover, Wright suggested that the Milky Way is a huge stellar system in which stars circulate around a common center. The Sun is not at the center of this system, nor is the Milky Way at the center of the universe – another modern assertion.

Wright explained the appearance of the Milky Way in the sky such that there is a huge sheet of stars, and that the Sun is inside it. When we look along the direction of the sheet, we necessarily see plenty of stars. When our line of sight is at a large angle

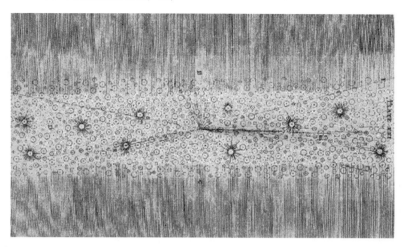

Fig. 20.3 Thomas Wright used the drawing to explain how the belt of stars of the Milky Way arises when we ourselves are in the middle of a flat distribution of stars

to the sheet, there are fewer stars in the sky in this direction. The visual appearance will be a ring of stars in the sky – the Milky Way (Fig. 20.3).

In fact, Wright did not actually propose a disk-like Milky Way. He preferred something more beautiful, an immense spherical shell of stars. At short distances its surface is almost like a plane, so the apparent sheet of stars was only a local feature. He imagined that there was a massive body at the center of the shell which made the stars rotate around this center which was supposedly not visible to observations.

Without having heard of Wright's and Kant's ideas, the German scientist Johann Heinrich Lambert (1728–1777) wrote about his cosmological views in his *Cosmological Letters* in 1761. He described the Milky Way as a rotating flat system of stars and assumed that the massive center of the system lies in the Orion nebula. Today we know the direction of the center is quite different, in the constellation of Sagittarius, and as Wright had guessed, invisible to the eye and even to an ordinary telescope.

It could be that Lambert was the first to get the idea that the Milky Way is a flat stellar system. In a letter to Immanuel Kant in 1765 Lambert tells that the idea occurred to him in 1749 when "contrary to my custom, I went to my room after the evening meal and looked at the stars and in particular the Milky Way." In *Cosmological Letters* he wrote:

> I wondered about the multitude of small stars in that arc (Milky Way)... I thought that those stars cannot be so close that they almost touch each other. They should be located one behind the other and the rows of stars must be many times deeper in the direction of the Milky Way than outside it. If the rows were equally deep everywhere, the whole sky would be as bright as the Milky Way. Rather outside the Milky Way I see almost entirely empty regions. To summarise, the structure of the fixed stars is not spherical but flat, even very flat.

William Herschel's Milky Way

The views of Wright, Kant, and Lambert about the Milky Way resulted simply from a visual impression of the distribution of stars in the sky. The first proper survey of the Milky Way using a telescope was started by William Herschel. Herschel moved from Germany to Britain at the age of 19 where he earned his living as a musician (later he obtained the position of organ player in the chapel of Bath). In 1773, at the age of 35, he happened to buy a book on astronomy. "When I read about the many enchanting discoveries made using a telescope, I became so fascinated by the subject that I wanted to see the sky and the planets by my own eyes through one of those instruments."

Herschel learned how to grind mirrors working on building telescopes day and night. He was assisted by his musician brother Alexander and sister Caroline. In the following years, music was left more in the background while Herschel learnt how to build bigger and better telescopes. He also started making systematic notes of what he saw in the sky. We already told about the discovery of the planet Uranus in 1791 (Chap. 11). This brought Herschel fame as an astronomer. The hobby turned into profession - he started receiving a salary from the King as the first Astronomer Royal. In scientific circles, Herschel was still rather unknown. For example J. E. Bode, the leading German astronomer, wondered if his name was Mersthel, Hertschel, Herrschel, or Hermstel.

In the study of the Milky Way, Herschel pioneered statistical methods. His brilliant idea was to chart its outline using *star counts*. Herschel trusted that his 47 cm telescope was powerful enough to see the edges of the Milky Way. The number of stars seen in the telescope tells how far the edge is in that particular direction: the more stars, the further the edge. Figure 20.4 shows the result of the star counts when translated into the outline of the Milky Way. This cross-section perpendicular to the plane of the Milky Way resulted from the study of 683 regions placed on an arc of a great circle in the sky. It agrees with the visual conclusion that the Milky Way is a flat stellar system.

Later Herschel became suspicious of the correctness of his picture. It was questionable whether the telescope was actually powerful enough to see the stars at the edge of the system. His new telescope of 120 cm diameter showed many more stars than what he had seen with the 47 cm tube. Otherwise this large telescope was not entirely satisfactory: it was clumsy to handle, and the assistant was injured several times while operating it. Also his studies of star clusters made Herschel believe that his initial assumption of uniform distribution of stars in space was far from the truth. However, the idea of the disk-like Milky Way survived until the twentieth century when it became possible to confirm it using more advanced methods.

Herschel made also good progress in the study of stars and nebulae. He discovered binary stars and his systematic "sweeps of the sky" revealed about 2,500 star clusters and nebulae (Chap. 21). Previously, only about one hundred such objects were known.

Fig. 20.4 (**a**) A portrait of William Herschel, painted at the time of his discovery of Uranus. (**b**) Cross-section of the Milky Way based on Herschel's star count using his telescope equipped with a 47 cm mirror (an illustration from the year 1785). This great astronomer spent numberless nights observing the starry sky

Great Star Catalogs and Kapteyn's Universe

What is needed to chart the distribution of stars in space? Clearly the directions to the stars, but one also needs to know their distances. Then one can define the outline of the Milky Way. But astronomers have to be satisfied with much less. As Herschel found, one can never see all the stars; some are definitely too dim even for modern telescopes. Moreover, it is impossible to measure the distances to all stars. There were very few parallax (that is, distance) measurements in the nineteenth century, and even today we are limited to our local neighborhood in the Milky Way. Only when the space telescope Gaia of the European Space Agency (ESA) is launched in the next decade will we start to get distances to a representative sample over the disk of our Milky Way. Gaia's goal is to make the largest, most precise map of the Milky Way by surveying an unprecedented number of stars – more than a billion.[1]

A much easier task than the distance measurement is the measurement of the brightness, or magnitude, of a star (Chap. 8). It gives some indication about the distance. In the nineteenth century the estimation of magnitudes became a routine operation, and they were included in great star catalogs. The most famous one is *Bonner Durchmusterung* (The Bonn General Survey). It was compiled by Friedrich Argelander (1799–1875) and his associates. After working in the Turku and Helsinki observatories in Finland, Argelander became director of the Bonn observatory in 1836. While in Bonn, he studied all stars brighter than magnitude 9.5, determined their coordinates in the sky, and measured their magnitudes. The Survey was completed in 1859 and it contained 324,000 stars. This huge catalog has been useful up to recent times in the study of the Milky Way.

The German astronomer Hugo von Seeliger (1849–1924) developed Herschel's star count method further. He realized that it is better to study the *change* in the number counts of stars going to successively dimmer stars, rather than the total count number. The great star catalogs had exactly the right kind of material for this line of research.

What can the change in the star counts tell us? Let us make the assumption that all stars are equally bright, and consider looking at a uniform spherical star system from its center. We would find that there should be four times as many stars of magnitude 7 than of stars of magnitude 6. The same difference would apply to any increase of magnitude by one unit. This follows simply from the way the magnitude scale has been defined, together with the decrease of brightness and increase of available space with increasing distance. However, when we meet the stars at the edge of the system, the number count at the next magnitude level suddenly drops to zero. By finding the magnitude after which the counts drop suddenly, we can identify the edge of the system.

In studies like this, von Seeliger found in 1884–1909 that the successive number count ratio is not 4, but more like 3. Thus the star density is not uniform around us, but seems to decrease with distance. At the faintest magnitudes the number counts dropped even below 3. He concluded these faint stars are close to the edge of the system. He found that the overall shape of the Milky Way is much like what Herschel had found previously.

The first proper model of the Milky Way, including the distance scale, was constructed by the Dutch astronomer Jacobus C. Kapteyn (1851–1922). He was elected to the professorship of astronomy at Groningen University at the age of 27. After arriving there, he found that the university did not have an observatory. This lack redirected his efforts to the study of catalogs compiled by others. He also became a spokesman for international collaboration.

---

[1] Gaia will be placed in an orbit around the Sun, at a distance of 1.5 million km further out than Earth. This special location ("L2") will keep pace with the orbit of the Earth; *Gaia* will map the stars from there. Its predecessor Hipparcos exceeded all expectations and cataloged more than 100,000 stars to high precision.

Kapteyn wanted to determine the structure of the Milky Way. Its shape was already known, but what about the scale? How far is the edge of the Milky Way that shows up in star counts? From the star counts astronomers had already identified a faint star at Milky Way's edge. If this star were as bright as the Sun, we could calculate its distance and thus determine the system size. But stars are not of equal brightness. Kapteyn studied nearby space and determined how the star brightness is distributed. For this, distances are needed. The parallax method was inadequate and Kapteyn used *proper motions*.

The distance of a star is revealed from the direction and the rate it moves across the sky, its proper motion (Chap. 8). These motions arise, not only from the stars' true space motions, but from the reflection of Sun's motion in space as well. Imagine driving at night through a snowstorm with snowflakes representing stars. Ahead of you, the flakes appear as dots as they come straight toward you with zero "proper motion." A similar view is seen through the rear window. However, to the sides, one sees streaks as the nearby flakes appear to move backward showing significant "proper motion."[2]

Today we know that the Sun moves at 20 km/s relative to nearby stars toward the constellation of Hercules. As discussed in our snowstorm example, depending on the size of the proper motion and the angle from the direction of the Sun's motion, we can estimate how far a star is from us. The tinier the motion appears, the greater is the distance likely to be. By using an ingenious analysis, Kapteyn derived statistical distance values and the brightness distribution of stars. Then he could derive the distance scale of the Milky Way. According to Kapteyn, the Milky Way is a disk with a diameter of 50,000 light years where the star density diminishes toward the edges (Fig. 20.5).

The problem with this model was that the Sun was only 2,000 light years from the center of the Milky Way, which seemed suspicious. As Kapteyn wrote himself in 1909:

> This would place the Sun at a very exceptional position in the stellar system, i.e. where the stars are at their densest. – On the other hand, if we suppose that the decrease in density is only apparent and is caused by absorption of light, then the apparent decrease in density in all directions is perfectly natural.

Kapteyn realized that if space is not transparent, but filled with some medium that dims the light enough, then the star counts can give an incorrect picture of the Milky Way – what appears as the edge is the effect of absorbing dust. He studied the possibility of absorption in space by various methods, but was unable to prove its existence. Thus his model was for years the dominant view of the Milky Way. A change began in 1918 when Harlow Shapley studied the distribution of globular star clusters in space, much less affected by absorption. He concluded that the Milky Way is much bigger than the "Kapteyn universe" and that the Sun is situated 50,000 light years from its center. To see how Shapley came to his radical conclusion, we now discuss a new way of estimating distances using variable stars.

---

[2] William Herschel estimated which way the Sun moves among other stars by looking at proper motions of just 13 stars. The first precise solar motion study, by Argelander, relied on 560 stars observed at Turku, Finland.

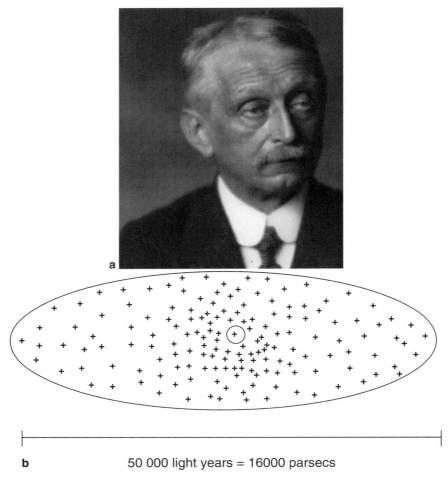

Fig. 20.5 (**a**) Jacobus C. Kapteyn studied the Milky Way using star counts. (**b**) "Kapteyn's universe" was the first model of the Milky Way with a distance scale. The Sun appeared to be almost at the center of the system

Cepheid Variable Stars: Standard Candles to Measure Large Distances

Next to the well-known constellation of Cassiopeia, there is the constellation of Cepheus. Using Fig. 20.6 it is easy to locate the fourth brightest star in the constellation, delta Cephei. Its magnitude is about 4 and thus it is visible to the naked eye. It is actually a very luminous giant star which varies its brightness in a regular manner in a 5 day cycle. Some stars may vary irregularly or even explode. Here, we will

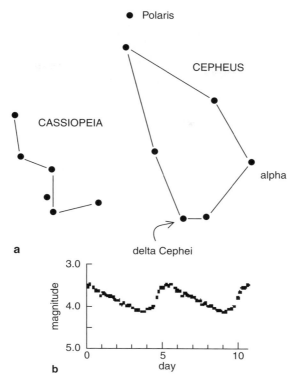

Fig. 20.6 The delta star of constellation Cepheus (*top*), the prototype of Cepheid variables. Its brightness varies in the cycle of a little over 5 days (*bottom*). The variability of this star was discovered by John Goodricke in 1784. This English astronomer died at the early age of 21 after having caught cold during observing nights

focus entirely on stars like delta Cephei where the brightness varies continuously and regularly, with a constant period. Cepheids can have periods ranging from a day to tens of days.

What is the reason for their variation? At the end of the nineteenth century, it was observed by the Russian astronomer Aristarkh Belopolski (1854–1934) that the wavelengths of the spectral lines vary in unison with the brightness. Using the Doppler effect one can determine that the surface of the star is in constant motion, in and out, with a typical speed of 100 km/s. This pulsation became the commonly accepted explanation when Arthur Eddington formulated the mathematical theory of pulsating stars.

In 1908 and 1912 Henrietta Swan Leavitt published at Harvard College Observatory her studies of variable stars in the Small Magellanic Cloud (SMC). This collection of stars, star clusters, and nebulae had been photographed at the Harvard observing station in Peru. Leavitt found 2,400 variable stars in these photographs.

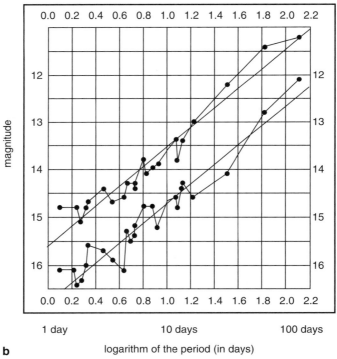

Fig. 20.7 (**a**) Henrietta S. Leavitt (1868–1921) (credit: American Association of Variable Star Observers) discovered the connection between the brightness and the variability period of Cepheids: a bright Cepheid pulsates more slowly than a faint one. (**b**) In the graph from her 1912 study we have added the explanations of the axes. Note that the range of the period is from a few days to more than hundred days

For a part of her sample Leavitt was able to determine the period of variation by graphing brightness as a function of time (Fig. 20.7). She noticed that the longer the period was, the brighter the star was at its average state. Since all the SMC stars are at practically the same distance to us, the period and true brightness of the Cepheid stars had to be also very closely connected.

Such a correlation opened a new way to determine distances: by measuring the period of a Cepheid one finds out its true brightness, that is, how strongly it radiates. Then it is a simple matter to compare it with the apparent magnitude and to calculate the distance.[3]

The distance to the Small Magellanic Cloud was not known. Thus Leavitt could not determine how bright the Cepheids really were. The calibration of the new distance determination method was done first by Ejnar Hertzsprung in 1913. He used a few Cepheids in our Milky Way for which he could calculate the average distance by Kapteyn's method. Harlow Shapley carried out similar research. Cepheids are very bright stars, giving off from 100 to 10,000 times more energy per second than the Sun. These "standard candles" have provided a new important method of estimating distances and studying the structure of the Milky Way as well as more distant stellar systems.

Shapley's Second Copernican Revolution

The American astronomer Harlow Shapley (1885–1972) moved the Sun from the central position in the Milky Way where star counts had put it. Shapley's path to science was not straightforward. In his memoirs Shapley tells that he went to University of Missouri to study journalism, but the beginning of the course had been shifted till the following year. He decided to study something in the mean time and thumbed through the university syllabus. The first subject in the syllabus the name of which he was able to pronounce was astronomy. So that is how it was decided.

In 1914 Shapley was employed at Mount Wilson observatory which had the world's biggest (1.5 m mirror) telescope. He started to study the Cepheids in *globular star clusters*, and their use in distance determinations. What are globular star clusters? Mostly the clusters are loose collections of stars, with hundreds of members, like Pleiades in the constellation Taurus. Globular star clusters are clearly different: they are spherical in shape, and the number of stars could be over a million. In their centers, the images of stars appear to merge together to form a smooth luminous nebula (Fig. 20.8).

The globular clusters are rather rare in the Milky Way, only a little over a hundred systems are known. But they are important objects of research for several reasons. Since they contain so many stars, they can be seen from far away and it is possible to find even rare stars in them. At Mount Wilson observatory, Shapley

---

[3] For example, if the period is 10 days, the Cepheid is 2,000 times brighter than the Sun. A simple calculation shows that the same Cepheid, if it has magnitude 6 (just visible by plain eye), is at the distance of 800 parsecs (2,600 light years) from us.

Fig. 20.8 There are two kinds of star clusters. The open star clusters are more common. They are loosely bound gatherings of often young stars. We show the open cluster of Pleiades ("The Seven Sisters") in Taurus (**a**) and the globular cluster Omega Centauri (**b**) (photos by Harry Lehto and Tapio Korhonen, respectively)

discovered variable stars in globular clusters and used them to measure distances. After measuring the distances of a dozen of them, he realized that the cluster diameters were all almost the same size. He could then calculate the distances of the rest of the globular clusters using the apparent diameter in the sky as an indicator of distance.

In this way Shapley determined the distances of several dozen clusters, and then marked their positions in a plot with their known distances and directions. He found that the globular clusters were distributed in almost spherical manner around the Milky Way (Fig. 20.9). Shapley announced his conclusion in 1919: the Milky Way is much bigger than had been previously thought on the basis of star counts. Its center is not near the Sun, but far away in the direction of Sagittarius. "Kapteyn's universe" is only a small part of the much bigger Milky Way.

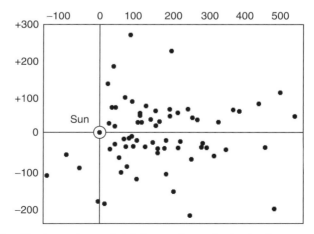

Fig. 20.9 Harlow Shapley made use of globular star clusters in charting the structure of the Milky Way. The chart shows clearly that the Sun is far from the center

It was a brave deduction, based on one class of celestial bodies. Also, Cepheids were a new distance indicator and people were suspicious of Shapley's great distances. Now we know the distances *were* too big for various reasons (e.g., the variable stars in globular clusters differ from "ordinary" Cepheids), but his general idea survived. A modern estimate of the diameter is 100,000 light years, near to one-third of Shapley's value.

Cosmic Dust Between the Stars

Two new results gave indirect support to Shapley. They were studies of motions of stars by Bertil Lindblad in 1921 and by Jan Oort in 1927, as well as the observation by Robert Trumpler in 1930 that the Milky Way has interstellar dust which dims light significantly.

Kapteyn himself was aware of the risk that absorption of light would spoil the results of star counts. The photographs taken by Edward Barnard (1857–1923) showed many dark patches in the Milky Way (see Fig. 20.10). It was believed that these could be made of some light dimming material, collected in clouds – but is there matter like this spread all over space? There was evidence of interstellar gas, but not of anything that would absorb light. In 1904 Johannes Hartmann had observed "extra" lines in the spectra of double stars, lines which did not take part in the Doppler shifts due to the stars' orbital motions. The lines thus must be formed in gas between the star and us. But was there dust along with the gas?

Fig. 20.10 In the Milky Way dust can appear in dense clouds which block the starlight from behind them. This picture was taken by the Hubble Space Telescope (credit: NASA/ESA/ STScI/AURA/P.McCullough)

At last in 1930 the Swiss astronomer Robert Trumpler (1886–1956), who worked at Lick Observatory, showed that interstellar space was far from transparent. He measured the distances of star clusters in two ways. One method used cluster angular diameters, just like Shapley had done for globular clusters. This method is unaffected by dimming of starlight. The other method was based on the apparent brightness of stars in star clusters. The calculation of distance from this method is affected by dimming of light, if it exists. By comparing the distance values Trumpler showed that the latter method gave incorrect results just in the way expected if there is an absorbing medium in the interstellar space.

We know now that for a distance of 3,000 light years the medium in the plane of the Milky Way weakens the light signal to one-sixth of the strength by comparison with a transparent space. The loss of light is caused by dust particles. Kapteyn's fears were legitimate. Even though the Milky Way is a stellar system, merely counting stars is not enough to determine its structure; one has to use more distant objects such as globular star clusters which are not concentrated in the dusty plane of the Milky Way.

The Milky Way Rotates

Immanuel Kant had suggested that the stars of the Milky Way circle around a distant center, not unlike planets that revolve around the Sun. But where is this mysterious center? Shapley's view of the Milky Way gained strong support when it was shown that (a) our stellar system does rotate around a center, and that (b) the center of rotation is in the direction which Shapley had pointed out, in the constellation of Sagittarius. How do we get such an excellent result?

William Herschel and Friedrich Argelander demonstrated using proper motion observations that the Sun moves through the field of nearby stars. But do the motions of the stars themselves display systematic trends? Early in the twentieth century there were proper motion determinations for several thousand stars, and also a large number of measurements of their radial motions were collected since 1890s using spectroscopy. Putting these two data banks together led to a peculiar discovery.

It was found there are a small number of *high-velocity stars* which pass by us at speeds 60–80 km/s. For the "ordinary" stars the speeds are smaller. The correct solution to this puzzle was discovered by the Swedish astronomer Bertil Lindblad (1895–1965). In 1921 he proposed that the Milky Way is composed of overlapping subsystems. These rotate about the common center, but with different speeds. This explains the high-velocity stars right away: they belong to a different subsystem than the "ordinary" stars. As a matter of fact, the "high-velocity stars" together with the globular clusters belong to a *slowly* rotating subsystem. The stars that move with high speed are actually the "ordinary" stars, the Sun among them. They are part of the flat disk of the Milky Way. We are overtaking the stars of the slowly rotating subsystem, and it appears as if they are passing us at high speed to the opposite direction.

Fig. 20.11 The center of the Milky Way is between the constellations of Sagittarius and Scorpius, just over the horizon in the right-hand part of this splendid panoramic Death Valley photograph. Note the dust clouds elongated along the Milky Way band. The picture was taken by Dan Duriscoe (U.S. National Park Service) as part of a program studying light pollution and methods that can protect the remaining dark skies on Earth

The final proof of the rotation of the Milky Way was provided by the Dutch astronomer Jan Oort (1900–1992) in 1927. Oort was a student of Kapteyn, and following the footsteps of his famous teacher, he started also to study the Milky Way. Lindblad's explanation of the high-velocity stars inspired Oort, and he wanted to figure out in greater detail how the stellar motions should appear in a rotating Milky Way in different directions and at different distances, as seen from the our position.

How do we expect a flat system of stars to rotate? If a large fraction of stars is concentrated at the center of the system, then stars further out will feel a gravitational pull as if it is coming from a giant single body. They revolve around the center roughly as planets revolve around the Sun, the nearer stars orbiting in a shorter time, the more distant stars orbiting in a longer time. Oort derived the formulae which tell us how the observed speeds of stars should vary in different parts of the Milky Way. It was remarkable to find that the radial speeds along the line of sight followed Oort's theory very well. The speed should go to zero in four directions 90 apart; one of them is the direction to the center of the Milky Way which Oort found to coincide with Shapley's model. Oort was also able to calculate the distance to the center. He found the value of 20,000 light years, about one-third of Shapley's value. According to later studies the distance is somewhat greater, about 26,000 light years (or 8,000 parsecs). You may see in Fig. 20.11 a fine view of our Galaxy, with a glimpse of its center.

The Sun in a Spiral Arm

It seems that the American Stephen Alexander (1806–1883) was the first to suggest that the Milky Way and its stars form spiral arms, in 1852. At that time several spiral nebulae had been observed in the sky. He did not have any proof for his contention, but the proposition was later repeated by some others. After failed attempts to see the spiral structure by charting the numbers of stars in different directions, the way of observing the spiral arms came from a surprising quarter in the 1940s.

German astronomer Walter Baade who worked at the Hamburg Observatory had migrated to the United States in 1931 (Fig. 20.12). When war broke out, he did not have American citizenship so he was not able to take part in the war effort, but he was allowed to carry out research at the Mount Wilson 2.5 m telescope. Wartime blackouts of nearby Los Angeles resulted in excellent conditions for photographing galaxies.

Baade studied the nearby spiral galaxies such as the Andromeda galaxy. He concluded that spiral galaxies consist of two different populations of stars: Population I which is in the flat subsystem and makes the spiral arms, and Population II which surrounds the flat disk almost spherically. The total brightness of all the stars in Population II is not so high, so the existence of the Population II subsystem in the neighborhood of the Sun is easily missed since the disk stars outshine it by a wide margin. Fortunately, the globular star clusters are prominent in this spheroidal subsystem. There are over 100 globular clusters in orbit about the center of the Milky Way. The orbits are elongated, and they cross the plane of the Milky Way from time to time. But mostly they are found far above or below the plane of the Milky Way.

Fig. 20.12 Walter Baade (1893–1960) showed that spiral galaxies consist of subsystems with different stellar populations. For example, the spiral arms belong to the young population I. This photograph is from the year 1923 (courtesy of Hamburg Observatory)

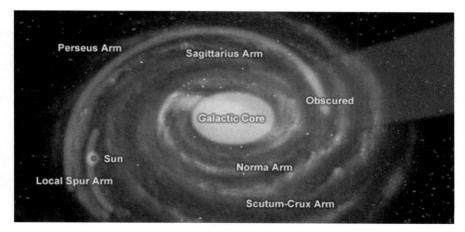

Fig. 20.13 A present-day schematic view of the spiral arms of our Milky Way galaxy (credit: NASA/JPL-Caltech)

The populations differ from each other also by rotation. As mentioned before, the Population II subsystem is in slow rotation in comparison with the disk subsystem.

Thus there was a clue that one should look for the spiral arms of the Milky Way by using Population I stars. Extreme examples of such stars are the bright, hot, and blue stars of spectral classes O and B. These stars are often associated with bright gaseous emission nebulae. William Morgan (1906–1994) and his colleagues reported their work on O and B stars and emission nebulae in the 1951 Christmas meeting of the American Astronomical Society. They had measured distances and plotted the positions of these objects on a graph representing the plane of the Milky Way. This map showed the spiral structure at the local neighborhood of the Sun for the first time. The Sun seems to be at the inner edge of one of the spiral branches.

Morgan's talk ended with cheers and stamping of feet by the enthusiastic audience. For a long time it had been speculated that we live in a spiral nebula; now it was proven. Unfortunately, dust prevents the extension of this method to much greater distances. The arm which we are part of is simply called the Local spiral arm. In fact, it may not be a major arm, but perhaps a "bridge" between two arms as sketched in Fig. 20.13.

Among the stars in the Local spiral arm are Capella, Sirius, and Betelgeuse in Orion; Deneb in Cygnus; and the well-known W stars of Cassiopeia. The direction across the arm is more or less in line with Capella. In the direction of the "head" of Cygnus (the Swan; Deneb is its "tail") we are looking more along the Local arm. Thick dust clouds limit visibility in that direction, and it appears as if the belt of the Milky Way is divided in two "tongues" along this line of sight.

The Milky Way spiral arms are much more than strings of bright stars. They are also condensations of gas and dust, and sites of the birth of new stars. New methods of charting gas clouds were developed in the 1950s using radio telescopes. Dust does not stop the long wavelength radio waves which travel right past the grains. Over the rest of the disk long branches of spiral arms have been mapped, but it is still difficult to connect them in an overall pattern, because we cannot look at the Milky Way from outside! In a recent study collecting different evidence on the question the Canadian astronomer Jacques P. Vallée concluded that the most likely number of spiral arms is four – incidentally the same number as suggested by Stephen Alexander in 1852, with no evidence whatsoever!

Chapter 21
Entering the Galaxy Universe

Early in ancient times it was realized that stars are not the only fixed lights in the sky. The haze of the Milky Way was known. There are other nonstellar objects which were called nebular stars or nebulae – Ptolemy's *Almagest* mentions seven of them. Until the riddle of nebulae was solved in the last century, there were many different kinds of objects under the nebula label. One did not know how far they were, nor was there any idea whether the nebulae were actually "foggy." Galileo's telescope revealed that the Milky Way was composed of a multitude of stars. Later, with bigger and better telescopes, many more new nebulae were found which appeared to be truly nebular.

The first catalogs of nebulae were published in the eighteenth century. The list of Edmond Halley from year 1716, called *An Account of several Nebulae or lucid Spots like Clouds, lately discovered among the Fixt Stars by help of the Telescope*, included six objects, illustrating the modest role that nebulae had at that time in astronomy. The most famous catalog of the century was made by Messier. Its origin was actually related to – comets!

Messier's Catalog of Nebulae

Edmond Halley demonstrated in 1705 that what is now known as Halley's Comet is in an elongated orbit and predicted its return in 1758. After the return was verified, searching for new comets became popular. To be the first discoverer, one would have to spot the comet when it was still a faint smudge in the telescope, not yet possessing a tail. The many other kinds of nebulae led to unpleasant false alarms.

To facilitate comet hunting, Charles Messier (1730–1817) made a list of nebulae that he and his colleagues had accidentally spotted during their comet searches. Messier had moved to Paris at the age of 21, where he was fortunate to have astronomer Joseph Delisle hire him as an assistant. The young man became a skillful observer detecting the return of Halley's Comet in 1759 (to his disappointment, he

Fig. 21.1 The first known description of the Andromeda galaxy was given by the Persian astronomer Al-Sufi (903–986) in his *Book of the Fixed Stars*. It is the object near the mouth of the fish, described as "small cloud"

was not the first one). During his life Messier discovered about twenty comets which made him internationally famous. Messier worked for years in Paris in a building called l'Hôtel de Cluny where he lived and made observations in an observatory maintained by the navy. The building was initially built as a monastery, and it still exists today as a museum with fine collections of medieval artifacts.

Messier was elected a member of the French Académie Royale des Sciences in 1770. His first report there was the beginning of his catalog of nebulae. The final 1781 version included 103 nebulae. Messier himself had discovered 38 of them. We still use the number in his list to designate brighter objects, e.g., the Andromeda galaxy is M31 (Fig. 21.1 shows an old drawing of this nebula). The Messier catalog has a short description of each object, and its number and coordinates. This helped users to be sure that it was the correct object that they saw with their telescope.

Messier would have been surprised to know that his name would be remembered especially for this work. He showed no interest whatsoever on the nature of the nebulae – comets were the only things that mattered. Fortunately, he sent a copy of his list to William Herschel who studied all of them through his telescope and decided to complement the list by carrying out a systematic search. And complement he did: in the following 19 years he found 2,500 new nebulae and star clusters.

Fig. 21.2 William Herschel's 47 cm telescope which he used for his "sweeps of the sky"

Herschel's new, powerful telescopes were most suitable for his "sweeps of the sky" (Fig. 21.2). The telescope tube was kept pointed to a fixed direction while the sky rotated through the field of view. Herschel made the inventory of the sky by dictating a description of each nebula passing through his view to his sister Caroline. She wrote about practical problems:

> My brother began his series of sweeps when the instrument was yet in a very unfinished state, ... every moment I was alarmed by a crack or fall, knowing him elevated fifteen feet or more on a temporary cross-beam, ..., and one night, in a very high wind, he had hardly touched the ground before the whole apparatus came down. Some laboring men were called up to help in extricating the mirror, which was, fortunately, uninjured ...

The Garden of Nebulae

William Herschel was interested in the nature of the nebulae, initially thinking that all such fuzzy objects were really star systems that a bigger telescope might resolve into stars. He was able to do so for many nebulae in the Messier list through his own telescope. He agreed with the view of Kant that faint nebular patches are really distant "island universes," systems like the Milky Way. However, he could not resolve the Orion nebula into stars even though this is a rather large nebula in the sky. He thought that it is a very big system of stars, much bigger than the Milky Way, but so far away that its stars were not seen separately.

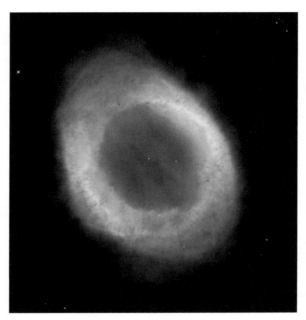

Fig. 21.3 A planetary nebula made William Herschel conclude that some nebulae are really "foggy" and not distant star systems. Here is another planetary nebula M57 (credit: Hubble Heritage Team (AURA/STScI/NASA))

Herschel's confidence in the "island universe" theory suffered considerably in 1790. He discovered a nebula which could not possibly be a stellar system, a "planetary nebula" (now known as NGC1514), where a central star is surrounded by a circular cloud of gas (Fig. 21.3). If the foggy part was really made of stars, then the central star would have to be incredibly luminous in comparison to them. Or if the central star is ordinary, the foggy ring must be made of incredibly small stars. Thus Herschel concluded that in this case the "fog" was real, not an impression created by innumerable distant stars close together. From now on he could not be sure of the nature of any other "unresolved" nebula.

Herschel was not only a skillful telescope maker and observer, he was also a thinker. He was fascinated by the possibility that nebulae evolve from one shape to another. But cosmic evolution is slow (or our life is too short!) – we cannot follow a star or a nebula from its birth to its death. He compared the situation with a garden where plants of the same species are observed in different stages of their life (seed, seedling, mature plant, and so on), and this can be used to reconstruct the lifecycle of the plant:

> The Heavens . . . resemble a luxuriant Garden which contains the greatest variety of productions, in different flourishing beds; and one advantage we may at least reap from it is that we can, as it were, extend the range of our experience to an immense duration. For, to continue the simile I have borrowed from the vegetable kingdom, is it not almost the same thing, whether we live successively to witness the germination, blooming, foliage,

fecundity, fading, withering, and conception of a plant, or whether a vast number of specimens, selected from every change through which the plant passes in the course of its existence be brought at once to our view.

However, just as there is more than one kind of plant in the garden, each with different life stages, there may be heavenly bodies of quite different kinds, which cannot be put in the same evolution sequence together. And even if we are able to single out a class of objects with no difference but their age, it is still not easy to infer the order of events from the frame of the cosmic film that we can inspect.

Initially Herschel thought that all nebulae are stellar systems and that their appearances reflect different stages of evolution. In their youth the nebulae may have been vast collections of stars far apart, and later they may have contracted under the force of gravity. The dense globular clusters would represent the end points of this evolution. When he later realized that there are also "foggy" nebulae, he concluded that a star can be born out of this "fog" by contraction. Herschel's speculations sound modern, but they were not discussed much at that time. Astronomers were more interested in matters of our Solar System, and anyhow nobody else had similar observational material to instigate any criticism.

William Herschel's only son John Herschel entered St John's College at Cambridge in 1809 and following his graduation was elected a fellow of the same college. Also in 1813 he became a fellow of the Royal Society of London, having written an important paper in mathematics. He decided to enter the legal profession, much against the advice of his father. "How many have miserably failed to obtain honest living in this way. A priest would have time to various cultural activities," complained William. John went to London in 1814 to begin legal training, but after 18 months he gave up and returned to Cambridge as a tutor and examiner in mathematics.

John Herschel Goes into Astronomy

John spent a holiday with his father in the summer 1816. He seems to have decided during these days to turn to astronomy; at 78 years of age his father's health was failing and there was nobody to continue his work. He wrote to a friend: "I shall go to Cambridge on Monday where I mean to stay just enough time to pay my bills, pack up my books and bid a long – perhaps a last farewell to the University.... I am going under my father's directions, to take up the series of his observations where he has left them (for he has now pretty well given over regular observing) and continuing his scrutiny of the heavens with powerful telescopes ... "

His first major work in astronomy, a catalog of double stars, was highly praised. In 1833 Herschel decided to go to the Royal Observatory at the Cape of Good Hope in South Africa to catalog astronomical objects which could not be seen from the northern hemisphere. Herschel took with him his family and his own 20 foot long refractor telescope. Their ship reached South Africa in January 1834.

Fig. 21.4 John Herschel (1792–1871) was William Herschel's only child. He lived at the time of discovery of photography and was one of the pioneers of this new technique: the word "photography" comes from him. In astronomy he discovered among other things, 2,200 nebulae. This is a photograph of Sir John taken in 1867

In the years 1825–1838 John Herschel discovered 2,200 new nebulae and star clusters. He spent much time studying the Large and Small Magellanic Clouds, two nebulae which are visible even to the naked eye in the southern sky. By his telescope Herschel saw that the Magellanic Clouds contain many stars, star clusters, and nebulae. It was only much later that other astronomers took interest in the Clouds and made major discoveries. As already stated, the key to measuring large distances came from the study of Cepheid stars in the Small Magellanic Cloud.

In 1838 Herschel returned to England. The next year he heard of Daguerre's work on realistic photography from a casual remark in a letter. Without knowing any details, Herschel began to take photographs himself within a few days. He was able to achieve this rapid breakthrough due to work he published in 1819 on chemical processes related to photography (Fig. 21.4).

John Herschel became rector of Marischal College in Aberdeen in 1842. In 1850 he accepted the post of Master of the Mint, with a major reform under way. This took all Herschel's time and energy, and he could no longer pursue his scientific interests. However, in 1864, he did publish all his and other available observations of nebulae in the *General Catalogue*. It contains more than 5,000 items.

In the meanwhile, the 3rd Earl of Rosse, William Parsons, at Birr Castle, had started working with the six-foot aperture telescope nicknamed the Leviathan, at the time world's largest telescope. A member of British Parliament at the age of 21,

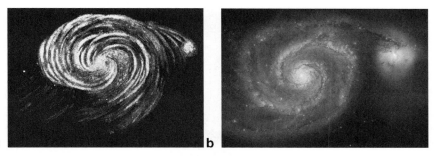

Fig. 21.5 Earl of Rosse William Parsons (1800–1867) built the biggest telescope of his time. (**a**) The nebula M51, in the direction of the Big Dipper in the sky, was seen to have spiral structure when viewed through the telescope. (**b**) Compare Parsons' drawing of 1845 with a photograph taken by the Hubble Space Telescope (credit: HST/STScI/AURA/NASA/ESA)

Parsons inherited the title of Earl of Rosse after his father in 1841. With an interest in mechanical devices and plenty of free time and money, he decided to try his skills in making telescopes. Parsons experimented for years in casting of metallic mirrors. Like Herschel before him, Parsons used an alloy of tin and copper as the raw material. This makes a good reflecting surface, but it is difficult to cast such a mirror blank; the blank breaks easily. In grinding the mirror to its correct shape, Parsons used a steam engine for the first time. After starting with a smaller telescope, Parsons finally finished his big telescope in 1845. Its mirror had a diameter of 183 cm.

The giant telescope collected much more light than Herschel's instruments, and it became possible to see more structure in the nebulae. One of Parsons' most important discoveries was the spiral structure which he saw first in the nebula M51. Soon after the telescope started operations, Parsons reported that he "saw the spirality of the principal nucleus very plainly; saw also spiral arrangement in the smaller nucleus." His drawing of the nebula was circulated in the meeting of the British Association for the Advancement of Science at Cambridge. It was sensational news, and since then the focus of the discussions shifted from the question whether nebulae could be resolved into stars to the question of their form. You may compare a modern photograph of M51 and Parsons' drawing of it in Fig. 21.5. Many years earlier John Herschel had looked at the same nebula through his smaller 48 cm telescope, but he was able to see only "a very bright round nucleus surrounded at a distance by a nebulous ring." Parsons saw "spirality" in other nebulae, too, and by 1850 fifteen examples were known; the number reached thousands by the end of the century. Spiral nebulae were a notable component of the universe.[1]

---

[1] Danish J.L.E. Dreyer (1852–1926) worked as an assistant to Lord Rosse's son. With the big telescope he began observing nebulae. Later Dreyer became director of Armagh Observatory in Northern Ireland. His *New General Catalogue* included 7,840 objects. Even today the NGC numbers are in use. So the Andromeda galaxy, M31, is also known as NGC224.

Fig. 21.6 William Huggins
(1824–1910) – the founder of
astrophysical spectroscopy.
He was the first one to measure the radial velocity of a
star from its spectrum. He
also discovered that the spectrum of a planetary nebula
resembles the emission line
spectrum of a cloud of gas

Astrophysics Is Born

As mentioned in Chap. 15, Gustav Kirchhoff and Robert Bunsen studied the spectrum of the Sun and identified several lines of known elements. A few years earlier
a wealthy amateur William Huggins had built an observatory by his house near
London. When he heard of Kirchhoff's work, he got the exciting idea of extending
the spectroscopic studies from the Sun to stars and nebulae. He built a spectroscope
for his telescope and started observations (Fig. 21.6).

He spent the first year studying of the spectra of stars and then moved on to nebulae. The first one he looked at was a planetary nebula in the constellation Draco.
He was surprised to see an emission line spectrum. According to Kirchhoff's laws
this means that the source of light is gaseous. Thus Huggins had proven what Herschel had conjectured. But when he pointed his telescope at the Andromeda nebula,
the result was quite different: a continuous spectrum – the light was spread over
all colors rather smoothly, just like in stars. Andromeda nebula was thus a collection of stars, a galaxy, which appears foggy only because it is so far away. Huggins
had found a way to separate gaseous nebulae from stellar systems! He studied the
spectra of sixty nebulae, and found that a third were gaseous, the rest stellar.

Toward the end of the nineteenth century it became apparent that spiral nebulae
are distributed in the sky in a peculiar way. There were few of them along the belt of
the Milky Way, but their numbers increase the more the further from the Milky Way
we look. The greatest density of nebulae is observed in directions perpendicular
to the plane of the Milky Way (Fig. 21.7). What does it mean? Most astronomers
had the opinion that the anticorrelation of stars and nebulae in the sky shows that

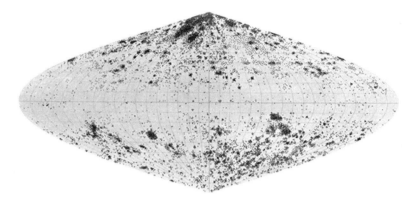

Fig. 21.7 The distribution of 11,475 spiral nebulae in the sky map made by Carl Charlier in the beginning of the twentieth century. Very few of the nebulae are found at the belt of the Milky Way (horizontal in the middle)

nebulae belong to the Milky Way. If the nebulae were distant "island universes," why would their numbers have anything to do with stars in our galaxy? Only later it was realized that there is a lot of dust in the Milky Way, and visibility through it is poor. The nebulae exist in all directions, but we cannot see them through the dust.

As another case against "island universes," a new star was seen in the Andromeda nebula in 1885. At its brightest this single star was one-tenth of the brightness of the whole nebula. If the nebula was really made of millions of stars, it appeared impossible that any single star could be so much brighter than the rest (supernovae were not yet known). It was simpler to view Andromeda as a gaseous nebula inside the Milky Way, in spite of its continuous spectrum. At the end of the century the majority opinion was that nebulae are part of the Milky Way. There were a few dissenting views. For example, Julius Schneider photographed the spectrum of the Andromeda nebula and found dark spectral lines (just like in the spectrum of the Sun). This was in favor of stellar composition.

"Island Universes" Gain Support

In 1911 the American astronomer F. Very calculated the distance to the Andromeda nebula assuming that the nova of 1885 was as bright as another nova inside the Milky Way, for which the distance was known. He obtained the distance of 1,600 light years. For some reason Very thought that the diameter of the Milky Way was only 120 light years. Reaching the right conclusions for the wrong reasons Very concluded that nebulae with continuous spectra are outside the Milky Way.[2]

---

[2] A few years earlier the Swedish Karl Bohlin reported a parallax for M31. He inferred that the nebula's compact nucleus shows an annual parallax shift of about 0.14 arcsec leading to a very short distance of $1/0.14 = 7.1$ pc or 23 light years. The true parallax would be about 0.000001! Later Lundmark suggested that there was a technical problem in the telescope.

By 1917 many novae had been discovered in other spiral nebulae. The newly discovered novae were about ten magnitudes fainter than the novae of the Milky Way which means that they are a hundred times more distant. That was a clear clue that the novae's host nebulae are independent "island universes," similar to our Milky Way. This chain of deduction assumes that the nova explosions in the nebulae and in the Milky Way are intrinsically the same in brightness; but this had not yet been proven.

By 1912 it had become apparent that dark spectral lines were common in the spectra of spiral nebulae, further evidence that these nebulae were stellar systems. Beyond this, the spectral lines could be used to measure the radial velocities (thanks to the Doppler effect, Chap. 12). The first velocity measurement of a star (Sirius) was carried out by Huggins in 1868. It took quite a few years before a similar measurement was made for a spiral nebula.

The director of the Flagstaff observatory Percival Lowell (1855–1916) was interested in the theory that spiral nebulae are a stage in the formation of planetary systems. He asked Vesto M. Slipher, one of the staff members, to use their 61 cm telescope and a new spectrograph to study the rotation of the nebulae. The task was a challenge, but Slipher had experience in the study of rotation of planets. In 1912 he managed to measure the faint spectrum of the Andromeda nebula. The result was most unexpected: the nebula speeds toward us at 300 km/s! Such a high speed was unheard of; speeds of stars and gas clouds in the Milky Way are more like 10 km/s. Today we know that most of the speed is contributed by the motion of the Sun as it carries us around the center of the Milky Way, a smaller part arises from the real motion of the Andromeda nebula relative us.

Slipher announced this and the results of 14 other radial velocity measurements in the meeting of the American Astronomical Society in 1914. The news was received with much acclaim. Slipher himself viewed his measurements as lending support to the island universe theory; spiral nebulae cannot belong to the Milky Way, they moved just too fast. For most of the nebulae, the lines in the spectrum were redshifted so they were escaping away from us, the largest speed in his sample being 1,100 km/s. This talented but unpretentious astronomer had discovered what is now called the cosmological redshift (Fig. 21.8).

Slipher found also what he had looked for: spiral nebulae do rotate and the typical rotation speed was 200 km/s. Francis G. Pease measured the rotation of the Andromeda nebula at Mt. Wilson Observatory in 1918. The Estonian astronomer Ernst J. Öpik (1893–1985) seized on the result right away and derived the distance to the nebula. He realized that the rotation speed gives an indication of the mass of the nebula in terms of the Sun's mass, from which one may infer the true brightness of the nebula, assuming it is made of sun-like stars or approximately so. When he compared the true brightness with the observed brightness as dimmed by distance, he obtained a huge distance: 2.5 million light years. Öpik reported the result in an astronomy meeting in Moscow in 1918, amid the Bolshevik Revolution. It was finally published in the *Astrophysical Journal* in 1922 (now the result was 1.5 million light years). If the method was correct (and it was, more or less), the Andromeda nebula was far outside our Milky Way.

Fig. 21.8 Vesto Slipher (1875–1969) measured the speed of the Andromeda nebula from its spectrum and discovered the cosmological redshift in the light of more distant galaxies

This result was in direct conflict with the measurements of the Dutch Adriaan van Maanen, who claimed that he can actually see the rotation of the spiral nebula M101 in the sky by following the changes in the photographic images from one year to another. If this were true, the nebula would make a complete turn around its axis in only 100,000 years (a short time, cosmically speaking). Such a nebula would have to be rather small and definitely inside the Milky Way!

"The Great Debate"

The most prominent centers for the study of nebulae in the early twentieth century were in California: Mount Wilson Observatory and Lick Observatory. The latter was famous for its 90 cm reflecting telescope, named for the British amateur Edward Crossley who donated the instrument. It started operation in 1895 and was used for photographing of nebulae from the very beginning. Mt. Wilson Observatory had a 1.5 m telescope since 1908, and a new 100-in. telescope, biggest in the world, saw first light in 1918. The name "Hooker telescope" refers to the businessman John Hooker.

Harlow Shapley worked at Mount Wilson while another leading astronomer, Heber D. Curtis (1872–1942), carried out observations at Lick observatory. Curtis had photographed spiral nebulae, looking for evidence of their rotation, but found none (unlike van Maanen). The staff of Lick favored the "island universe" theory,

Fig. 21.9 The plane of a spiral nebula shows a layer of dust when seen close to edge-on. Heber D. Curtis concluded that the odd distribution of spiral nebulae in the sky is a result of a similar layer of dust in the Milky Way. This figure shows an edge-on spiral M104 called "Sombrero" (credit: NASA/Hubble Heritage Team)

as did Curtis. While looking at photographs of spiral nebulae, Curtis noticed that the plane of the nebula often has a layer of dust which shows up as a dark lane in edge-on views (Fig. 21.9). If the Milky Way is also a spiral nebula, it should have a similar dust layer at its midplane. This would limit our view so that we could not see the distant stellar nebulae except away from the Milky Way belt, exactly as is observed. Curtis also argued that the high speeds of spiral nebulae and novae brightness favored the island universe theory.

Earlier Shapley had favored the island universe idea. But after finding that the Milky Way is 300,000 light years across, he found it easier to place the nebulae inside this huge system. He did not believe in cosmic dust, except for occasional clouds. To him, the distribution of spiral nebulae led to the opposite conclusion as compared with Curtis. The measurements of rotation by van Maanen, Shapley's good friend at Mount Wilson, confirmed this view.

A debate took place between Curtis and Shapley at the 1920 meeting of National Academy of Science in Washington. It was first planned that there would be a debate on relativity theory; however, this new topic was regarded as possibly incomprehensible for the majority of the participants and, instead, "the scale of the universe" was chosen. The "Great Debate" was less of an event than had been expected. More exactly, the two gentlemen read a prepared speech each, outlining the arguments for their own point of view. Shapley held on to his Milky Way of 300,000 light years across, while Curtis was willing to accept only 30,000 light years. Today, we talk about 100,000 light years for the diameter.

Hubble Finds Cepheids

Both Shapley and Curtis claimed that they had won the debate. Both were unaware of Öpik's work which in a way had already solved the dispute in Curtis' favor. Also the Swedish astronomer Knut Lundmark (1889–1958) had in his Dissertation in 1919 derived a large distance to the Andromeda nebula on the basis of its novae. But the decisive evidence came soon from Edwin Hubble (1889–1953). He was born in Missouri, the son of an insurance executive who moved to Chicago nine years later. At his high school graduation in 1906, the principal said: "Edwin Hubble, I have watched you for four years and I have never seen you study for ten minutes." He paused, before continuing: "Here is a scholarship for the University of Chicago." There he obtained a degree in Mathematics and Astronomy in 1910.

A tall, powerfully built young man, Hubble loved boxing and was on the University of Chicago championship basketball team. The combination of athletic prowess and academic ability earned him a Rhodes scholarship to Oxford. There, a promise made to his dying father, who never accepted Edwin's infatuation for astronomy, led him to study Roman and English Law rather than science.

He returned to the United States in 1913. After passing the bar examination, he practiced law half-heartedly for a year in Kentucky, where his family then lived. He later said that "I chucked the law for astronomy, and I knew that even if I were second-rate or third-rate, it was astronomy that mattered." Thus in 1914 he returned to the University of Chicago for graduate work leading to his doctoral degree in astronomy. While finishing his doctorate in 1917, Hubble was invited by G. E. Hale[3], the head of Mt. Wilson Observatory, to join his staff. Although this was a great opportunity, it came while the United States entered World War I. After sitting up all night to finish his Ph.D. thesis, and taking the oral examination the next morning, Hubble enlisted in the infantry and telegraphed Hale, "Regret cannot accept your invitation. Am off to the war." He returned to the United States in the summer of 1919, was mustered out in San Francisco, and went immediately to Mount Wilson Observatory.

At first Hubble studied reflection nebulae, dust clouds which reflect light from a nearby star. After this he started using the 100-in. telescope for the study of spiral nebulae. In 1917 Hubble had concluded that they are "island universes," based on their high speeds, but now he started looking for individual stars which could indicate their distances. With a large telescope it is possible to see individual stars separately. But, at a great distance one can see only a "porridge" of stars. Therefore it is good to look for variable stars, in order to identify possible distance indicators. The best case would be a Cepheid.

As a matter of fact, Hubble was hunting novae when he discovered a Cepheid in the Andromeda nebula in 1923. Its faintness already told him that it must be very far away. More exactly, he found that the period of the Cepheid was 31 days; then using the dependence between the period and brightness, derived by Shapley, he was able to calculate its distance. This turned out to be one million light years. Thus he confirmed that the Andromeda nebula is clearly outside the Milky Way. Later in the same year Hubble found nine more Cepheids in Andromeda, all of which agreed with the same large distance. He also found Cepheids in the spiral galaxy M33 (near

---

[3] George Ellery Hale (1868–1938) was a remarkably influential figure. He established three observatories: the Yerkes, Mount Wilson, and Palomar Observatories. Mount Wilson dominated astronomy in the first half of the twentieth century. It was here that astronomers found out the cosmic significance of galaxies. The nature of quasars was discovered at Palomar.

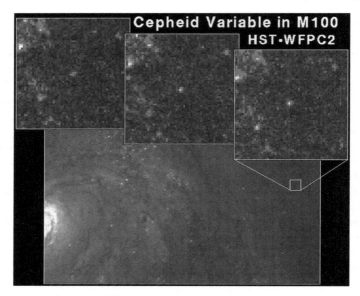

Fig. 21.10 Edwin Hubble found Cepheids in the Andromeda nebula and used them to measure its distance. The picture shows the about twenty times more distant galaxy M100, whose distance was measured seventy years later by the same method; while normal stars appear the same at different times, variable stars change in brightness (credit: HST, NASA, W. Freedman (CIW), R. Kennicutt (U. Arizona), J. Mould (ANU))

to the Andromeda nebula M31 in the sky, in the constellation Triangle); calculations showed that it is at the same distance as M31 (Fig. 21.10).

The official announcement of Hubble's results was made in the January 1925 meeting of the American Astronomical Society in Washington, where it received plenty of attention. The whole society knew that the Great Debate was over. Curtis had been correct in that stellar spiral nebulae are really outside the Milky Way, forming the new realm of *galaxies*. By the way, this term was suggested by Shapley – for example Hubble spoke to the end of his life about extragalactic nebulae.

Hubble's Classification of Galaxies

Recognition of classes of objects has been an important aim in science since Aristotle. He understood that if one can put natural phenomena into different classes according to their *essential* properties one obtains knowledge about the world. There is much more to a galaxy than meets the eye; however, even their appearances – the first thing we discern – have offered useful keys for understanding galaxies. Already William and John Herschel started classifying nebulae based on their observations through the telescope. Later photographs showed many kinds of nebulae, not only spirals. In 1926 Hubble concluded that most galaxies can be divided into two major classes, spiral galaxies and elliptical galaxies.

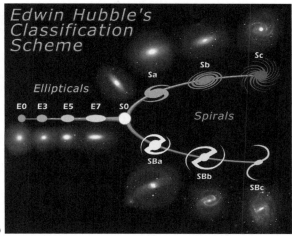

Fig. 21.11 (**a**) Edwin Hubble at the telescope (With permission of The Huntington Library, San Marino, California). (**b**) Hubble's "tuning fork" diagram with photographs of corresponding types of galaxies (NASA)

Box 21.1 Some members of the Local Group and some other nearby galaxies

Galaxy	Distance 1,000 light years	Type[a]	Luminosity Milky Way $= 1$	Diameter 1,000 light years
Milky Way[b]	–	SBc	1	100
LMC[b]	160	SBm	0.09	30
SMC[b]	200	Im	0.02	28
Sculptor dwarf	290	dE3	0.0009	8
Fornax dwarf	460	dE0	0.0001	9
Leo dwarf	820	dE3	0.00004	5.7
Draco dwarf	260	dE0	0.000,004	3.5
Barnard's	1,600	Im	0.006	10
IC1613	2,200	Im	0.004	17
IC10	2,200	Im	0.003	3
Andromeda[b]	2,500	Sb	1.71	140
Triangulum (M33)[b]	2,800	Sc	0.17	70
M32	2,500	E2	0.006	8
NGC205	2,500	E5	0.008	16
NGC147	2,500	E5	0.003	13
NGC185	2,500	E3	0.005	11

[a]The modern classification system is a modified version of Hubble's system. Intermediate cases are shown by two letters: Sab is between Sa and Sb. Sd describes the extreme end of the spiral sequence beyond Sc, "I" refers to irregular galaxies, "m" means "Magellanic," and "dE" is a dwarf elliptical galaxy
[b]Visible by eye

Some nearby galaxies beyond the Local Group

Galaxy	Distance[c] Millions of light years	Type	Luminosity Milky Way $= 1$	Diameter 1,000 light years	Group
NGC55	7	Sc	0.20	70	Sculptor
NGC300	7	Sc	0.08	40	Sculptor
NGC253	12	Sc	0.58	65	Sculptor
NGC247	10	Sc	0.10	46	Sculptor
NGC7793	11	Sd	0.09	30	Sculptor
Circinus	13	Sb	0.26	60	
M81	12	Sb	0.75	55	Ursa Majoris
Cigar (M82)	12	I	0.10	27	Ursa Majoris
NGC2403	10	Sc	0.16	46	Ursa Majoris
IC342	11	Scd	0.41	60	Maffei
Maffei 1	10	E3	0.28	33	Maffei
Maffei 2	9	Sbc	0.22	30	Maffei
Centaurus A	14	S0	1.71	70	Centaurus
NGC4945	12	Sc	0.26	50	Centaurus

[c]Some of the cited distances may have considerable uncertainty which is also reflected in the values for luminosity and diameter

COLOR PLATES

Fig. 1 The Very Large Telescope (VLT) of the European Southern Observatory in Chile is currently the largest optical ground based telescope. It actually consists of four telescopes, each with a 8.2-m mirror, which can be used either separately or together (credit: ESO)

Fig. 2 The Nançay radio telescope in France operates in the decimetric wavelength range, studying a wide range of subjects from comets and pulsars to galaxies and cosmology. For example, it can measure the 21-cm line radiation emitted by the neutral hydrogen gas which is abundant in our Milky Way and in many other galaxies. Courtesy of I. Cognard, CNRS

Fig. 3 The Hubble Space Telescope orbits the Earth at a height of 600 km above the atmosphere and can thus make sharper images than ordinary telescopes (credit: NASA)

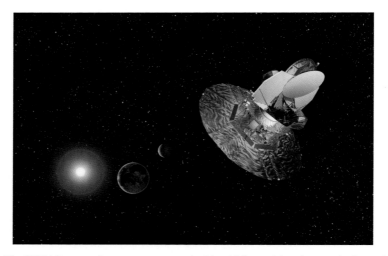

Fig. 4 The WMAP space observatory measured with a high precision the cosmic thermal background radiation, allowing cosmologists to study the geometry and composition of our universe. The European Space Agency PLANCK observatory will complement and improve these observations. This artist's depiction shows the WMAP's location a million miles from the Earth in the direction away from the Sun (credit: NASA)

Fig. 5 An ultraviolet image of our Sun taken by the SOHO space observatory in 1999. Our Sun is an ordinary star, about five billion years old. Despite short term activity such as the eruptive prominence shown, our Sun's long stable phase of development has allowed life to exist on the Earth for a large fraction of its age. The huge mass of the Sun (over 300 000 earth masses) holds its planetary system circling around it (credit: SOHO-EIT Consortium, ESA, NASA)

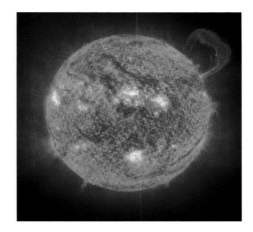

Fig. 6 Valles Marineris (the Valleys of Mariner) on the Tharsis plateau in Mars. This huge (200 km wide and 4,500 km long) canyon is a spectacular example of former geologic activity (credit: NASA)

Fig. 7 A close view of a rough portion of the Martian landscape was obtained by the Spirit Mars Exploration Rover. The nearby dark volcanic boulder is about 40 cm high (credit: NASA/JPLCaltech/Cornell/NMMNH)

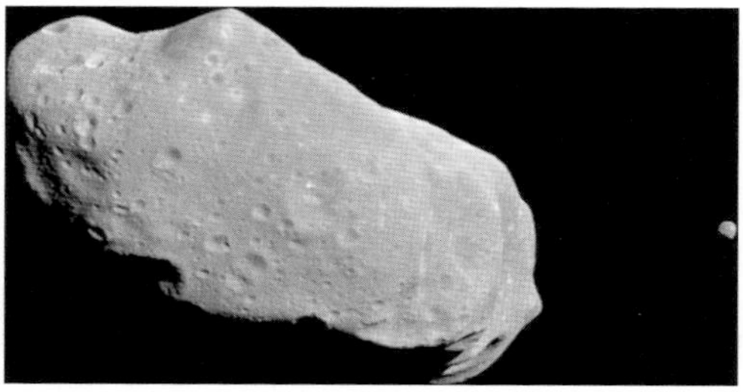

Fig. 8 On its way to Jupiter in 1993 the Galileo Probe took this photo of the asteroid 243 Ida orbiting about 440 million km from the Sun. Being only about 50 km in length, Ida's weak gravity cannot even pull it into a round shape making it an asteroid rather than a dwarf planet. In this picture, we see also Ida's small moon Dactyl (credit: NASA/JPL)

Fig. 9 The comet Shoemaker- Levy 9 broke into over 20 pieces before it collided with Jupiter in 1994. This Hubble Space Telescope image shows gigantic dark spots where four pieces of the comet penetrated the atmosphere. Similar impacts like this may have affected the early environment of life on Earth and may be a hazard for life today (credit: HST Comet Team & NASA)

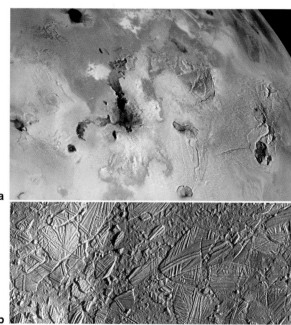

Fig. 10 Images of two different moons of Jupiter from the Galileo Probe show: **a** Mountains and volcanic calderas on the geologically active Io. **b** The icy world of Europa where liquid water oceans beneath the ice may provide a habitat for life (credit: NASA/JPL)

Fig. 11 This 2004 Hubble Space Telescope combined visual and ultraviolet image shows a huge auroral display in Saturn's southern polar region (credit: NASA, ESA, J. Clarke (Boston University) & Z. Levay (STScI))

Fig. 12 This photograph of the Sagittarius star cloud illustrates the huge amounts of stars inhabiting our Milky Way galaxy. All stars are not like our Sun, but may differ considerably e.g. in mass, temperature, and luminosity. In this picture you can easily discern *red* (cool), *yellow* (sun-like), and *bluish* (hot) stars (credit: Hubble Heritage Team (AURA/STScI/NASA/ESA))

Fig. 13 The region of the Eagle nebula in the constellation Serpens, 7000 light-years away, offers striking scenes of cold gas and dust where new stars are actively born in the Milky Way. Energy from massive, hot, and young stars works as sculptor carving ghostly shapes from interstellar matter. This "tower" has a length of about 9.5 light-years, roughly a million times the diameter of the Earth's orbit around the Sun (credit: NASA, ESA, & The Hubble Heritage Team STScI/AURA)

Fig. 14 This beautiful planetary nebula (NGC 6751) in the constellation Aquila is a shell of gas ejected thousands of years ago from the hot star visible in the middle. Such glowing nebulae tell about the forthcoming death of stars roughly similar as our Sun. The "planetary" nebulae have actually nothing to do with planets or planetary systems. In fact, this nebula's diameter is almost 1 light year, 700 times the size of our Solar System (credit: NASA, The Hubble Heritage Team STScI/AURA)

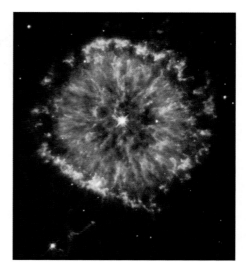

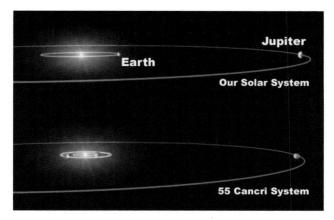

Fig. 15 Extrasolar planetary system 55 Cancri as sketched on the basis of observations and compared with the Solar System. It consists of five known planets on orbits around the central, sun-like star (credit: NASA)

Fig. 16 Artist's view of the multitude of earth-like planets expected to be in our Milky Way, each individuals in their detailed surface structure, water cover, and atmosphere. It is a great question of astrobiology, whether a fraction of them carry some kind of life (credit: NASA)

Fig. 17 Fig. 21.15. Our neighboring large "island" in the universe, the Andromeda galaxy M31, and its small companion galaxies M32 and M110. This spiral galaxy, which can be dimly seen with naked eye, lies at a distance of 2.5 million light-years (credit: John Lanoue www.bedfordnights.com)

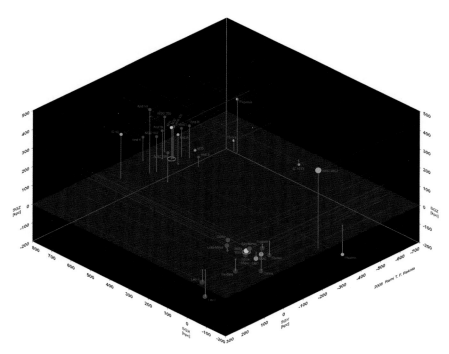

Fig. 18 The structure of the Local Group: The MilkyWay and the Andromeda galaxy, surrounded by their smaller companion galaxies. Courtesy of Rami Rekola

Fig. 19 The spiral galaxy
M81 is a member of a nearby
galaxy group in the constellation Ursa Major. It is about
five times farther away than
the Andromeda galaxy. At this
distance it recedes from us at
a speed of about 250 km s^{-1},
participating in the expansion
of the universe (credit: NASA,
ESA & The Hubble Heritage
Team STScI/AURA)

Fig. 20 Many spiral galaxies have a bar-like structure with the spiral arms starting out of its ends.
This beautiful barred spiral is called NGC1300. It is impressive to think that what we see as starlight
and glowing gas is just a fraction of mass within a larger lump of invisible mysterious dark matter
(credit: NASA, ESA, & The Hubble Heritage Team STScI/AURA)

Fig. 21 This rare system, known as IRAS 19115-2124 and dubbed "The Bird" or even "The Tinker
Bell" by astronomers, consists of two large spiral galaxies and one irregular galaxy which are
merging together. Observations using an adaptive optics system on the Very Large Telescope (ESO)
at near-infrared wavelengths revealed this dramatic cosmic collision. The image combines near-
infrared data with optical images from the Hubble Space Telescope (credit: ESO & Henri Boffin
and Petri Väisänen & Seppo Mattila)

Fig. 22 The supernova that exploded in 1604 (and was observed e.g. by Kepler) left behind it a shell of gas expanding at a speed of 2,000 km s^{-1}, shown here in a composite picture based on different wavelengths from infrared to x-rays. Located 13,000 light-years away in the constellation Ophiuchus, this was the last supernova thus far observed in our own Galaxy (credit: NASA, ESA/JPLCaltech/R.Sankrit & W. Blair (John Hopkins University))

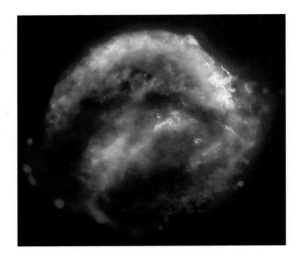

Fig. 23 A supernova explosion in the outskirts of a galaxy called NGC 4526. Supernovae occur about once in a century in a typical galaxy. Certain types of supernovae are "standard candles." Their observations at very large distances have revealed that the expansion of the universe is accelerating due to a mysterious antigravitating "dark energy" (credit: NASA/ESA, The Hubble Key Project Team, and The High-Z Supernova Search Team)

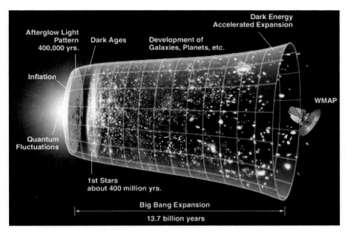

Fig. 24 Our current view of the development of our universe during its about 14 billion years of existence starting from the mysterious Big Bang. When the temperature decreased, during the first second various elementary particles including hydrogen nuclei (protons) were formed, through the first minutes helium nuclei were created, and about 400 000 years later first atoms were formed and the thermal background radiation started wandering in space. During billions of years stars and galaxies were gathered by gravitation from the expanding cosmic matter (credit: NASA)

Fig. 25 An all-sky picture of the infant universe as observed in the thermal cosmic background radiation. This is based on data gathered by the WMAP space observatory. The effect of the motion of the Earth, moving in space at a speed of about 350 km s^{-1}, has been cleaned away from this map. The 14 billion year old slight temperature fluctuations (shown as color differences) are caused by the seeds that grew to become the galaxies (credit: NASA/WMAP Science Team)

Large Scale Structure in the Local Universe

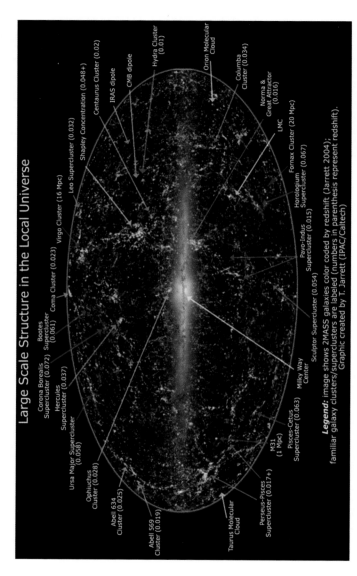

Fig. 26 An all-sky view of the local galaxy universe based on about 1.5 million galaxies from the 2MASS galaxy survey. The MilkyWay is shown in the middle. The redshifts (distances) of the galaxies are coded by color (the *red* ones are most distant). Some galaxies, galaxy clusters, and superclusters are identified in this picture where one can discern complex large scale structures. Courtesy of Tom Jarrett (IPAC/Caltech)

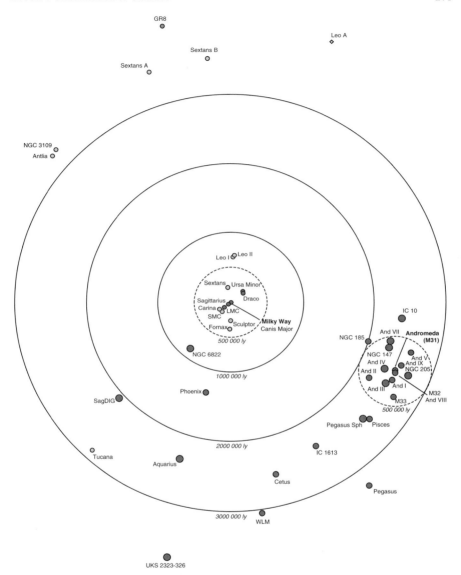

Fig. 21.12 Main galaxies of the Local Group. Note the swarms of companions near the Milky Way and Andromeda galaxies (courtesy of Rami Rekola)

Elliptical galaxies (E) appear as rather smooth spherical or flattened patches of light, with the light concentrated to the center and fading off toward the edges. A number tells the degree of flattening (E0: spherical, ..., E7: very flattened). Spiral galaxies (S) come in two categories: normal or barred. In normal galaxies the spirals start from the center of the galaxy, in barred galaxies they start from the ends of a bar. Based on the tightness of winding of the spiral structure, the spirals are further divided into subclasses Sa, Sb, and Sc (for barred galaxies SBa, etc.). The winding is tightest in the subclass Sa, and the least tight in Sc. Hubble identified also an intermediate class S0; this looks smooth like an elliptical galaxy, but is flat like a spiral. He presented all these types in a "tuning fork" diagram (Fig. 21.11).

Our own Milky Way is either an SBb or SBc galaxy; the exact structure is hard to determine without being able to go outside to look at it. Observations at infrared wavelengths (which are much less affected by dust than visual light) indicate the presence of a bar in the center of our galaxy. You may see a nice example of a barred galaxy (NGC1300) in the color supplement.

Almost all galaxies can be placed in one of the Hubble classes and it is a very useful way to classify galaxies even today. Like Herschel, Hubble thought that his sequence of galaxies might represent different stages in the life of a galaxy. We now know that it is not the case. Nevertheless, in addition to its simplicity, the classification is valuable, because the appearance of a galaxy is correlated with its physical properties which one cannot see from its photograph, such as its rotation speed and mass.

Box 21.1 is a list of part of galaxies in what Hubble named the Local Group, the system to which the Milky Way belongs (Fig. 21.12). It shows that most galaxies are of very low mass and low brightness compared to our home galaxy. However, most of the total mass of this group is in two big galaxies, the Milky Way and Andromeda. Box 21.1 lists also other examples of nearby galaxies of a variety of Hubble types.

The Hubble Law of Redshifts

When Slipher started to measure the radial speeds of galaxies in 1914, it came as a surprise that almost all galaxies have lines shifted to the red end of the spectrum. If this *redshift* is due to motion (cf. the Doppler effect) then it seems that the galaxies are escaping away from us. Already in 1917 Willem de Sitter had developed a model based on General Relativity which predicted a redshift for distant objects in the universe. It was a competing model for the Einstein's static universe (which did not predict a redshift). Actually, it was a peculiar model since it assumed that the universe has no matter. But if the real world is "thin" on matter, then one might expect a "de Sitter effect" of redshifts which should be larger for more distant light sources. This inspired people, including Edwin Hubble, to study if the redshift of nebulae depends on distance.

Hubble had an able assistant, the legendary Milton Humason (1891–1972), who used the large 100-in. telescope to photograph the spectra of galaxies (earlier these measurements came mainly from Slipher and the 61 cm telescope at Lowell observatory). Having dropped out of school, Humason became a mule driver for the pack trains that traveled the trail between the Sierra Madre and Mount Wilson during construction work on the Observatory. In 1911 he married the daughter of the Observatory's engineer and became a foreman on a relative's ranch, but in 1917 he joined the staff of the observatory as a janitor and was soon promoted to night

Fig. 21.13 A group of eminent scientists in 1931 in front of the portrait of G.E. Hale. Left to right: M. Humason, E. Hubble, Ch. St. John, A. Michelson, A. Einstein, W. Campbell, and W. Adams. Of those not discussed elsewhere we note that St. John showed in 1922 that the atmosphere of Venus is almost devoid of water and oxygen, Adams (director of Mt. Wilson Observatory) identified Sirius B as a white dwarf star and Campbell, head of Lick Observatory, was also famous for spectroscopy studies (with permission of The Huntington Library)

assistant. In 1919 George Hale, the director, recognized Humason's unusual ability as an observer and appointed him to the scientific staff. Thus he finally became a self-educated astronomer. During his career Humason measured the redshifts of 620 galaxies (Fig. 21.13).

In 1929 Hubble published his fundamental discovery on how the redshift of a galaxy depends on its distance. A loose correlation between these two quantities had been noticed by Knut Lundmark, but it was Hubble's work that revealed for the first time that the redshift is directly proportional to the distance. The result has been since then amply verified by using more distant galaxies. In terms of the speed derived from the redshift, it is expressed in the form of the famous *Hubble law*:

Speed of escape = Hubble constant × Distance.

The widely accepted interpretation of this important law is that in the world of galaxies the distances are really increasing, or, as stated more commonly, that the universe is expanding. Note that we do not actually "see" galaxies moving, but infer this from the slight displacements (redshifts) of spectral lines (Fig. 21.14).

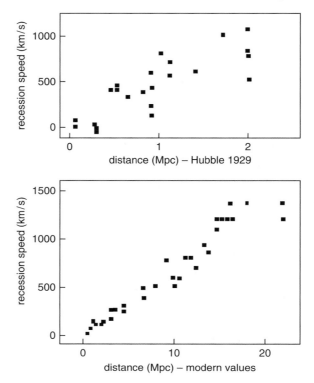

Fig. 21.14 The galaxy's speed of recession and its distance. The points represent observations of certain galaxies, and the straight line represents the Hubble law. The upper panel is based on Hubble's 1929 work. Below we show a modern Hubble diagram for about the same distance interval. Note that modern distances are almost ten times longer than those by Hubble. This is due to the large systematic error in the old measurements

The Hubble constant is a very important quantity for cosmology, related to both the size and the age of the universe. Furthermore, for most galaxies we only know the redshift. If the Hubble constant is known, then it is simple to divide the speed of recession by this constant, and we have the distance. But to infer the value of the Hubble constant, we first need reliable distances for sufficiently many galaxies.

How to Measure Cosmic Distances?

An astronomer started his review on distances of galaxies: "As a matter of fact, the determination of distances of galaxies is an impossible task." The pessimistic phrase has some truth to it, because the measurement of cosmic distances is based on a complex chain of methods having its weak links. The chain starts from the Sun, ties together nearby and distant stars in our Milky Way, jumps to nearest galaxies, and stretches out to more and more remote galaxies, forming the *cosmic distance ladder*.

The distances to the nearest galaxies are mainly measured using Cepheid stars, but at larger distances Cepheids are too faint to be observed with ordinary telescopes on Earth. The Hubble Space Telescope, working above the Earth's atmosphere, has recently been very helpful in extending the range of the Cepheid method to greater distances, up to 30 times the distance of the Andromeda galaxy.

Supernovae are much brighter than Cepheids (the 1885 supernova in M31 was one-tenth of the brightness of the whole galaxy). In recent years there has been a dramatic increase in the detection of distant supernovae, as well as understanding their behavior. Thus certain types of supernovae have become the "standard candle" that can be used to measure the distances in truly universal scale.

Many other methods have been applied over the years. One could assume that a galaxy of certain kind has a known brightness. If it were true, one could calculate its distance. Unfortunately, no such good "standard candle galaxies" are known. For example, take a look at the Andromeda galaxy and its companions M32 and M110 in the color supplement. If one believes that all galaxies are equally big, then one would put the companions far behind Andromeda. Thus estimating distances of galaxies by their size or brightness is very uncertain. One can do better by going back to Öpik's method of determining the distance to the Andromeda galaxy by using the rotation speed of the galaxy to estimate its true brightness. In its modern form it is called the Tully-Fisher method, and it can give distance values at better than 30% accuracy. The measurement of the speed of rotation is easiest using a radio telescope. Why does the rotation indicate the true brightness of a galaxy? This is because more massive galaxies rotate faster than light ones, and more mass means also more stars and thus more star light. Of late the Tully-Fisher distance has been measured for thousands of galaxies.

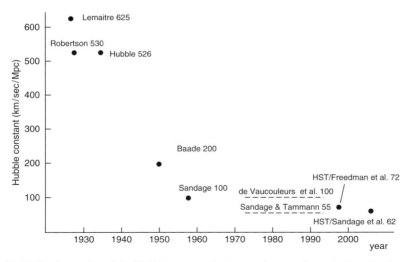

Fig. 21.15 The best value of the Hubble constant in the past decades. Its evolution tells about the difficulty of measuring distances of galaxies. The last two points come from the Hubble Space Telescope observations

Box 21.2 The Hubble constant, the distance scale, and the age of the universe

The Hubble constant (H) is intimately related to the distance scale and the age of the expanding universe. The distance R to a galaxy with escape speed V is V/H (from the Hubble law). Hence, the smaller the Hubble constant, the larger the distances calculated from the speed (redshift). The link to the age is also easy to see. Let us assume that the expansion speed V between two galaxies has been constant throughout the history of the universe. Then these galaxies (and all others) were very close together a finite time ago. This time, or "the age of the universe," is obtained by simple school mathematics, dividing the present distance R between the galaxies by their mutual speed V. On the other hand, this ratio R/V is just equal to $1/H$. Hence the inferred age of the universe is inversely proportional to the assumed value of the Hubble constant.

For example, if the Hubble constant equals 70 km/s per Mpc, the galaxies escape from each other with the average speed of about 70 km/s if they are one million parsecs (or 3.26 million light years) from each other. The speed of one kilometer per second corresponds to one parsec per million years; thus with the speed 70 km/s the same distance takes 14,000 years, and the million parsec distance takes 14 billion years. Thus the age of the universe, since the time when the galaxies were "on top" of each other, is about 14 billion years. The exact age depends on whether the expansion has accelerated or slowed down since the Big Bang (this question arises in Chap. 23).

When many decades ago it was thought that the Hubble constant is about seven times larger (see Fig. 21.15), the corresponding age of the universe was only $14/7 = 2$ billion years.

One should mention a tricky problem in the methods using standard candles. It illustrates the inconvenient fact that an astronomer cannot travel between the galaxies, but is stuck in one point, the Solar System. When galaxies are collected for study, there is an inevitable tendency to include fainter galaxies in nearby space and brighter galaxies from the far end of the distance range. It brings an error to the final result called *Malmquist bias*, initially discussed by the Swedish Gunnar Malmquist (1893–1982) in connection with studies of stars. This bias tends to creep into the evaluation of many astronomical data bases. It makes distances too small, and thus the estimates of the Hubble constant (= speed/distance) become too big.

Figure 21.15 illustrates how the ideas of the value of the Hubble constant have changed in the past decades. It tells about the difficulty of measuring distances of galaxies. The division into two schools of thought around 1980 had quite a lot to do with the Malmquist effect, and how it was corrected. Gérard de Vaucoulers and his associates favored a value of around 100, while "the grand old man of modern observational cosmology" Allan Sandage and his longtime European associate Gustav Tammann derived values around 55 (the unit of the Hubble constant is km/s per Mpc; Mpc = one million parsecs). In general, the large differences in the derived

Table 21.1 Measured distances to the Andromeda galaxy

1907	Bohlin[a]	23 light years
1910	Strömberg	65
1911	Very	1,600
1919	Lundmark	650,000
1918/1921	Öpik	2,500,000
1922	Öpik	1,500,000
1925	Hubble	900,000
1948	Lundmark	1,300,000
1952	Baade	1,600,000
1982	Sandage/others	2,150,000
2007	Various methods	2,500,000 light years $= 770,000\,pc$

[a]Some methods used by the cited authors are discussed in the text.

value of the Hubble constant reveal how difficult the art of cosmic distance measurement is. Today there are also a few special methods which skip the usual distance ladder and are free of the Malmquist bias. These also give values in the range 60 to 80. See Box 21.2. for a brief discussion on how the value of the Hubble constant is related to the size and age of the expanding universe.

Fig. 21.16 The rotation of the Triangulum galaxy M33 in the sky was measured by observing the tiny motions of giant water vapor masers. The arrows show the directions of measured shifts. This galaxy in the constellation of Triangle is sometimes called the Pinwheel (courtesy of Travis Rector)

Another example of the difficulty of distance determinations is given by the "changing" distance to the Andromeda galaxy. Table 21.1 shows some results from the last one hundred years.

And Yet It Moves!

We have mentioned van Maanen's belief that he could detect the rotation of some spiral galaxies from photographs taken at different times. These results played a role when the distances of spiral nebulae were debated. However, now we know that it was not at all possible for him to detect the rotation; it takes a galaxy 100 million years or more to make only one rotation. Van Maanen's error was certainly not intentional; he was a careful measurer of stellar motions. More likely we have here an example of so called personal bias. When one tries to see very small effects, actually impossible to perceive, one's unconscious expectations of "what should be seen" may take the lead and influence the measurements.

On the other hand, spectroscopic Doppler shifts show that galaxies do rotate, and in 2005, some 80 years after van Maanen's measurements, an international team of astronomers led by Andreas Brunthaler and Mark Reid could indeed detect the rotation of one of the galaxies in van Maanen's sample, M33. The rotation was observed not from a photograph but with the help of giant water vapor masers in

Fig. 21.17 Allan Sandage continued the work of Hubble in the field of observational cosmology, painstakingly investigating with large telescopes the fundamental problems of age, size, and geometry of the universe. He has been awarded the prestigious Crafoord and Gruber prizes for his achievements (courtesy of The Library of the Observatories of the Carnegie Institute in Washington, Pasadena)

the gas clouds of M33. These natural masers (like lasers) radiate intensively into one direction and within a narrow range of radio frequency. Their position (and its changes) can be very accurately measured by radio astronomers using "very long base line interferometry," in this case ten large radio antennas situated all over the USA and operated by the National Radio Astronomy Observatory (NRAO). The team could measure the very slight shifts in the sky of two water vapor masers in the spiral arms of M33 (just several *millionths* of second of arc per year). These shifts revealed the rotation as expected from independent spectroscopic (Doppler effect) evidence (Fig. 21.16). The measurements also allowed calculation of the distance to M33, in agreement with other methods which tell us that this Local Group galaxy is about as distant as the Andromeda galaxy.

Allan Sandage (Fig. 21.17) likes to say: "What seems so simple is often complex in its dreadful detail". We have spared the reader the dreadful details of the strong struggle to establish the cosmic distance ladder. The distances found using this ladder have enabled humanity to reach new heights of understanding ranging from an appreciation of our insignificance in the scale of the universe, the existence of multitudinous stars like our sun, and the great antiquity of the universe. Still another application is cosmic cartography, a "geography" of the universe, which we will discuss in the next chapter.

Chapter 22
Large-scale Structure of the Universe

By naked eye we can easily detect only three galaxies: the Andromeda galaxy in the northern sky, and the Large and Small Magellanic Clouds in the southern sky. Pictures taken via large telescopes contain millions of galaxies, and it is estimated that there are hundreds of billions of faint galaxies all over the sky. Also spectra of millions of galaxies have been obtained in recent years; hence, the distances to these galaxies are known with the help of the Hubble law (distance proportional to redshift). Thus it is possible to study how galaxies are distributed three dimensionally, indicating matter distribution in the universe. Previously, before the current "redshift industry," astronomers could study only the two-dimensional distribution of galaxies on the celestial sphere.

Galaxy Clustering in Our Neighborhood

William Herschel pointed out that nebulae were social creatures: they tend to be in pairs, groups, and clusters. More recently, the cosmologist James Peebles, a leading student of the distribution of galaxies, has said that "the best place to find a galaxy is next to another galaxy." This tendency is so strong that there are few isolated galaxies. As already stated, our Milky Way is no exception; it belongs to the Local Group whose two dominant members are the Milky Way and the Andromeda galaxy 2.5 million light years apart. Most others are much smaller galaxies than these two (as you can see from the previous Box 21.1).

The Local Group is a rather modest grouping of galaxies, surrounded by similar other groups (Fig. 22.1). A much greater *cluster of galaxies* is found in the direction of the constellation Virgo about 60 million light years from us. It has hundreds of galaxy members, the brightest of which may be dimly seen by good binoculars. In the sky they are spread out over a circle of 10 degrees in diameter, i.e., 20 times the diameter of the full Moon (Fig. 22.2). It is an example of a somewhat small irregular cluster of galaxies. Further away much bigger *rich clusters* are found.

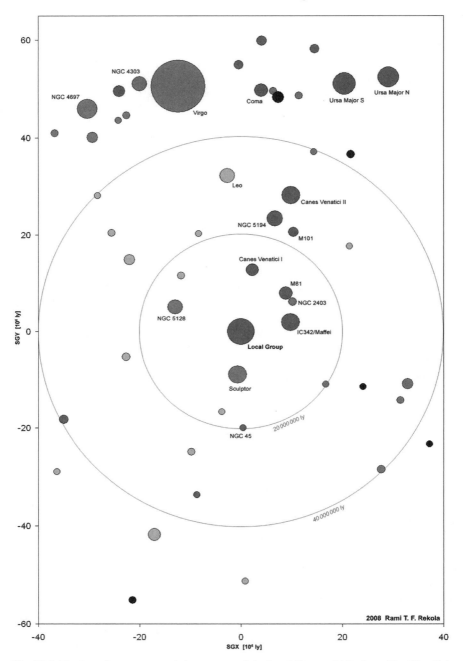

Fig. 22.1 Nearby galaxy groups and clusters around the Local Group within about 60 million light years and having at least ten member galaxies. The number of members is roughly indicated by the size of the symbol. The Virgo cluster is the largest single system here (courtesy of Rami Rekola)

Fig. 22.2 The Virgo cluster of galaxies is the nearest large collection of galaxies, and the center of the Local Supercluster. This picture shows its central parts populated by galaxies. The brightest galaxy in this photograph is the giant elliptical M86 (credit: Chris Mihos)

To find a rich cluster of galaxies, where the membership may rise to 10,000 galaxies, we have to go 300 million light years to the direction of Coma Berenices. There we find a rather regular-looking cluster which contains mostly elliptical and S0 galaxies. In contrast to the smaller and less dense Virgo cluster which has many spiral galaxies, it appears that spiral galaxies cannot survive in the more extreme environment of the Coma cluster. The tidal forces arising from the overall gravity field of the cluster strongly perturb the disk galaxies and make them lose their gas content and thus the ability to form spiral arms, as has been demonstrated with computer calculations.

In the 1950s the French astronomer Gérard de Vaucouleurs (1918–1995) presented evidence that our Local Group of galaxies is a member of a "Local Supercluster" of galaxies. It consists of the Virgo cluster at its center and of outlying smaller clusters and groups of galaxies. The whole system is rather flat, resembling our Milky Way in this respect. But unlike the Milky Way which rotates around its center, the Local Supercluster does not rotate, and neither is it a structure held together by gravity. This large system of galaxies is expanding like the galaxy universe in general, though the gravitation of the central Virgo cluster has somewhat slowed down the recession velocities. Clusters themselves thus cluster, and a cluster may have extensive galaxy neighborhoods where the gravity of the cluster is felt, but not strong enough that we would have a bound system of clusters.

Toward Larger Scales: Mapping Three-Dimensional Structures

The Schmidt telescope can take large-angle photographs (Fig. 22.3). Using the "Big Schmidt" of the Palomar Observatory the whole northern sky and part of the southern sky was photographed, in the 1950s. Each picture, nine hundred altogether,

Fig. 22.3 Schmidt telescopes were important in early studies of the distribution of galaxies in the sky. (**a**) The telescope is named after its 1931 Estonian inventor Bernhard Schmidt (1879–1935) (courtesy of Hamburg Observatory). (**b**) Many years earlier Yrjö Väisälä (1891–1971) in Finland presented the same telescopic principle, but did not publish his results, and his idea was not known to Schmidt. (**c**) The 70-cm Schmidt telescope of Tuorla observatory in Finland (photo by Rami Rekola)

covers an area of 6×6 degrees of the sky. This Palomar Sky Atlas formed a basic tool for astronomy at observatories all around the world for decades. It became possible to study faint galaxies and how they cluster. For example, American George O. Abell (1927–1983) identified 2,700 clusters of galaxies. He realized that clusters form superclusters, just as de Vaucouleurs had found locally. However, because at the time the studies were mainly confined to inspection of how galaxies were scattered across the sky, without good distance information, a long debate ensued about the reality of such superclusters – many astronomers viewed these celestial patterns as chance superpositions, perhaps caused by the extinction of light by the lumpy cosmic dust.

Sometimes one phenomenon discovered in the universe offers us a way to investigate quite other important things. Such was the case with the Hubble law. Thanks to this profound cosmological regularity, the redshift of light from a galaxy can be used as a distance indicator, which is relatively easy to measure and which can then be used to map the real three-dimensional distribution of galaxies. This was pioneered by Mihkel Jõeveer (1937–2006) and Jaan Einasto in Estonia in late 1970s, when the number of measured redshifts had increased to about 2,000. They started making 3D maps of the galaxy distribution and pronounced their startling discovery in the first international conference devoted to the large-scale structure of the universe, held in Tallinn, Estonia, in 1977. The maps showed wonderful structures around us in space, in the form of long filaments and giant walls marking out a kind of honeycomb. Between these formations made of groups and clusters of galaxies there were huge voids that contained practically no galaxies. The sizes of the "cells" were as big as one hundred million light years across (about 30 megaparsecs), something like the size of the Local Supercluster.

Not all were at once convinced about the reality of such unexpected structures. Perhaps the redshift data gathered by various people for different purposes did not yet provide a good view of the galaxy distribution? It was necessary to produce well-planned redshift surveys, uniformly covering large regions of the sky, in order to check the Estonians' findings. The first such project was carried out at the Harvard-Smithsonian Center for Astrophysics (CfA) in the USA. This effort

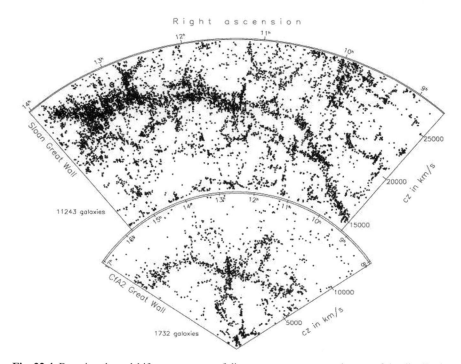

Fig. 22.4 By using the redshift as a measure of distance one can create pictures of the distribution of galaxies in space. The smaller map is based on the CfA work published in 1986 (see the text). Many redshifts were measured in a narrow strip of sky, and the galaxies were plotted in the figure using the velocity (distance) scale. The bigger figure is the SDSS map which goes deeper in space. Now similar structures tend to appear in view, though having larger sizes. The Great Wall of CfA and the Sloan Great Wall extend across these maps. Notice the large voids (courtesy of J. Richard Gott III and Mario Juric)

required measuring redshifts for 1,900 new galaxies. In 1986 in an article titled "A Slice of the Universe" the astronomers V. Lapparant, M. Geller, and J. Huchra confirmed the existence of shell-like galaxy clustering and found still more variety in the realm of galaxies. Their map (shown as part of Fig. 22.4) became a symbol of the complexity of the distribution of galaxies in space.

Inspired by these early results, several extensive redshift surveys have been performed for the new 3D cartography of the universe. The special technique of multi-object spectrographs is used to measure redshifts of many galaxies simultaneously. The largest such program is the Sloan Digital Sky Survey (SDSS) that is measuring the redshifts of one million of galaxies over a quarter of the sky, reaching the depth of about one and half billion light years. It uses a special telescope at Apache Point Observatory in New Mexico. The telescope itself is not a giant (its mirror is 2.5 m in size), but its state-of-the-art spectrograph can measure the redshifts of 640 galaxies in one exposure.

The Novel Realm of Large-Scale Structures

The new 3D maps have extended our view of the galaxy universe from the Local Supercluster up to scales tens of times larger. Our cosmic milieu outside the Milky Way has turned out to be surprisingly complex. One would be willing to characterize this structure by some familiar manner. Could it be just random variation? In fact, a smooth random distribution of points has a constant density, on the average, but with small variation from one place to another. The amount of variation should follow a law found by Siméon-Denis Poisson (1781–1840), physics professor at Sorbonne in Paris. "Poisson's distribution" is generated, e.g., when with eyes closed one sows grains over graph paper and then counts the grains in the squares. A majority of squares will have the number of grains close to the expected average (the total number of grains/the number of squares), a little more or less, while only a tiny fraction of squares have numbers much deviating from the average.

Such a distribution may well be true when one looks at very large volumes of space. It is clear, however, that on scales of tens and even hundreds of millions of light years galaxies are not "sown" in space just randomly – voids, superclusters, and giant walls are too conspicuous for that (Fig. 22.5).

Possibly there is no apt analogy on Earth for the structures of the galaxy universe, made by gravitation in unfamiliar cosmic conditions on huge spatial scales during very long times. Originally the Estonians spoke about "cell structure," while the CfA team compared the revealed landscapes with "soap suds," where between the bubbles there are plane surfaces. Still another model comes from the same familiar circle of life: one has also compared the realm of galaxies with a bath sponge. Its empty places are interconnected so that one can usefully squeeze air and water out of the sponge. If one lays sponges side by side, their insides are complex, but from one sponge to another the amount of matter is about the same. In this way one could represent the crossover from lumpy to uniform spatial distribution.

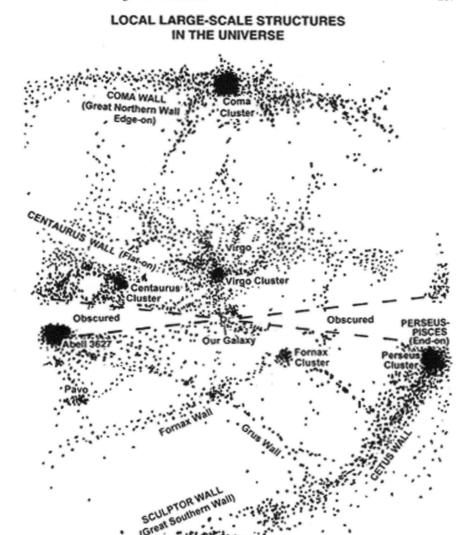

Fig. 22.5 A schematic map of our neighborhood in the galaxy universe within about 250 million light years (with permission from Anthony Fairall & Praxis Publishing Ltd, Chichester, UK)

Hierarchies and Fractals

The trend that the observed structures, both superclusters and voids, tend to increase when deeper maps are made is similar as in hierarchically distributed systems. Also the organization of galaxies into pairs, groups, clusters, and superclusters reminds one of a hierarchy. Such things were pondered already by the eighteenth-century thinkers (e.g., Wright, Kant, Lambert). Galaxies were not known, but the planets circling the Sun and the moons around them helped one to imagine larger systems. Even now we may describe our location relative to different levels in a cosmic hierarchy: We live in the Solar System that is located in the Local Spiral Arm belonging to a galaxy called the Milky Way. This galaxy is a member of the Local Group that in its turn is a part of the Local Supercluster which belongs to the Pisces-Cetus hypercluster that... Well, here we have reached scales of hundreds of millions of light-years, beyond which little is yet known in detail.

The old cosmic hierarchies imagined before galaxies were known were overly simple and stiff to describe the true complexity of the galaxy universe. However, they have a modern descendant that may give a more realistic picture. This is fractal, a mathematical entity introduced by Benoit Mandelbrot in the 1970s, which has now many applications in natural and humanistic sciences.[1] Fractals are systems, the parts of which are similar to the whole. A magnifying class reveals in such a *self-similar* system a new structure that looks similar to what one can see by plain eye. In other words, from a picture of a part of a fractal structure one cannot tell its real size! This is something like suggested by Fig. 22.4, and fractal analysis is now often applied to study how galaxies are distributed in space.

An interesting property of a fractal (and of the old hierarchies, too) is that when one looks at increasingly larger volumes, the average density progressively decreases. The rate of this decrease defines a numerical quantity characterizing the fractal: the fractal dimension. The quicker is the decrease, the smaller is the fractal dimension. In fact, the fractal dimension 3 corresponds to the situation when the average density remains the same, irrespective of the scale – this is a smooth random distribution, like that of the molecules of ordinary gas. A genuine fractal has the fractal dimension less than 3.

Though it is generally agreed that in some sense the spatial distribution of galaxies has a resemblance to fractal, its exact nature is still intensively studied. For example, there are different results on the value of the fractal dimension and on how deep in space fractality extends before there is a crossover to the expected smooth distribution. Some astronomers have concluded that the fractal dimension is about 2, and one really does not yet know where uniformity begins. However, many astronomers regard that on scales of a few tens of millions of light years the distribution is already almost uniform. Such differences in opinion tell

---

[1] Mandelbrot, the father of fractal, coined this term from the Latin adjective "fractus" that derives from the verb "to break" or to create irregular fragments. The fractal properties of the spatial distribution of galaxies have been especially much studied by the Italian Luciano Pietronero of the University of Rome and his team.

that it is not quite easy to study the organization of galaxies even from the large three-dimensional galaxy maps (such as SDSS).

Where Uniformity Begins

In 1934 Edwin Hubble finished his deep survey where he counted galaxies from 1,283 areas of the sky. Recall that there appear to be fewer galaxies in the Milky Way belt purely as the result of the cosmic dust blocking away light from distant galaxies in these directions. Hubble avoided those directions in his survey that elsewhere reached the depth of 6,000 million light years. The result was that whatever direction we look at, there is the same number of galaxies. This means isotropy around us and is very important for understanding the large-scale distribution of galaxies. If our location is typical – as told by the Copernican Principle – then every observer, no matter where he/she is, will also see similar isotropic landscapes. If so, then a mathematical theorem proved by the British mathematician Goeffrey Walker (1909–2001) in 1944 tells that on large scales the distribution of galaxies is uniform. In his book *The First Three Minutes* the Nobel-physicist Steven Weinberg gives a delightfully simple geometric proof that from "isotropy everywhere" follows "uniformity everywhere" (see Fig. 22.6).

For cosmology, the thermal background radiation is a key observation. We discuss it later, and here we just note the remarkable fact that its intensity is almost the same from every direction. It is believed that this radiation has its origin in the hot gas at a very ancient epoch, and its isotropy thus agrees with the view that the universe was on large scales uniform. This is why modern cosmology utilizes the world models of Friedmann, which describe a uniform and isotropic matter distribution (Chap. 23). Why then do we see the very lumpy distribution of galaxies around us?

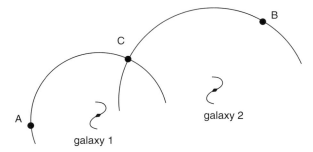

Fig. 22.6 From isotropy around each point it follows that the density on the circle around galaxy 1 is the same on each point of the circle, and on the circle around galaxy 2 the density is also constant. Because of the common point C, the density in fact is the same on both circles. Adding more circles around different points, one may conclude that the density is the same at any point (or the matter is uniformly distributed)

As we will discuss later on, the structure of the present-day galaxy universe is determined by tiny fluctuations of hot gas soon after the Big Bang. These served as seeds out of which gravitation gradually built the current structures small and big. A law, suggested around 1970 by Edward Harrison (1919–2007) at Massachusetts and Yakov Zeldovich at Moscow implies that there should be initial perturbations on all scales, with no end in sight. But it also asserted that the larger structures become less prominent with increasing size (on the very largest scales the relative density contrast should be inversely proportional to the square of the scale in question). Measurements of the background radiation have now revealed this spectrum, showing that its shape is close to what Harrison and Zeldovich predicted. Thus true uniformity may actually begin nowhere! However, on very large scales gravitation has not had enough time to disturb the original matter distribution much, and we expect that *strong* inhomogeneities and voids should end after some large, though still uncertain distance.

In the old CfA map one can see a long sheet-like formation called the Great Wall. It is 750 million light years long and 250 million light years wide. A still larger structure was found in a map from the SDSS survey, now simply termed the Sloan Wall after Alfred P. Sloan, whose foundation has financed the redshift survey. This is perhaps the largest known cosmic structure, a one and half billion light years long sheet of superclusters, clusters, and groups (see Fig. 22.4).

Recently radio astronomers may have found the biggest hole ever seen in the universe. It was discovered by Lawrence Rudnick and colleagues of the University of Minnesota in Minneapolis, who studied the distribution of radio galaxies and quasars in the direction of a cold spot in the cosmic background radiation. They saw little or no radio sources in a volume that is about a billion light years across. This means that there are no galaxies or clusters of galaxies in that volume. Typical voids seen in optical surveys are usually less than 100 megaparsecs (or about 300 million light years) wide.

Although there is strong indirect evidence that at some great distance the uniformity must set in, the discovery of such big structures as the Sloan Wall and Rudnick's void means that we do not yet know for sure on how large a scale the universe starts to look smooth, that is, where one can ignore the lumpy cellular, sponge-like or fractal structure as tiny ripples on the surface of an immense ocean.

Chapter 23
Finite or Infinite Universe: Cosmological Models

Now it is time to recall, from the first chapter, that the first cosmologist may have been the mysterious Pythagoras who originated the word *cosmos* as a term for an ordered universe. Geometrical forms and numbers became a part of the attempts to describe the whole world. If the cosmos is ruled by mathematics, it is possible to make models for our universe and to learn about its structure. As a matter of fact, one of the first applications of General Relativity was to the universe as a whole. This marked the beginning of modern cosmology. Before that our means for creating world models were limited, although different views of the structure of the world have always been considered.

Ancient Views

The universe envisaged by Aristotle was finite. Everything was inside the sphere of fixed stars. Its size was unknown, even though Ptolemy estimated that the sphere is 20 000 Earth radii distant. In this view the world above the Moon's sphere was different from the world below. A human made of ordinary matter had no place in the upper world. Outside this outer sphere there was nothing. If we try to visualize a universe like this, we fail; instead we mentally put it inside an even larger empty volume (Fig. 23.1).

Another answer to the riddle of the edge given in antiquity was that there is no edge since the universe is infinite in extent. This was the view of the atomists for whom everything depended on the complicated interaction of atoms and the resulting evolution of structure – even humans. These processes required plenty of time and space and were easiest to vision in an infinite universe. Lucretius, who lived in the first century BC (see Chap. 2), describes infinity in his book *On the Nature of Things* as follows: "*There is therefore a limitless abyss of space, such that even the dazzling flashes of the lightning cannot traverse in their course, racing through an interminable tract of time, nor can they even shorten the distance still to*

P. Teerikorpi et al., *The Evolving Universe and the Origin of Life*
© Springer Science+Business Media, LLC 2009

Un missionnaire du moyen âge raconte qu'il avait trouvé le point
où le ciel et la Terre se touchent...

Fig. 23.1 This famous woodcut which appeared in 1888 in a book by Camille Flammarion shows a man in his finite world peering through the celestial sphere, the enigmatic edge of the universe. The text in French tells about a medieval missionary who has found the point where the sky and the Earth meet

be covered. So vast is the scope that lies open to things far and wide without limit in any dimension."

Giordano Bruno was aware of Lucretius' thoughts, and was one of the first supporters of infinite space and countless stars in the Renaissance. His vision was that there are celestial bodies like the Earth in infinite number. In this respect he was well ahead of Copernicus, Kepler, and even Galileo, though it should be said that the nonastronomer Bruno did not have observational evidence for his ideas.

The third line of reasoning in antiquity was that the world was partly finite, partly infinite. In this view, our material world is like an island in an infinite universe. This was the idea of stoic philosophers who followed the teachings of Zenon (336–246 BC). The popular view in the nineteenth century of the Milky Way containing everything in it had some resemblance to the stoic ideas. On the other hand, the competing "island universe" theory was like the atomist view. The latter, of course, was found to be correct. But can we follow the atomists all the way and say that our universe of galaxies is infinitely large?

Newton and the Infinite Universe

Newton's law of gravity provided a mathematical starting point for cosmological considerations, but it also spelled trouble. Interesting letters between Newton and theologian Richard Bentley in winter 1692–1693 show the first elements of the new thinking. Bentley was looking to science for weapons in his fight against atheism. Science reveals rational laws of nature (such as gravitation), but do these imply the existence (or intervention) of a supernatural being? Newton, who himself was a deeply religious man and at the same time the greatest expert in physical science, was for Bentley the natural choice to ask for comments about deep origins.

Bentley asked Newton sharp questions, among them being how matter distributed uniformly in space would behave. Newton answered that matter could be in equilibrium if the gravitational pulls from different directions on every particle cancelled out. But this balance is extremely unstable. Newton compared the situation with needles (in fact, an infinite number of them!) standing on their points. Even a tiny imbalance would lead to catastrophic collapse. Thus the past and continuing existence of the universe of stars together with gravity appeared a finely tuned mystery. Newton admitted that it may be possible, at least by a divine power. This was what Bentley sought, and it also agreed with Newton's wish to see God's fingerprints in nature. Today we are less inclined to think that the existence of God could be argued from specific provisional enigmas of the physical world. In this respect many modern scientists (religious and non-religious alike) may rather feel more at home with mathematician Blaise Pascal (1623–1662) in his deep *Pensées*. For him, God is a hidden God, and he prefers not to look at "the heavens and birds" for a strict proof of His presence.

In 1895 Hugo von Seeliger concluded that under Newton's law of gravity, an infinite Euclidean universe with stars distributed uniformly throughout cannot be at rest. In fact, one cannot calculate a unique value for the force affecting a particle in any given point in space. Nature cannot remain in such an undetermined state. This new problem in the old world model inspired von Seeliger to propose a small modification into Newton's force law, a small extra weakening in the gravity force in addition to the inverse square law. The modification has a parallel in Einstein's later proposal to add the so called cosmological constant in his equations of General Relativity, in order to make possible his own finite universe model in a state of rest.

The Uniform Universe

The discovery of non-Euclidean geometries in the nineteenth century revolutionized this subject (Chap. 15). One could have a finite universe, and at the same time, avoid the embarrassing question of the edge of the universe. Thus the universe of galaxies may be either finite or infinite. A special case is a homogeneous and isotropic universe. Since we observe isotropy from our Milky Way (equal galaxy numbers in opposite directions), it is likely that the universe is also homogeneous on

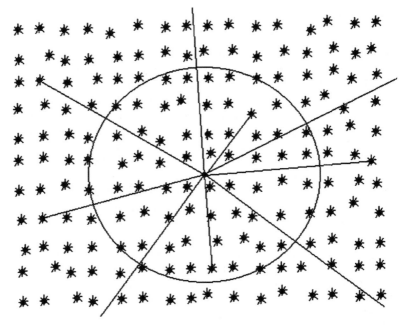

Fig. 23.2 According to Olbers' paradox, the night sky should blaze as brightly as the disk of the Sun, if the universe is infinitely large and infinitely old, because then every line of sight would encounter a star. In this scheme we are supposed to be at the center of the circle

sufficiently large scales, unless it is centered on us. The latter is against the principle of Copernicus.

In regard to the finiteness of the universe, there is one cosmological observation which can be made by naked eye and can be easily understood. Namely, it is obviously dark at night. However, if the universe were infinite in extent and full of stars, then every line of sight would sooner or later meet the surface of a star. And if we are looking at the surface of the star everywhere, then the sky would be as bright as the surface of the Sun, day and night. The fact that this is not the case is called *Olbers' paradox*.[1] So, what does the dark sky at night tell us?

There is a hidden assumption in the aforementioned reasoning, hidden in the phrase "sooner or later..." When we look far away, we look into past. It means that in order that every line of sight meets a star, there must be sufficient time in the past. In a universe of short age Olbers' Paradox does not exist. Thus the universe may well be infinitely large as long as it is of finite age. The night sky is illuminated by only a finite number of stars so that actually very few lines of sight meet the surface of a star (Fig. 23.2). The current view is that indeed the universe is "only" about 14 billion years old. This would be a "time edge" of the universe (Fig. 23.3). Aristotle's universe had the intriguing edge of the universe in space. A time-like edge in the past for some scientists is a conceptual problem of similar magnitude.

---

[1] Heinrich Olbers (1758–1840) was a German physician and astronomer. The paradox that he stated in 1823 was noted even earlier by some other astronomers (Kepler, Halley, Chéseaux)

Fig. 23.3 This "Ultra Deep Field" photograph taken by the Hubble Space Telescope shows, beyond the stars of our Galaxy, galaxies all over the sky and between them holes with little or no fainter galaxies. We can see only a finite (though big, over 100 billion!) number of galaxies, because the universe has a finite age and there has been not enough of time for the light to arrive from very distant galaxies (credit: NASA, ESA, S. Beckwith (STScI) and the HUDF Team)

It is interesting that the finite age solution for Olbers' paradox was already hinted at by the poet and writer Edgar Allan Poe in his cosmological prose-poem *Eureka*, published in 1848. He wrote: *"Were the succession of stars endless, then the background of the sky would present us a uniform luminosity, like that displayed by the Galaxy – since there could be absolutely no point, in all that background, at which would not exist a star. The only mode, therefore, in which, under such a state of affairs, we could comprehend the voids which our telescopes find in innumerable directions, would be by supposing the distance of the invisible background so immense that no ray from it has yet been able to reach us at all."*

Einstein's Finite Unchanging Universe

In 1917 Einstein extended the concept of curvature of space from the application to a single star to the whole universe. Gravity dominates in cosmological considerations. The view of gravity, space, and time as given by the theory of General Relativity is quite different from the previous concepts. Thus it is not surprising that since General Relativity came on the scene "the universe is not what it used to be." One of the most impressive monuments of this change was Einstein's model of a static, finite size but still boundless universe. How did Einstein arrive at this model?

In General Relativity theory, "matter determines the geometry of the spacetime while geometry determines how matter moves." Einstein and Karl Schwarzschild first applied the theory to the Solar System, and made the natural assumption that at large distances the influence of the Sun on the overall geometry vanishes. When we go far enough from the source of gravity, space assumes the same form as in the Special Theory of Relativity, i.e., becomes flat. This assumption was adequate for the description of spacetime around a single star. But what about the whole universe? In 1917, Einstein published a radically new world model. Prior to this there had been speculation about the curvature of space in the manner of a sphere, e.g., by Schwarzschild, but only now it became connected to physical reality. In his model Einstein wanted to bypass the difficulties associated with infinity. But in addition, the model was simple, and this appealed to Einstein whose thinking was guided by the need to see beauty in the fundamental simplicity of nature.

Einstein used *Mach's principle* as a guiding idea of his theory. According to Ernst Mach (1838–1916), the property of matter which opposes motion, called *inertia*, is due to the interaction of matter with the rest of the universe. Einstein thought that if a particle is very far from all other matter, its inertia or inertial mass, which appears in Newton's laws of motion, will vanish. He tried to build a cosmological model where inertia disappears far from the Milky Way. The task turned out to be overwhelming. Thus Einstein decided to bypass the problem of what happens to inertial mass infinitely far away by removing the concept of infinity from cosmology altogether. The geometry of his universe became bounded and closed, neatly curved about itself.

In developing his theory Einstein gave up the idea that the Milky Way is the only island universe, and assumed that matter is distributed uniformly, on average, throughout a much bigger cosmos. He compared himself with a geodesist who describes the average shape of the Earth by a sphere, neglecting all the details of hills and valleys. In the universe, stars and their clusters form the landscape, but Einstein chose to ignore the fine details. He assumed that the stars (galaxies were not known yet) are distributed uniformly in space and that they curve space equally everywhere, resulting in the finite "spherical" space.[2] The assumption that matter is uniformly distributed in space, at least when one looks at sufficiently large volumes, is now called the Cosmological Principle.

The other important property of Einstein's model, besides being finite in volume, is that it is static: on average stars are at rest relative to each other and the geometry is unchanging. At that time, astronomical observations did not contradict the static assumption. The recession speeds of some nebulae had been observed, but the

---

[2] As explained in Ch. 15, the three-dimensional spherical universe of stars has a 2D analogy of stars on the surface of a sphere. Do not confuse this with an actual sphere inside which stars are contained.

discussion about their significance had hardly begun. Einstein must have intuitively preferred the universe to be unchanging.

Einstein had to pay a price for an at-rest universe. Just as von Seeliger adjusted the Newton's theory of gravity in order to make an infinite static universe possible, similarly, Einstein had to add a so called *lambda term* (or the cosmological constant) to his equations. It can be interpreted as a universal *repulsion* which is unnoticeable over short distances such as the scale of the Solar System but which becomes important in the scale of the universe.

Einstein was not happy with this generalization of his theory, later remarking that the lambda term "was the biggest blunder in my life." Without it, he might have predicted the expanding universe well before its discovery by Hubble. Furthermore, the model was not an adequate solution to the demand of keeping the universe static. Arthur Eddington later demonstrated that the universe in Einstein's model was unstable and would start to contract catastrophically or to expand. Both Newton and Einstein had to admit that it is not easy to make a universe that is permanently at rest. In our days, the idea of cosmic repulsion is again a part of our cosmological world view, as will be discussed later.

Friedmann World Models

The models of the universe in standard use today were derived by the Russian Alexander Friedmann (1888–1925). Friedmann was a professor of mathematics at St. Petersburg University and a specialist in the newly founded General Relativity. He published his investigations in the leading scientific journal *Zeitschrift fur Physik* in 1922 under the title "On the curvature of space." Two years later his second article on the same topic "On the possibility of a universe with a constant negative curvature" appeared. These researches form a turning point in the study of cosmology, but they remained almost without notice. Friedmann became ill and died a year after the publication of his second article. In 1927 Georges Lemaître rediscovered the world models, now known as *Friedmann universes* (Fig. 23.4).

Friedmann showed that Einstein's equations possess nonstatic solutions which may describe the real world. He assumed, as Einstein had done, that matter is distributed uniformly throughout space, but he did not require that the density of matter should remain constant. Therefore, even though the curvature of spacetime is everywhere the same at a given universal time, it changes with time: the universe either contracts or expands. One of the Friedmann models has a special name, the Einstein–de Sitter universe after Einstein and the Dutch astronomer Willem de Sitter who discussed it in a 1932 publication. The density of matter in this model is such that the space of this universe is flat (Euclidean) at all times.

The "just right" density of the Einstein–de Sitter universe is called the *critical density*. When matter is spread out uniformly through space, a cube with sides of a million kilometers would contain just 9 kg of matter at the critical density. The real density of matter made of all massive celestial bodies is probably about one-third of

Fig. 23.4 (**a**) Alexander Friedmann and (**b**) Georges Lemaître founded the theory of the expanding universe in the 1920s

the critical density, so this number gives a good idea of the emptiness of universe. After all, the million kilometer cube would contain 10^{27} kg if filled with the "thin" air we breathe!

The Gallery of Possible Worlds

There are four basic types of Friedmann universes. For the first three types the cosmological lambda term is zero, and thus there is no universal repulsion. The types are: universes of spherical geometry, of hyperbolic geometry, and between them, the flat Einstein–de Sitter universe. In addition, universes where the lambda term is not zero form the fourth broad category. In reading the following description the reader may wish to refer to Fig. 23.6 and Table 23.1 where these are summarized.

For zero lambda, if the mean density of the universe is greater than critical, the geometry is spherical or *closed*. However, if the amount of matter is below the critical level, space is hyperbolic. The closed Friedmann model differs from the Einstein model in that it is not static. As a matter of fact, General Relativity theory tells us that a static space where galaxies are at rest relative to each other is not possible even in principle. The whole system of galaxies is either in a state of contraction whereby galaxies approach each other or in expansion whereby they recede from each other (Fig. 23.5). The situation is like when you throw a stone up in the air: the stone moves either up or down, but it cannot remain floating at a constant height.

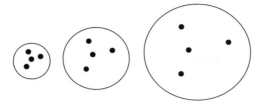

Fig. 23.5 The expanding universe may be likened to the surface of a balloon that grows constantly bigger. The dots represent galaxies that have been distributed over the surface more or less uniformly. When the surface expands, the distances between galaxies grow. Even though the dots are fixed to the surface, it appears that all other dots are escaping away from any given dot

A nonzero lambda can compensate for the matter density. A special case is the model where the lambda term exactly compensates for the matter density that is short of the critical level. Then the overall geometry is flat. The *concordance model*, which according to our current knowledge is closest to the reality, belongs to this category. The concordance, hyperbolic, and Einstein–de Sitter universes all extend to infinite distance, and thus they are called *open* universes. These universes contain infinitely many galaxies. The closed Friedmann model has a finite (though changing) volume just like in Einstein's 1917 static model, and it contains a finite number of galaxies.

Einstein was initially doubtful about Friedmann's results. In the same issue where Friedmann published his models Einstein inserted a five-sentence critique. There he claimed that in fact Friedmann had proved that the static model was the only possible one. In spring 1923 Einstein added four more sentences in the same journal where he admitted that his critique had been wrong – Einstein had made a small error in his calculation – and he now regarded "Mr Friedmann's results correct and clear."

The Hubble law mentioned previously was exactly the observational result that was needed to confirm the Friedmann models. The universe seems to be expanding. If the closed Friedmann universe happened to be the correct model (even though it does not appear be so) then the expansion would turn into contraction one day. Galaxies would fall on top of each other and finally all structures in the universe would be destroyed. In the open world model and in the Einstein–de Sitter model the expansion continues forever, even though it becomes gradually weaker. In the concordance model the expansion not only goes on forever, but it is also accelerating in pace (Table 23.1).

Table 23.1 Friedmann models of the universe

Model	Volume	Matter density	Lambda	Geometry	Future evolution
Open	Infinite	<critical	Zero	Hyperbolic	Decelerating expansion
Einstein–de Sitter	Infinite	=critical	Zero	Flat	Decelerating expansion
Closed	Finite	>critical	Zero	Spherical	Expansion–contraction
concordance	Infinite	<critical	>zero	Flat	Accelerating expansion

Since the galaxies now escape from each other, they must have been closer in the past and long ago right next to each other. It means that an expanding universe must have a finite age. There must have been an initial event, a *Big Bang* which has put the matter in the universe in the state of expansion.

The Accelerating Universe

The reason why the universe may be accelerating is in the possibility of the cosmological lambda term of Einstein's equations. Whether there actually is a nonzero lambda, in other words whether the general repulsion of gravity (or antigravity as it is sometimes called) exists, is a question of observational determination. For a long time the evidence for repulsion was considered weak or nonexistent, so the nonzero lambda term was usually neglected.

This changed in the late 1990s when it had become possible to study very distant stellar explosions, supernovae (these are discussed in Chap. 19). Observationally, it was found that the highest brightness of one type of supernovae (SNIa) is rather constant from one explosion to another, meaning that the supernovae can be used as standard candles, like light houses in the vast sea of galaxies. Since the power of each light house is known, we can estimate its distance based on how brightly it shines in the sky. Then we can construct a Hubble diagram like the one shown in Fig. 23.7. The way the line curves at very large distances tells us about the correct world model.[3]

Several teams worked on the supernova standard candle problem in 1990s. One is the Supernova Cosmology Project, led by Saul Perlmutter of the Lawrence Berkeley National Laboratory in California. Formed in 1988, it included both astronomers and physicists associated with the laboratory. To discover supernovae, the group used wide-angle cameras attached to large telescopes.

Another group was the High-z Supernova Search Team led by Brian Schmidt from the Harvard-Smithsonian Center for Astrophysics. To get detailed brightness observations after discovery, both teams used the Hubble Space Telescope and the biggest ground-based telescopes, including the 10-meter Keck Telescope in Hawaii. By 1997 the team had discovered 16 high redshift supernovae, and they gave the first indication of an amazing discovery: the universe is accelerating. The supernovae appeared fainter, and thus they were more distant than they should have been in a decelerating universe. The most obvious explanation was that the correct model for the universe must contain Einstein's positive lambda term, that is, there is antigravity after all!

The second group reported their result at the January 1998 meeting of American Astronomical Society. At the same meeting, also the first group gave tentative evidence for the cosmic acceleration. Both results were immediately reviewed in

---

[3] The maximum brightness of the supernovae of type Ia depends on how quickly the brightness fades away. This property, first noticed by Yu. Pskovskii of Moscow University in 1977, serves to increase their accuracy as standard candles.

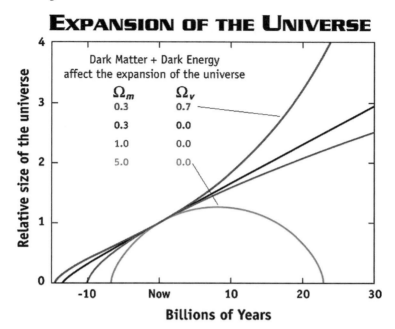

EXPANSION OF THE UNIVERSE

Fig. 23.6 Different "universes" plotted as a function of time. The vertical axis can be thought of as average separation between typical galaxies versus time. The top plotted line is the currently popular nonzero lambda model with accelerating expansion. The lower curves have zero lambda and no acceleration. The next lower line is a "hyperbolic" model with only the deceleration due to gravity which has no important effect on the expansion. The third lowest curve is a critical density model where the expansion and the deceleration balance. Finally, the bottom curve corresponds to a model whose density is sufficiently large, so gravity overcomes the expansion and it reverses bringing the galaxies together again (credit: NASA)

Science magazine, and later in the year the second group with Adam Riess of University of California at Berkeley as first author reported the work in the *Astronomical Journal*. Among the many model parameters they were able to determine was the age of the universe, close to 14 billion years (Fig. 23.7).

The work of the Supernova Cosmology Project appeared in 1999 in the *Astrophysical Journal*. It was based on an independent set of 42 high redshift supernovae and confirmed the findings of the High-z Supernova Search Team. It is a rare event when an important scientific discovery is made and verified "beyond reasonable doubt" within one year. Even members of the teams did not expect their results. Brian Schmidt said: "*My own reaction is somewhere between amazement and horror; amazement because I just did not expect this result, and horror in knowing that it will likely be disbelieved by a majority of astronomers who, like myself, are extremely sceptical of the unexpected.*"

Continuing high redshift supernovae observations have confirmed the acceleration. The clinching evidence for the accelerating universe has come from totally

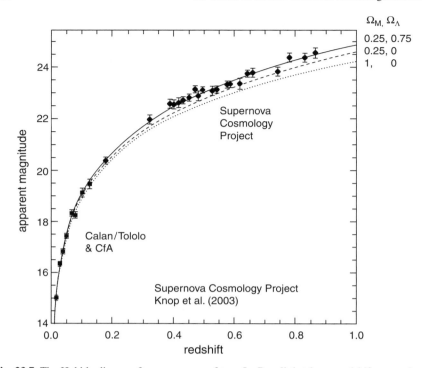

Fig. 23.7 The Hubble diagram for supernovae of type Ia. Recall that larger redshifts mean larger distances and dimmer supernovae. The top plotted line is the concordance model with accelerating expansion due to vacuum energy. The lower curves have no acceleration. Especially, the bottom curve is a model where the matter has the critical density and there is no vacuum energy. It does not agree with the observations (credit: Supernova Cosmology Project)

different observations, cosmic microwave background measurements by a satellite called the *Wilkinson Microwave Anisotropy Probe (WMAP)* in 2003, named after one of the pioneers of this field, David Wilkinson (1935–2002) of Princeton University. This has led to acceptance of the concordance model as the leading picture of our universe. In 2007 Samuel Perlmutter and Brian Schmidt, together with their teams, received the prestigious Peter Gruber cosmology prize for their discovery of the accelerating universe.

Redshift and Cosmic Distances

When we talk about the distances of galaxies we usually mention their redshift (symbol z) which for nearby galaxies is proportional to their distance (Hubble law). The redshift of a very distant object is directly measurable from the spectrum, but

Table 23.2 Redshift, light travel time distance, and "distance now"

Redshift	r_{light} (billions of light years)	r_{now} (billions of light years)
0.2	2.4	2.66
0.5	5.1	6.24
1	7.89	11.04
2	10.56	17.51
3	11.74	21.60
4	12.38	24.44
6	13.03	28.21
10	13.51	32.39

The distances are valid in the Friedmann model where the Hubble constant = 70 km/s/Mpc, space is flat, matter fraction = 0.24, dark energy fraction = 0.76.

its distance must be inferred in some other way. Even the concept of distance can be defined in different ways.

One possibility is to use the light travel time which tells how long the light has been in the way to us. This gives the distance in light years. In order to calculate the required "look-back time" we have to use a Friedmann model which assumes the exact age of the universe as well as its composition (one often uses the current best model, the concordance model).[4] The light travel time distance tells us of the remoteness of the time when the object emitted the light that we now receive.

Another way to define distance is to give it instantly at the moment when the light arrives at us (in the balloon analogy of Fig. 23.5, the distance measured with a measuring tape along the surface between two points). This distance can be conveniently compared with the distances between galaxies today. It gives us an idea of the depth of space at which the object lies. For example, it tells us how many our Galaxy-Andromeda galaxy distances there are between us and the object. This kind of distance is rather close to our usual concept of distance. However, we cannot actually measure such a distance (we cannot stretch a tape between us and a distant galaxy!) – it can only be inferred from the redshift of the galaxy together with an adopted Friedmann world model. Also the calculation of the light travel time distance requires both the redshift and the world model. This reminds us that in cosmology the world model is not only a theoretical construction for understanding the universe, but also a practical instrument without which we cannot tell the distance to a distant celestial body (and without the distance we cannot derive its size or radiation power).

Let us take a galaxy at redshift $z = 2$ as an example. From Table 23.2 we can read that light left the galaxy about ten billion years ago. We can also calculate that

---

[4] Let t_1 be the time when light leaves a distant object and t_0 the present time when we receive the light. Then the light travel time distance is $r_{light} = c(t_0 - t_1)$, where c is the speed of light.

at the present cosmic time this galaxy lies at about 7,000 times farther from us than the Andromeda galaxy (whose distance is 2.5 million light years). The table gives values of the light travel time distance and the "distance now" for certain redshift values. The unit is in both cases one billion light years, and we use the standard world model where the age of the universe is 14 billion years.

Astronomers can now make rather routine observations of galaxies up to redshift of about 0.5 which corresponds to the universe at 64% of its present age. With greater difficulty one observes now galaxies up to about $z = 3$ when only 16% of the current age of the universe had passed. At redshift $z = 10$ merely 3.5% of the 14 billion year age of the universe had passed.

Topology of Space: Still Another Cause of Headache

The flat infinite Friedmann models seem to work well. However, we will end by discussing intriguing conjectures about whether the universe might be flat but finite, and with a finite number of galaxies.

Alexander Friedmann wrote that "distorted ideas have spread about the finiteness, closeness, curvature and other properties of our space which are supposedly established by the Relativity Principle . . . I mean the notorious question of the finiteness of the universe, i.e., of the finiteness of our physical space filled with luminous stars. It is claimed that having found a constant positive curvature of the space one can conclude that it is finite, and above all that a straight line in the universe has a finite length, that the volume of the universe is also finite, etc." He wanted to emphasize that even though in General Relativity the curvature of space is a key quantity, its measurement does not necessarily tell about the global shape and volume of space. The topology[5] of space is a separate question, and it cannot be derived from General Relativity: there is no simple one-to-one relation between the overall design of space and its curvature.

In the quoted book *The World as Space and Time* which was published in Russian in 1923, two years before his untimely death, Friedmann gives a pedagogical example. The two-dimensional geometry of the surface of a cylinder and the geometry of a plane are identical: both are 2D Euclidean spaces. A cylinder can be glued out of a plane, and if one draws a triangle on the plane, nothing special will happen to it when the sheet edges are glued together. The sum of the angles of the triangle will remain equal to two right angles, and the Pythagorean theorem which holds for the Euclidean plane preserves its validity for the cylinder.

But topologically they are different: on the cylinder, there are "straight lines of finite length," while there are no such lines in a plane. The cylinder has a finite size in the directions normal to its axis, so it is finite and closed in these directions. It is infinite in the directions parallel to its axis. Using the plane and the cylinder,

---

[5] Topology is a branch of mathematics studying, among other things, the properties of geometric figures or solids that are not changed by homeomorphisms, such as stretching or bending. In this sense doughnuts and picture frames are topologically equivalent, for example.

Friedmann guides the reader to the conclusion: "Thus, the geometry of the world alone does not enable us to solve the problem of the finiteness of the universe. To solve it, we need additional theoretical and experimental investigations."

After Friedmann's remark early in the last century about "additional investigations," it may be said that up to now no general theory is available that would relate the topology of spacetime to its matter content (mathematicians can tell that 18 different versions of topology are possible for Euclidean flat geometries!). Nevertheless, there are ways to approach the problem observationally. For instance, multiple "ghost" images of the same object might be observed in the sky in a topologically closed space of a finite volume, because the light from a luminous object may reach the observer in various ways. It may come, say, from the opposite direction in space passing around the world, so that we shall see the same object in two diametrically opposite directions. No such effect has yet been observed.

A closed topology of space will also leave its ghostly fingerprint in the cosmic background radiation. The first observational indications about spatial topology were discussed in this way by Jean-Pierre Luminet and his associates in Paris in 2003. They studied the topological information contained in the largest angular scale variations of this radiation. The very largest scale of variation is the dipole. With the angular scale of 180° it is unobservable because the Doppler effect due to our motion relative to the universe (to be discussed in Chap. 24) creates a dipole effect which is 100 times stronger than what is expected from topology. The largest observable scale of variation is the quadrupole with an angular scale of 90°. The recent WMAP data show that the quadrupole variation is only about one-seventh as strong as would be expected in an infinite flat space. The octopole with the angular scale of 60° is about 70% of that expected in an infinite space. For smaller angular scales no weakening effect was found.

The small variation power on the angular scales wider than 60° could mean that the broadest spatial scales are missing, and Luminet suggests that this is because space itself is not big enough. The case may be compared to a vibrating string fixed at its two ends: the maximum wavelength of an oscillation is twice the string length. Luminet studied a concrete model of a finite-space universe. This model has the exotic name the Poincaré dodecahedral space, but it is familiar for topologists. To get a rough mental image of this space, one may notice first that any ordinary sphere may be completely covered by 12 regular spherical pentagons which fit together snugly. Each of them is a pentagonal part of a sphere. An ordinary Euclidean pentagonal dodecahedron is a solid with 12 equal flat faces (see Fig. 6.4); here the faces are pieces of a spherical surface.

Let us turn then to a hypersphere (as Einstein's finite, but borderless 3D world). To cover a hypersphere, 120 regular spherical dodecahedra are needed. They can be fitted together snugly, and each of them is a dodecahedral piece of a hypersphere. The Poincaré dodecahedral space is made of those spherical dodecahedra. It is not at all easy to imagine such a space, but technically speaking the Poincaré space is a positively curved space with multiply connected topology.

Our simplified two-dimensional model of the universe would now be, instead of an expanding balloon, an expanding football (soccer!) where "our world" is one

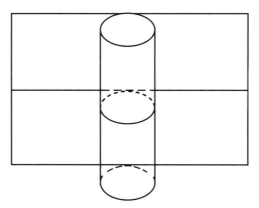

Fig. 23.8 A cylinder can be made from a flat rectangle. The surface of a cylinder and the plane have the same intrinsic Euclidean geometry, but their global structure or topology are quite different

of the 12 pentagonal patches. Now we might think that we can cross the border and visit the neighboring "world." Not so in the Poincaré space! The opposite faces of a dodecahedral block are connected to each other so that when light goes out through one face, it strangely re-enters from the face on the opposite side. It is like the example of a sheet of paper rolled into a cylinder (Fig. 23.8) so that the opposite edges of the sheet are glued together. We are aware only of one block of the dodecahedral space (see Fig. 23.9).

Calculations for the model by Luminet and his associates using a complex computer program show that the observed pattern of cosmic microwaves is well reproduced, if the present-day cosmological curvature radius has a special value. Such a finite universe would contain a finite amount of energy, and the total number of

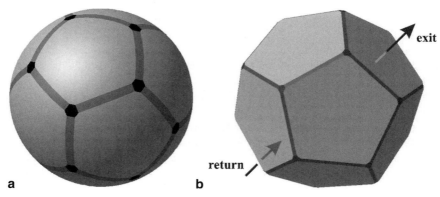

Fig. 23.9 (**a**) The football model of our expanding universe, with 12 pentagonal parts representing separate "worlds." (**b**) One dodecahedral block in the much more complicated Poincaré space: the light ray going out through one face, immediately re-enters from the opposite face (credit: Jean-Pierre Luminet)

stars and galaxies would be finite as well. However, this exciting idea is still conjectural, and to prove that we actually live in a flat, but topologically finite space it is important to verify the behavior of the background radiation on the angular scales wider than 60 degrees. This requires more accurate measurements by future space probes, such as the forthcoming *Planck Surveyor* of the European Space Agency, to be launched in 2009. In any case, Luminet's idea shows that modern cosmology may offer surprises even in those regions of space where the good old infinite Friedmann model at first sight seems to do its job quite well.

Chapter 24
When it all Began: Big Bang

What was the Big Bang like? This question puzzled Georges Lemaître (1894–1966) as early as 1931. As an ordained priest (and the professor of astronomy at the Louvain Catholic University in Belgium) the deep origins and creation of the universe had natural fascination to him, though he saw science and faith as separate things. In 1927 in a fine theoretical study, he pointed out the expected redshift in the light of remote galaxies and its dependence on the distance (the Hubble law). He talked about *l'atome primitif*; it was like a big radioactive nucleus that started to decay. He surmised that "the very origin is not reachable even by thinking, but that one could approach it in some asymptotic way." At that time most astronomers did not consider it proper even to try think about problems of the first origins.

Deducing the Existence and Properties of the Hot Big Bang

Actual research on processes during the Big Bang was originated by George Gamow who studied under Friedmann at St. Petersburg University and first obtained fame with his studies of quantum physics (tunneling and alpha-decay). In the 1930s he "tunneled" away from the Soviet Union and moved to the United States and worked at George Washington University. Younger colleagues Ralph Alpher and Robert Herman worked with Gamow. They asked what the initial, very dense matter was like and reached two significant results: the original matter had to be very hot and thus must have produced strong radiation, and secondly, this radiation should be still around us, even though it should have weakened into a faint afterglow of the Big Bang.

These ideas can be roughly understood by extrapolating backward trends seen today. Stars are forming from gas clouds. Thus in the past there was a greater proportion of gas versus stars in galaxies. Far back in time, the galaxies must have been all gas. Today, the galaxies are all seen to be flying away from one another, hence the early all-gas galaxies must have been jammed next to one another. Still farther back in time, the gas must have been hotter before it expanded. At some time in the past

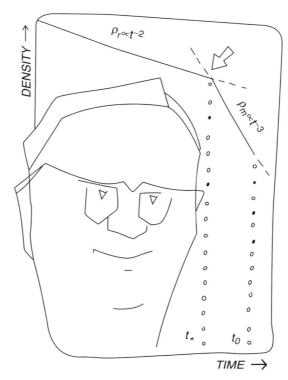

Fig. 24.1 George Gamow (1904–1968), the founder of the hot big bang theory. As time goes by, space expands and the density and temperature of the universe decrease (drawing by Arthur Chernin)

this gas was so dense and hot that it was opaque. After that era space has been transparent. The radiation emitted at the time of transition has been traveling through space until now, though greatly cooled by the expansion of the universe (Fig. 24.1).

Creating Light Elements in the Big Bang

As mentioned earlier, Cecelia Payne-Gaposchkin discovered that the matter in stars is mostly hydrogen with some helium and tiny amounts of heavier atoms (interstellar gas shows similar proportions). How did these elements come to be? Gamow's goal was to explain the origin of all elements in the Big Bang. In 1946 he suggested that initially all matter was composed of neutrons. When two neutrons collide, a deuterium nucleus can form, and after two more neutron collisions with the nucleus, a helium nucleus is born. Gamow assumed that under suitable conditions the process can go on until nuclei of up to 250 atomic mass units arise. The calculations showed how high densities and temperatures are required for this process. Alpher and Herman inferred that the leftover radiation from the Big Bang should today be like radiation emitted by a body at the temperature of $-268°C$ (or 5 K).

After a few years it became clear that the elements beyond helium cannot be created via neutron capture, because heavier nuclei break up back to light nuclei. Moreover, the observed abundance of elements heavier than helium varies as much as 100-fold from one star to another. If the heavy elements were born in the beginning, they should appear in the same proportions everywhere in the universe in all stars. One had to find some other furnace for making them.

In 1956 Fred Hoyle, working with his American colleague William Fowler (1911–1995) and with English astronomers Margaret and Geoffrey Burbidge, demonstrated that elements heavier than helium are naturally born in nuclear reactions in the hot interiors of stars. They found out how much of each element is formed in various stages of stellar evolution, and what fraction of it is carried back to the interstellar gas clouds. The exciting result was that in such a process the fractions of elements came out correctly, corresponding to the abundances of elements as observed in nature.

Furthermore, Hoyle showed together with Roger Tayler that all helium could not have been born in stars. Namely the production of all the helium, about one-fourth of all matter, in stellar fusion reactions would have produced far too much radiation to agree with the brightness of galaxies – at least 90% of the helium must come from somewhere else. With the Big Bang taken into account, the calculations agreed well with the observed amount of helium. We have already told about these processes inside stars and the expulsion of the elements into interstellar space via supernova explosions (Chap. 19) (Fig. 24.2).

Fig. 24.2 Fred Hoyle (1915–2001) during his visit to Finland in 1982 (photographed by Markku Poutanen)

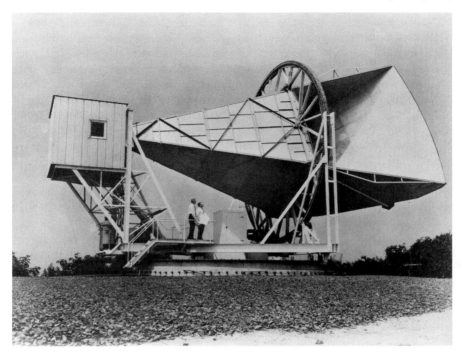

Fig. 24.3 The horn antenna with which Penzias and Wilson discovered the cosmic background radiation (credit: NASA)

Cosmic Background Radiation

At the same time that Hoyle and Tayler calculated the consequences of the Big Bang in England, at the other side of the ocean Robert Dicke and others at Princeton started to look for the remnant radiation. His younger colleague Jim Peebles figured out theoretically the nature of the radiation, while other team members built measuring instruments. But even before the search started, quite by chance the radiation was discovered elsewhere. Arno Penzias and Robert Wilson of Bell Laboratories were studying the radio noise that disturbs telephone links. They found that some of the noise comes from outside the Earth and possibly even from outside the Milky Way. Thus there was not much that the Bell telephone company could do to get rid of this noise, but what was its cause? Penzias happened to hear about a seminar in which Peebles had a short time earlier described the expected afterglow of the Big Bang. Its properties agreed well with their observed radio noise. Thus in 1965 the *cosmic background radiation* had been discovered. Penzias and Wilson received a Nobel Prize for the discovery (Fig. 24.3).

The cosmic background radiation is distributed among different wavelengths following the blackbody spectrum (Fig. 24.4). As we learned earlier, such a spectrum is described by a single number: the temperature. The higher the temperature, the

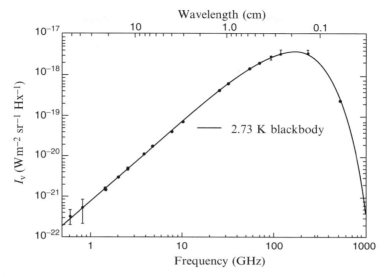

Fig. 24.4 The intensity of the blackbody radiation of temperature 2.73 K (or $-270.42°$C) versus the wavelength (*solid line*), as compared with the COBE observations of the cosmic background radiation

shorter the peak wavelength of emission. The observed peak of the background radiation is in the microwave region corresponding to 2.7 K temperature. The black body law was reliably proved in 1992 by the American space laboratory COBE (Cosmic Background Explorer). In 2006, John Mather and George Smoot shared the Nobel Prize for this result.

The spectrum is as expected from the hot gas existing after the Big Bang. Another clue to the origin of the radiation is given by the fact that its distribution is almost entirely uniform all around the sky (i.e., isotropic). The radiation is slightly enhanced (its temperature highest) in the direction of the constellation of Leo, while the lowest temperature is measured on the opposite side of the sky. Such a distribution reflects the motion of the Earth through a uniform radiation field. The Doppler effect makes the oncoming radiation brighter and warmer than the radiation coming from behind. Its measurement allows us to determine the motion of the Earth through the universe. More exactly, we can measure the speed relative to this radiation that in itself was born with the same intensity in different parts of the universe and forms a natural unique frame for measuring motions (do not confuse this with the failed attempts to measure our absolute motion relative to the ether in the nineteenth century).

From the motion of our Earth through the background radiation we can infer the motion of our Local Group of galaxies: it appears to travel toward the southern constellation of Hydra with the speed of about 600 km/s. In fact, we take part in a wider collective galaxy stream toward that direction. Part of the stream is apparently caused by the nearby massive Virgo cluster, but the gravity of some more distant and larger masses has been pulling our surroundings for the whole age of the universe, thus giving it the high speed. As we saw earlier, the large-scale structure is characterized by superclusters of galaxies. Beyond the Local Supercluster centered on the Virgo cluster, among the nearest large masses is the Hydra-Centaurus supercluster not far from the direction of our motion. It or other complexes beyond it may have given rise to the galaxy stream we are taking part in.

Temperature, Matter, and Radiation

Earlier in cosmic history the background glow was warmer than the very cool 2.7 K today. When the universe expands, the wavelength of any radiation is stretched along with it. It is a remarkable fact that during this process the blackbody nature of the spectrum is preserved, while the temperature is lowered. In fact, the temperature is inversely proportional to the scale of the universe.

The temperature of the ancient gas from which the background glow originates has been estimated to be about 3,000 K. Since the departure of the radiation toward us, the universe must have expanded by a factor of $3,000/2.7 = 1,100$ as a round figure. The age of the universe was about 400,000 years at that time. There was also another important event around the same time. Up to those times, radiation had been the dominant cosmic "element." A little earlier the matter component had taken over. The cosmic microwaves carry messages from the times when the changeover had just happened.

Using Einstein's formula ($E = mc^2$) we may calculate the energy content of the matter in a cube and compare it with the energy of radiation in the same volume. These two dissimilar forms of energy react to the expansion of space differently: radiation dilutes faster than matter. It might seem of little importance whether the cosmic energy was in the form of radiation or matter. It is not so. Only matter can make structures, radiation just spreads uniformly. In a world ruled by radiation, no real structures, including ourselves, could exist. Radiation would have blown apart any attempted gatherings of matter.

Astronomical Time Machine

Astronomical observations look straight into history. The further the light is coming from, the earlier is the history that it reports to us. Cosmic microwaves bring messages from the time 14 billion years ago. They tell about a major event in the universal history, about the *birth of the first atoms*. Before that, electrons and atomic nuclei had wandered about independently (i.e., the gas was ionized). After the temperature and density had dropped low enough, the electrons could take up their positions in orbits circling the nuclei in a stable manner. With the electrons tucked away into

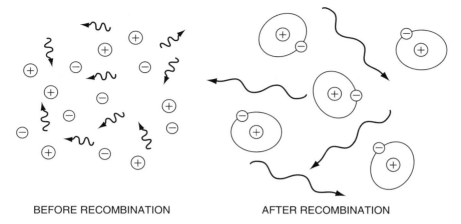

BEFORE RECOMBINATION   AFTER RECOMBINATION

Fig. 24.5 (**a**) Earlier in the history of the universe photons of the cosmic radiation prevented protons (*plus*) and electrons (*minus*) from forming permanent hydrogen atoms. (**b**) When the radiation weakened, atoms formed. At the same time the space became transparent since photons can travel past atoms relatively freely

atoms, the universe became transparent, and light could carry forward messages. At the present time the news about this ancient event causes a small part of the noise that disturbs radio and television programs (Fig. 24.5).

Besides telling about the birth of hydrogen atoms, the background radiation is a cosmic fossil record of the stage of evolution of structure at that point in time. Reading the record is not simple: the structural features are weak, at the level of 0.00001 of the radiation intensity. Their first detection required satellites orbiting the Earth. The pioneers in this enterprise were the Russian Relikt-1 space laboratory and the American COBE. The results from the COBE experiment were announced in April 1992 by the team led by George Smoot of University of California at Berkeley (later it was confirmed that Relikt-1 saw similar features, even though less clearly). A dramatic improvement to the measurements was provided by the American WMAP satellite in 2003, but there was a stream of discoveries even before that from balloon flights and from Earth-based observatories in suitable climates (such as Antarctica).

Measuring the Geometry of Space

Theoreticians expected that the most prominent patches of excess radiation in the microwave sky should appear at about the scale of the Moon in the sky. In fact, the sizes of these patches depend on the geometry of the universe. As we explained in Chap. 15, the angle at which a remote object is seen depends on the curvature of the space. In a spherical space the object appears larger than it would be in a flat, Euclidean universe, while in a hyperbolic space it tends to appear smaller. In this

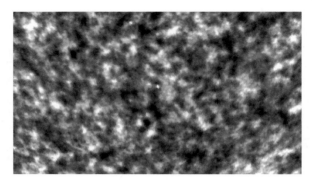

Fig. 24.6 Tiny temperature variations of the cosmic background radiation as measured by the Boomerang experiment inside a 10×20 square degree region of the sky. The typical angular size of the structures, about one degree, allows one to conclude that the spatial geometry of the universe is flat (credit: The Boomerang Collaboration)

way, by measuring the sizes of the microwave patches an accurate measure of the overall geometry is achieved (Fig. 24.6).

The first indication of the existence of such preferential size patches came from a telescope at Saskatoon in 1993–1995. The crucial measurements were made from

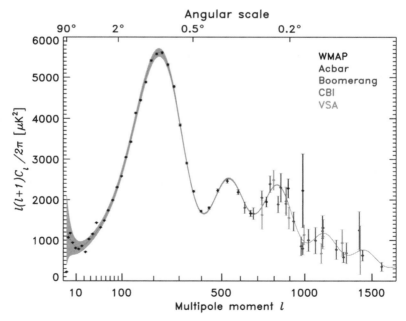

Fig. 24.7 A comparison of the best cosmological model of infinite extent and flat geometry (*solid line*) with the measurements of the cosmic microwave background radiation by the WMAP space probe and other instruments (points with error bars). The angular scale of the variation is given on top of the figure (credit: NASA)

balloon flights in 1998 in two programs.[1] They showed that the preferable scale of the background radiation patches corresponds to a flat universe. Finally, WMAP (a large collaboration headed by Charles Bennett from Goddard Space Flight Center and Johns Hopkins university) confirmed the previous results with much higher accuracy and measured the parameter omega $= 1.02 \pm 0.02$. The value of omega $= 1$ signifies a flat universe, values greater than 1 belong to spherical spaces, and omega less than 1 implies hyperbolic geometry. Thus our space may be exactly Euclidean, and any deviation away from flatness must be small (Fig. 24.7).

The Origin of Helium

A good part of the first 100,000 years of the cosmic history passed under radiation domination. The universe was different, but simple: it was filled with uniform gas at uniform high temperature. The temperature and the density decreased as the universe expanded. The chances of the radiation having universal influence have become less and less since the era when radiation ruled. But going back in time, a few minutes after the Big Bang the temperature was higher than in the center of the Sun. It is impressive to think that all over the universe nuclear reactions took place which were similar to those generating the energy of the Sun today. Protons and neutrons fused to make deuterium nuclei that became helium nuclei after further collisions with each other and with protons.

The amount of helium generated depends primarily on the relative number of neutrons in relation to protons. When 100 s had passed since the Big Bang and the temperature was a billion degrees, there were six neutrons to every 42 protons. Those six neutrons fused with six protons to make six deuterium nuclei which then turned into three helium nuclei. The end result was 36 hydrogen nuclei (protons) to every three helium nuclei. The relative proportions of helium and hydrogen (as mass fractions) are thus $4 \times 3/48 = 25\%$ for helium and $36/48 = 75\%$ for hydrogen (recall: a He nucleus is four times heavier than a H nucleus). When the reaction was over 200 s after the Big Bang, when temperature was 700 million K, these helium–hydrogen fractions remained in every part of the universe.

The First Second

Let us push further back. Since the Big Bang, our present time is about 14 billion years, the first atoms were born at 400,000 years, and helium was formed in its full amount at about 3 min. During the first second the universe is made of practically

---

[1] *Boomerang* (= Balloon Observations of Millimetric Extragalactic Radiation And Geophysics) led by A. Lange (California Institute of Technology) and P. de Bernardis (Univ. of Rome), and *Maxima* led by P. Richards (Univ. of California, Berkeley). The Saskatoon telescope was operated by the German Max-Planck-Institute and the American Princeton group.

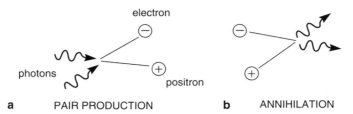

Fig. 24.8 (**a**) A sufficiently energetic photon may create a particle–antiparticle pair, for example an electron and a positron. (**b**) A photon of high-frequency radiation is born in a collision of a particle and an antiparticle. In the early universe these reverse processes were in equilibrium with each other

equal amounts of matter and antimatter. In contrast, the present world is made of matter, while antimatter particles are rare and short lived. When a particle and its antiparticle collide, both are lost and become radiation. In today's world it is not easy to be an antiparticle: there is a particle lurking around every corner ready to do away with the antiparticle.

How is it then possible that antiparticles could exist during the first second? The answer is that during that period the radiation was so bright and energetic that new particle–antiparticle pairs could be continuously created from the radiation quanta. The process is the reverse of the destruction of particle–antiparticle pairs. The reverse processes are possible since matter and energy are interchangeable according to Einstein's formula $E = mc^2$. When the temperature is high enough, both the destruction and creation of particle–antiparticle pairs happen at the same rate, and it is possible to maintain the equilibrium between them (Fig. 24.8).

When the universe expanded and the temperature decreased, at some point the creation side became impossible, and the destruction side caused a mass murder of the particle–antiparticle pairs. The fact that any particles survived at all is explained by a *small* asymmetry, by having just a small excess of particles over antiparticles. It has been estimated that for every 1,500 million antiparticles there was exactly 1,500 million and one particles. When the 1,500 million particles eliminated the same number of antiparticles, there was still one particle left that could later on become part of the structures of the universe. The remainder of the destroyed particles survived as radiation.[2]

Each kind of particle has its antiparticle, but it depends on their mass how long after the Big Bang they could exist in equal numbers. Light particle–antiparticle pairs can be created from less energetic photons and at lower temperature than heavy particle–antiparticle pairs. The temperature, above which the particle–antiparticle balance is possible, is the *threshold temperature* of this particle. The electron and its antiparticle, the positron, are the lightest particles (we discount the neutrinos whose masses are much smaller but uncertain). The threshold temperature of the electron is 10 billion degrees. The last time when the universal temperature was this high

---

[2] The tiny asymmetry between matter and antimatter is a problem. Apparently Nature does not differentiate between matter and antimatter; why then start out with the small asymmetry in favor of matter? Fundamental physics cannot yet give a reliable answer to this question. Of course, thanks to this favoritism we exist!

was 1 s after the Big Bang, giving this point in the time axis special significance. It was around this time or a bit later when the last annihilations between electron and positron occurred – now the cosmic stuff was basically free of antimatter.

Leptons were the most common particles in the universe during the period between $0.0001\,s(=10^{-4}\,s)$ and 1 s. In that interval or less, electrons and positrons were continually being created and destroyed but numerous. Therefore this period is called the *Age of Leptons*. (Recall the three main classes of subatomic particles: *leptons*, *hadrons*, and *photons*. The leptons include electrons, muons, and neutrinos. Hadrons (including baryons and mesons) consist of much smaller parts, quarks.)

From $0.00001\,s(=10^{-5}\,s)$ to $10^{-4}\,s$, the more massive hadron-antihadron pairs (especially pions) were still around in large numbers. This is called *The Age of Hadrons*. After this time, even the lightest hadron-antihadron pairs had by then annihilated each other and were no longer being created anymore. Photons after this did not have enough energy for hadron-antihadron pair production.

At earlier times quarks and antiquarks were dominant; the period from one millionth of a millionth of a second ($=10^{-12}\,s$) to about $0.00001\,s\,(=10^{-5}\,s)$ is the *Age of Quarks*. In that period the density of matter was so great that hadrons could not exist individually. There were only free quarks. When the Age of Quarks began the temperature was about 10^{16} ($=10$ million billion) degrees.

During the Age of Hadrons (between the Age of Quarks and the Age of Leptons), hadrons could appear as separate particles, but the annihilation of hadrons and antihadrons was not yet completed. The origin of protons (i.e., hydrogen) can thus be timed at the beginning of the Age of Hadrons, at 0.00001 s. At that time the density of matter was very high, comparable to the density inside a proton, or 10^{16} times the density of water.

Neutrinos deserve a special mention. According to the theory, they are the most common particles today; every cubic centimeter of space should contain 600 neutrinos which originated at the early universe. Unfortunately they interact so weakly with ordinary matter that we cannot easily detect them

The Mystery of the Big Bang

The history of the universe can be traced back in time to the epoch of the nucleosynthesis at the cosmic age of a few minutes. Basic physical theory and astronomical data are reliable for this purpose. The earlier epochs described here, however, are seen much less distinctly. And what is totally obscure for us is the very birth of the universe. We can say that the Big Bang is not much more than a metaphor. There was definitely no "explosion," like that of the H-bomb. But what was it that made the universe expand? There are also a few other more concrete questions related to the nature of the Big Bang:

- Why is the strength of the Big Bang just right to result in the critical density for the universe (flat space)?
- Why is the universe isotropic, the same in all directions?
- Why did the universe have small density seeds which later became galaxies?

Fig. 24.9 The cosmic horizon is at the distance from which light has had time to come to us during the age of the universe (in about 14 billion years). Light from further away is still on the way to us. When time goes on, the horizon widens and brings new regions to our field of view

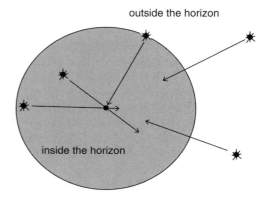

The Big Bang could have been too weak in which case the universe would have collapsed back to its original state, or it could have been too strong in which case galaxies would not have been born. The expansion is just right: there are regions where the expansion has been halted (galaxies), while in wide regions in between the expansion safely keeps the galaxies from not piling up on top of each other.

One popular answer to this fine-tuning problem is given by the *anthropic principle*. One may imagine that there are countless universes. Almost all are unfit for life as the needed long-lived structures do not arise. But there is at least one among them with the correct fine-tuning and expansion – ours! If no life-suitable universe had arisen, there would be nobody to miss it. We return to this aspect in Chap. 33.

Why is isotropy problematic? This is connected with *cosmic horizons*. A horizon is a distance beyond which we cannot see, at least so far. Inside the horizon there is the whole observable universe, outside the space continues, perhaps to infinity. The light that originated outside the horizon is still on the way to us. This border exists because light has a finite speed, and also because the universe has a finite age (Fig. 24.9). The horizon thus restricts how far in space we can see, but – as a remarkable compensation – there we see the birth of the universe, or more exactly, those events after the birth from which we can detect some radiation. The background radiation is currently the most distant message. If we learn to detect cosmic neutrinos one day, then we will get messages from within one light second of the birth.

Similarly as we have our own horizon, every point in the expanding universe has its horizon. If two points are sufficiently far away from each other, their horizons do not overlap at all. In this way the universe may be viewed as made of an immense number of separate regions that never could have been communicating with each other. At earlier times the size of the horizon was smaller than now, as light had less time to travel since the Big Bang. But even today, it is easy to find faraway regions in different directions, which have not known about each other. For example, take any two opposite directions in the sky. The cosmic background radiation from these directions originated in places millions of light years apart, when the age of the universe was much smaller than a million years. Theoretical calculations show that patches of the background radiation separated by more than about a couple of degrees in the sky originated in regions that never could have been in contact with each other. At the same time the properties of this radiation vary very little from one region to another. How is this possible? This is the isotropy problem.

Inflation and Cosmic World Periods

At the present time the horizon expands faster than the space itself, but has it always been like this? The universe may have expanded much faster in its youthful vigor, faster than horizons. If that is true, then the horizons will cover more and more of the universe the closer we go to the Big Bang. This is the idea behind the so called *inflation theory* that has been used to solve the isotropy problem. Then it is possible that practically the whole universe has once been inside the same horizon or at least such a large volume of it that it is much bigger than our current horizon. All regions within our field of view would have been in touch with each other in the past which explains the uniformity and isotropy of the observed universe. But what made the baby universe start expanding with the huge acceleration needed in the inflation model? This phase can be described by using a repulsive force that Einstein first introduced and then abandoned. Raise the strength of the repulsion by a factor of 10^{120} higher than in Einstein's static world model, and limit its period of influence to 10^{-32} s, and you have a model for the inflationary universe. It was not until 1965 that Erast Gliner of Ioffe Physical-Technical Institute in St.Petersburg (Russia) realized that the repulsive force may arise from cosmic vacuum. We describe this problem after a brief further excursion through the world periods as they are currently viewed.

Briefly said, in the inflation model the initially (almost) empty space started to expand rapidly and the universe remained relatively empty and cold. Then suddenly, at about 10^{-32} s, the universe was filled with matter and radiation at the high temperature of 10^{28} degrees. The energy for this creation of matter and radiation comes from the vacuum that subsequently falls to its current level. After this the expansion is "normal."

In this way ended the first period in the cosmic history, the *Age of Inflation*. The matter that was born was not anything like we know today, and interactions were different, e.g., electromagnetic forces and weak forces were not yet independent forces, but were parts of the electroweak force. Such particles as photons and W and Z bosons were intertwined with each other, and one could not talk about electrons, muons, or neutrinos in the same sense as today. There may also have been quite unfamiliar particles, such as the hypothetical X particle that cannot be produced even in big particle accelerators.

The time between the Age of Inflation and the later Age of Quarks can be divided into two parts. The first phase is called the *Age of Grand Unified Theories*, the latter the *Age of Weinberg–Salam Theory*. The names refer to modern theories of interactions. In the former era the color force and electroweak force were still united into a single force, in the latter period they have already separated (Box 24.1.)

Box 24.1 Cosmological world periods

When discussing the early universe, it is convenient to use a logarithmic time scale. Extremely short periods in the beginning may have important events, while later on long periods may pass and nothing very interesting happens. The logarithmic scale of time (in seconds) gives equal importance both to the initial short intervals and later longer periods. We give here the rough moments of time when different world periods happened or started.

World period	Logarithm of time	Events
Planck time	−43	Origin of spacetime? 10 dimensions?
Age of inflation?	< −32	Matter and radiation are born?
Age of grand unified theories	−32	Color force separates from electroweak force
Age of Weinberg–Salam theory	−12	Weak force separates from electromagnetic force
Age of quarks	−5	Protons are born
Age of hadrons	−4	Antipions annihilate
Age of leptons	0 (1 s)	Positrons annihilate, neutrinos become separate
Age of Big Bang nucleosynthesis	2	Helium is born
Age of radiation domination	13 (300,000 years)	Cosmic background radiation arises
Dark ages	16	First stars are born
Age of slingshot	16.5	Mergers of dark matter halos, slingshot ejections of black holes
Bright ages	17 (3 billion years)	Galaxies and quasars are born
Age of dominant dark energy	17.3 (7 billion years)	Present accelerating universe
Now	17.6	14 billion years

Though the hypothetical Age of Inflation is totally outside our range of observations, the inflation theory has, besides explaining isotropy, other interesting consequences that may throw light on the Big Bang and the origin of galaxies. At the end of the rapid inflation the universe settles automatically in the right rate of expansion: neither too slow nor too rapid. The theory predicts that space should be close to or exactly flat, and the studies of the cosmic background radiation support this claim.

The inflation has also been used to explain the origin of the tiny density seeds that later became galaxies. One may appeal to Heisenberg's uncertainty principle and say that the transition from the initial vacuum to the current vacuum state could not happen simultaneously everywhere. Matter and radiation were born a little earlier in some parts of the universe, a little later in others. This process could have made small ripples that survived through following ages as pressure waves (more in Chap. 27).

Antigravity, Cosmic Vacuum, and Dark Energy

Erast Gliner conjectured that the force that could give matter the huge initial velocities of the Big Bang expansion is a cosmic antigravity described by Einstein's cosmological constant. This is essentially the same physics as for the standard model in which the present-day observed accelerating expansion of the universe is ruled by antigravity (Chap. 23). But, to account for the initial Big Bang, one has to assume that the cosmological constant was initially much larger than now to be able to give rise to an initial cosmological expansion at a very high exponential ("inflationary") rate.

This approach was further studied by Irina Dymnikova of the Ioffe Institute in the 1970s. Since the early 1980s, it has become very popular in cosmology due to the efforts by Alan Guth of Massachusetts Institute of Technology, Andrey Linde of Lebedev Physics Institute in Moscow, Katsuhiko Sato of University of Tokyo, and others. They have suggested an impressive variety of inflation models to demonstrate that Gliner's conjecture is indeed promising in the search for the Big Bang physics.

Inflation models assume that General Relativity and "ordinary" physical theory are valid at the extreme early conditions in the Big Bang. This is a far-reaching extrapolation of our present knowledge. Hence the scientific status of the inflation model is still quite uncertain. In contrast, starting a few minutes after the Big Bang to the present day, the standard Friedmann cosmology is solidly accepted, confirmed by many astronomical observations with no extreme extrapolation of physical laws.

Gliner's basic idea is that the cosmological constant represents a cosmic medium with very special properties that may be described in terms of density and pressure. It is important to notice that different states of rest and motion do not exist relative to this medium. There may be two bodies moving relative to each other with a certain velocity, but this medium will be comoving with both of them! This means that the medium cannot serve as a reference frame. In mechanics, this special property is attributed to a vacuum which was usually understood as mere emptiness. Now we have one more example of a vacuum that has a definite density and pressure, and therefore carries definite energy. Gliner's vacuum is uniform in space, omnipresent, and unchanging in time.

Developing this idea, Yakov Zeldovich assumed in the late 1960s that the cosmic vacuum of the early universe was identical to the vacuum of quantum mechanics discovered by Paul Dirac of University of Cambridge in 1927. The quantum vacuum is also not emptiness, but is instead a field with so called zero-point energy – a consequence of the quantum nature of particles and fields. These matters are fundamental and difficult to understand. Zeldovich's assumption remains to be proved or disproved, in spite of efforts by many physicists during the last decades.

Gliner's vacuum has reappeared in the cosmology of the present-day universe in a new incarnation termed *dark energy*. This is not a hypothetical initial vacuum, but a real vacuum detected in cosmological observations. Dark energy is invisible and reveals itself only by its antigravity effects on the motions of galaxies. Its macroscopic properties as a medium are known thanks to Gliner, but its internal microscopic structure is still completely unclear.

As described in Chap. 23, the density of dark energy was first measured at very large distances of billions of light years, using supernovae as standard candles. But it is likely that its influence is also seen at much smaller distances of a few million light years in the vicinity of the Milky Way. This was pointed out by an international team, including some of the authors of the present book. In both cases, the Hubble flow of expansion serves as a Nature-given tool to detect the repulsive force of dark energy. In fact, the gravity of the mass of the Local Group and the antigravity of dark energy cancel each other surprisingly close to us around the outskirts of the Local Group (at about two times the distance to the Andromeda galaxy!). Studying the motion of galaxies at these distances, the "local" density of dark energy is found to be near its "global" density, or they may well be exactly equal. This suggests the remarkable fact that Einstein's antigravity is indeed an omnipresent universal phenomenon – in the same sense as Newton's gravity.

The Very Beginning

In antiquity, Plato stated that time came into being together with the heavens (or space). We have come a long way since then, but we always come back to the fundamental question: where did everything come from, how did it begin? Somehow the universe that we see around us emerged from the Big Bang, but we do not know how. One argument calls in question the common wisdom that there is no free lunch: if a vacuum can fill itself spontaneously with particles, even though with short-lived ones, why cannot the whole universe come out of nothing? Perhaps there is a free lunch after all, and not just a good meal but the whole material world.

These are ideas that theoreticians discuss in *quantum cosmology*. When the universe was very young, even younger than the previously mentioned 10^{-32} s, Heisenberg's uncertainty principle starts to rule. Quantum effects become dominant when we go back to so called *Planck time*, 10^{-43} s after the Big Bang. The concept of time becomes so confused at that point that it does not make sense to talk about earlier moments of time. Correspondingly, the energy has such huge uncertainties that the universe might have come "out of nothing." Perhaps it is Heisenberg's famous principle that offers a possibility to faintly understand how both space and time originated 14 billion years ago in their peculiar state from which they have evolved to what we see at present. The details are very much guesswork at present.

Chapter 25
The Dark Side of the Universe

The invisible came into astronomy in the nineteenth century when Friedrich Bessel concluded, from tiny motions of Sirius in the sky, that it is circled by a dark body. The companion of Sirius, a white dwarf star, was discovered later, in 1862, by the talented American telescope maker Alvan Clark when he tested a new half-meter size objective lens. Bessel did not live to see the discovery, but he was convinced that the universe has its dark secrets: "There is no reason to assume that brightness is an essential property of celestial bodies. Countless visible stars do not exclude numberless invisible ones." As we have discussed earlier, the faint high-density white dwarf companion of Sirius is not totally devoid of light, but modern astronomy speaks about genuinely dark substances.

Discovery of Dark Matter in the Coma Cluster

The modern period of dark matter studies began as early as 1933 when the Swiss astronomer Fritz Zwicky (1898–1974), who had emigrated to the United States in 1925, noticed that galaxies in the Coma cluster (Fig. 25.1) move surprisingly fast relative to each other. To keep the galaxies together in the cluster one has to have much more mass in the cluster than is seen in the galaxies. Zwicky suggested that the cluster is composed primarily of dark matter, and only in small part of visible matter. According to modern estimates the need for dark matter is ten times the visible matter.

How could one explain the large spread in velocities without dark matter? Armenian astronomer Victor Ambartsumian (1908–1996) of Yerevan observatory suggested in 1956 that galaxy clusters in general might be in a state of expansion. Then there is no need to postulate as much dark matter. The latest observations have revealed that the Coma cluster is composed of two subclusters near its center. These dark matter subclusters are marked by the two brightest galaxies and their companions. They apparently revolve around each other like stars in a binary star system, but at a much greater scale.

P. Teerikorpi et al., *The Evolving Universe and the Origin of Life*

Fig. 25.1 The cluster of galaxies at the constellation of Coma Berenices. It is five times further away than the nearby Virgo cluster. Some of the dots of light are foreground stars in our own Galaxy. Others that are fuzzier (not exactly points) are distant galaxies, mostly in the Coma cluster. Note the two large galaxies in the cluster center (NASA/JPL-Caltech/GSFC/SDSS)

Most of the acceleration of galaxies is caused simply by the pull of the dark matter in the two subclusters. But the situation is more complicated due to the binary nature of the cluster center. An ordinary galaxy which happens to wander to the central region of the cluster becomes a part of a three-body system, together with the two dark subclusters. As we learned earlier, a three-body system is generally unstable; sooner or later one of the bodies is thrown away. In this case it is the passing galaxy which acquires the escape speed, being the lightest member of the three bodies. It can get even enough energy to leave the cluster altogether. Some of the galaxies in the Coma cluster are undoubtedly in escape orbits, and to this extent Ambartsumian may have been right. However, the idea that clusters are held together by dark matter has gained strength in recent years. Overall Zwicky's idea was correct; most of the galaxies in the Coma cluster are bound to the cluster by dark matter.

Dark Matter in Spiral Galaxies

The study of the dark matter is somewhat simpler in spiral galaxies where stars and gas clouds orbit the galaxy center in a rather flat disk. When the motions of stars and gas far away from the center are measured, such high speeds of rotation are observed that the combined mass of ordinary stars inside the orbit cannot be responsible for the speeds.[1] It appears that in addition to stars and gas there is still something else in galaxies, matter which makes the greatest contribution to the total mass of the galaxy. It is usually assumed that the dark matter is in a more or less spherical *halo* around the galaxy.

The presence of massive halos in which galaxies are embedded was suspected by Estonian Jaan Einasto and his associates in the 1970s and at about the same time by Jeremiah Ostriker and Jim Peebles. The halos' existence was later concretely demonstrated by extensive studies of rotation of spiral galaxies by Americans Vera Rubin and Kent Ford. In 2002 Rubin was awarded the prestigious Gruber Cosmology Prize for her role in the discovery of dark matter (Peebles received his Gruber Prize in 2000 for his work in theoretical cosmology).

We do not know what this dark matter is made of. It evidently has mass and exerts gravitational force, but that is about all we know even after decades of studies. Consequently, there have also been assertions that dark matter does not actually exist. The large mass values are said to be a result of observational difficulties in "weighing" galaxies and their clusters or even new properties of the Newtonian gravity force. However, different kinds of independent evidence have appeared pointing to the reality of the dark substance.

New Methods of Detecting Dark Matter

Even if the nature of dark matter is not known, we can still make quantitative calculations about the total amount of dark matter in the universe. One of the best ways is through x-rays. Clusters of galaxies contain plenty of gas, which is typically so hot that it radiates x-rays. The gravity of the cluster must be able to hold the gas in the cluster, and this fact tells very effectively the total amount of gravitating matter in the cluster. It was a surprise that the gas itself has more mass than the galaxies in the cluster, but even when the mass of the gas and the mass of the galaxies is added together, there remains a wide gap which must be filled by dark matter. Generations of x-ray observatories following Uhuru (e.g., Copernicus launched in 1972 and Einstein in 1979 and more recently XMM and Chandra observatories) have done a thorough job in studying the x-ray gas and the amount of dark matter required to hold the gas together in the clusters (like the Virgo cluster in Fig. 25.2).

During the past decade a new efficient method of detecting dark matter has been developed: *gravitational lensing*. It makes use of the gravitational influence of dark

---

[1] The mass of stars can be estimated from the total amount of light and assuming a reasonable value for the ordinary star mass emitting a given amount of light. This is roughly the same as the light output of the Sun for a mass equal to the mass of the Sun (the mass to light ratio). The total light times the ratio gives the total ordinary star mass, a great deal less than the total mass from the orbital motions.

Fig. 25.2 The big galaxy M86 in the Virgo cluster as seen in x-rays. The x-ray emission is brightest in the center of the galaxy and weakens outward. Note the x-ray tail which is caused by the motion of the galaxy through the cluster and from the loss of hot gas out of the gravitational binding potential of the galaxy (credit: NASA/CXC/SAO/X-ray: C. Jones, W. Forman & S.Murray; Optical: Pal. Obs. DSS)

matter on rays of light. According to General Relativity, a ray of light is bent when it passes by a massive body. The bending of light was detected for light rays passing close to the Sun as early as 1919. The possibility of lensing effect was first pointed out in 1924 by Orest Chwolson (1852–1934), at the University of St. Petersburg. In 1936 Einstein independently made detailed calculations and concluded that among stars this effect must be hard to observe. Then Fritz Zwicky realized that galaxies, much more massive than stars, could produce observable size images of background objects (a few arc seconds) – such an image was indeed discovered decades later; a galaxy formed a double image of a distant quasar. We will return to such examples of gravitational lensing in Chap. 26.

Both visible matter and dark matter in a cluster of galaxies can gravitationally bend light from background galaxies forming a gravitational "lens." The bending

Fig. 25.3 Hubble Space Telescope image showing the cluster Abell 2218. Scattered over the picture there are arc-like images, galaxies 50 times more distant than the cluster. These are formed by the lensing of dark matter in the cluster whose mass is about 7×10^{14} solar masses (credit: NASA, ESA, Andrew Fruchter (STScI), and the ERO team (STScI + ST-ECF))

can be measured quantitatively to calculate the mass of the lens. This method indirectly makes the dark matter "visible." It is gratifying that the amount of dark matter thus detected agrees with the X-ray determinations. The measurements of the total amount of dark matter in galaxies and clusters of galaxies are now considered reliable (Figs. 25.3 and 25.4).

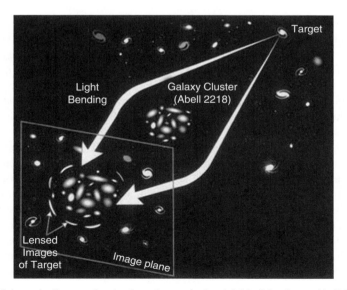

Fig. 25.4 Schematic diagram showing how the gravitational field of the cluster Abell 2218 forms arc-like images of distant background galaxies (credit: NASA, ESA, Andrew Fruchter (STScI), and the ERO team (STScI + ST-ECF))

What Could All that Dark Stuff Be?

What can the dark matter be? Is it ordinary matter or is it something totally different? By ordinary, we mean "baryonic" matter with different numbers of protons in different elemental nuclei and differing numbers of neutrons for isotopes of a given element. Recall, for example, that there is ordinary hydrogen with one proton in its nucleus versus deuterium which has both a proton and a neutron. Now we return to a subtle point in the calculation of the baryonic matter products of the Big Bang nucleosynthesis as described in Chap. 24. The resulting relative amounts of different isotopes of an element (as well as elements) depend sensitively on the total fractional amount of baryonic matter to total matter in the universe. This total amount is thought to be equal to the critical density. Recall that the critical density is the density that is required to make the overall geometry of the universe flat in Friedmann models. It is remarkable that from the present-day relative amounts of those elemental products, measured in the local environment of our Milky Way galaxy, one may infer the fractional amount of ordinary matter in the universe!

Deuterium and hydrogen, with similar chemical properties, are particularly useful for this estimate. In particular, there are leftover deuterium nuclei, the amount of which depends on the present-day matter density relative to the critical density of the universe. If the ordinary matter were to add up to the critical value then we would expect that the deuterium abundance would be only about a billionth part of the hydrogen abundance. In observations we find 10,000 times more deuterium than this value! This implies, according to the theory, that the density of ordinary matter can be only about 4% of the critical value. On the other hand, the total amount of matter in the form of gas and stars in galaxies is less than 1% of the critical density. Hence, we have actually two kinds of matter "missing": the ordinary baryonic kind and, in a much larger amount, the enigmatic nonbaryonic dark matter.

Some of the unseen baryonic dark matter in spiral galaxies may be in the form of neutron stars, white dwarfs, black holes, faint red stars, and planets. Neutron stars and black holes discovered through x-rays are related to rather brief periods in the evolution of close binary stars. Therefore x-ray stars on the whole are rare. However, this does not mean that neutron stars and black holes are rare. Yet, it is almost impossible to detect them when they are alone. In any case one expects additional unseen baryonic matter, but only a few percent of the critical value.

As to the remaining large amount of nonbaryonic dark matter, we do not know its nature. At present it is usually assumed that there are new particles, so far unknown to us, but which would affect our matter through gravity and through the weak interaction. The candidates have names (e.g., neutralino) and assumed properties, but they have not been detected in laboratory experiments. One of the suggested particles, the photino, is thought to be similar to a neutrino but much more massive. It is likely that new particle accelerators will bring some information on these particles in the next few years.

Previously it was thought that neutrinos are massless particles, but in recent years it has been shown that they have a tiny mass. Even if a neutrino would weigh as little as an electron divided by ten thousand, the mass of neutrinos in the universe would exceed the total mass of ordinary matter. But the neutrino mass is not even this much. In accelerator experiments at CERN it has been shown that the neutrino mass is less than the electron mass divided by 30,000. The result was confirmed in 1987 when a supernova exploded in our nearby companion galaxy, the Large Magellanic Cloud (Fig. 25.5). It was the nearest supernova that we have observed in the past 400 years, and when it was at its brightest, one could see it by naked eye in the sky. But what was most important, for the first time we had the opportunity to observe neutrinos born in a supernova explosion. Fortunately several neutrino telescopes were in operation, two of which (in Japan and United States) recorded the neutrinos. The travel time of neutrinos is about 163,000 years (the distance to the Large Magellanic Cloud is the same number in light years), but it varies a little for different neutrinos if they are particles with mass. The neutrinos were observed to arrive almost at the same time which means that their mass cannot be greater than the electron mass divided by 50,000.

Fig. 25.5 The supernova that exploded in 1987 in the Large Magellanic Cloud is visible as the very bright star in the middle right. (credit: European Southern Observatory)

In recent years it has even been suggested that there exists a shadow world made of particles which do not react at all to detectors and whose only influence is through gravity and the curvature of space associated with gravity. Since such particles are extremely difficult to observe, their existence or otherwise is still a matter of speculation. The shadow world may be there; even at this very moment blocks of matter from the shadow world may be traversing us, but we have no way of being aware of it. In some ways the shadow world is the last resort if plenty of dark matter is recognized through its gravity, but all other explanations fail.

Still Darker: Dark Energy

Ordinary matter contributes about 4% of the critical density, dark matter about 25%. If the overall geometry of the universe is flat, as we have good reasons to believe on the basis of the properties of the cosmic background radiation, then there must be a missing component which is neither dark ordinary matter nor dark matter particles. This new component is called dark energy, as we learned earlier. What is behind the word "dark" is currently quite unclear, except that in one way or another the dark energy makes the universal expansion accelerate, and at the same time provides the explanation for the flatness of the universe. Mathematically, dark energy was used already by Einstein when he introduced the lambda term in his equations to make the universe unchanging. It was not until 1965 that Gleiner realized that the natural explanation of the lambda term was a special kind of vacuum. The term "dark" aptly describes the fact that we do not understand where this level of vacuum energy comes from. The concept of vacuum energy itself is not so strange, but what requires an explanation most of all is why nature has chosen exactly this value for the dark energy filling our universe.

The Four Fundamental Elements: Some Concluding Thoughts

What is the universe made up of? Empedocles in the fifth-century BC, Aristotle, and other sages in antiquity believed that everything in the world consisted of four elements – earth, water, air, and fire. These correspond well to the four familiar states of matter (solid, fluid, gas, and plasma[2]). But also modern cosmology tells us about four basic elements, or cosmic energies as they are now called. At the present cosmic epoch, the dark energy of the cosmic vacuum is the dominant element, contributing to about three-quarters of the total energy of the universe. All bodies of nature are embedded in this uniform medium, but no structures are made of it – except the vacuum itself.

---

[2] In order to see plasma, hot ionized gas, just look at the candle flame, or the Sun, or those points of light of the starry sky. Stars are giant plasma balls, so most of the ordinary matter in the universe is in the form of plasma.

Three other energies are the dark matter (some 20% in round numbers), ordinary or baryonic matter (4%), and radiation whose fraction is now very small (0.01%). These shares were different in the past and will change in the future. For example, conditions were very different during the first three minutes of the cosmic expansion. Then the vacuum fraction was close to zero, while radiation formed almost 100% of the energy.

This cosmic recipe which changes with time may look accidental and complex, making the universe strange or even absurd. But this is so only at first glance. In fact, there is hidden regularity behind all this. It is a new kind of symmetry that – contrary to familiar geometrical symmetries (uniformity and isotropy) – does not involve space and time. Nongeometrical symmetries are usually called internal symmetries. A simple example of an internal symmetry comes from particle physics: the symmetry between the proton and the neutron. They are very similar, despite differences in mass, electric charge, lifetime, etc. Each of the two particles is equally able to participate in strong interactions inside the atomic nuclei, and this similarity unifies them into a set called the nucleon doublet.

In the same manner, cosmic internal symmetry unifies the four cosmic energies into a regular set – a quartet. Each of them has a permanent physical characteristic termed the Friedmann integral. This quantity has the dimension of length and it was introduced by Friedmann in his world models. The length is of truly cosmological size, comparable to the distance of the cosmic horizon of about ten billion light years. The values of the four integrals prove to be similar, to the order of magnitude. Since the identity is not exact, the symmetry itself is not exact and is said to be violated. Nevertheless, this set of cosmic lengths looks simple and natural. As the integrals are constant in time, they offer us an "eternal" recipe of the cosmic energy mixture, which remains the same at all times whenever the four energies exist in nature – at least since the first minutes of the cosmic expansion.

The cosmic internal symmetry gives regularity to cosmic energetics and suggests that there are deep internal links among the fundamental elements of Nature. Through this symmetry the universe manifests new order and beauty – the very qualities that were ascribed to our cosmos by the first cosmologists of antiquity.

Chapter 26
Active Galaxies: Messages Through Radio Waves

Through the ages the human eye alone was the most important means of observation. In the early seventeenth century the telescope was invented. Gradually the size of the biggest telescopes became larger, and, complemented with photography, they helped us to see ever further into the universe. In the 1930s this process was in full swing, and hardly anyone could imagine that there would be any other means of extending our vision except by further development of ordinary optical telescopes.

Early Years of Radio Astronomy

In 1933 an engineer employed by the Bell telephone company, Karl Jansky, studied Trans-Atlantic telephone connections and tried to figure out the sources of disturbing noise. He noticed that radio noise increased daily always at the same hour. After a while he realized that the increase was not exactly at the same hour, but that it appeared always 4 min earlier than during the previous day. Recall that our Sun, the basis of civil or solar time, moves eastward among the stars over the year; stars and galaxies rise in the sky 4 min earlier than the day before. The four minutes was a clear hint that the origin of the noise must be outside the Earth. Jansky identified the source of noise in the Milky Way but could not follow up this research since he became occupied with other duties (Fig. 26.1).

One of the few people who were aware of Jansky's discovery was American engineer and radio amateur Grote Reber (Fig. 26.2). He built a dish antenna 10 m in diameter in his backyard and started studying the cosmic radio noise in his free time. He confirmed Jansky's discovery and adding to it, he identified regions of strong localized emission in the sky called radio sources. One of them is the center of the Milky Way. It was more difficult to identify the other sources, but generally they were called radio stars. Years later it became clear that the radio sources were not stars; some astronomers had guessed this already in the 1940s since the Sun would not be easily detected by its radio emission if it were moved to the distance of other stars.

P. Teerikorpi et al., *The Evolving Universe and the Origin of Life*
© Springer Science+Business Media, LLC 2009

Fig. 26.1 Karl Jansky (1905–1950) with his radio antenna (Credit: NRAO/AUI/NSF)

Reber's antenna was not known to scientific circles. For his neighbors in Wheaton, Illinois, the purpose of the apparatus was a source of amazement and a talking point. The most popular view was that it was a rain-making machine since the dish obviously collected water which was drained out through a hole in the bottom. Reber's explanation that he was listening to radio noise from outer space

Fig. 26.2 Grote Reber (1911–2002) made first observations of radio sources (Credit: NRAO /AUI/NSF)

Fig. 26.3 Electrons spiraling around magnetic lines of force emit synchrotron radiation

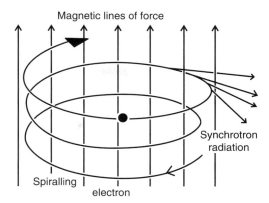

was just too incredible in 1937. When Reber wrote about his findings to *Astrophysical Journal*, an editor of the journal went to visit him to see that there really was a dish in his backyard. The editors had a nagging suspicion that the unknown engineer might have invented the whole story about cosmic radio waves.

However, the time was now ripe for the study of celestial radio waves. During World War II antenna technology had taken huge strides forward, and after the war a number of radar antennas had remained unused. Astronomers grasped the opportunity and so radio astronomy was born in late 1940s. In addition to Jansky and Reber, one must mention Martin Ryle (1918–1984) of Cambridge University as one of the pioneers of this new science and a Nobel Prize winner. Jan Oort – we already told about his Milky Way studies – also realized early the value of this form of radiation as a new tool for astronomy.

The radio emission was found to be roughly equal in strength in different neighboring frequencies – one speaks about a continuous spectrum. Reber suggested that the radiation is emitted by electrons when they move through a field of atomic nuclei in their curving paths in an ionized medium. Observations did not agree with this idea.[1]

The mystery of radio noise started to clear when Karl Kiepenhauer (1910–1975) realized a connection between cosmic rays and radio noise in 1950. In the same year also Hannes Alfvén (Sweden) and Nicolai Herlofson (Norway) proposed that the propagation of cosmic rays near to the speed of light causes the noise. This *synchrotron radiation* is observed also in particle accelerators where magnetic fields force charged particles to move in circles. In space, highly energetic electrons circle in magnetic fields, causing radio emission somewhat similar to that of electrons vibrating back and forth in the antenna of a radio station (Fig. 26.3). Vitali Ginzburg (Nobel Prize in 2003) and Josif Shklovskii (1916–1986) are among the scientists most noted for the development of the theory of sychrotron radiation.

---

[1] Such "bremsstrahlung" has a continuous spectrum but its characteristic shape and cut-off point did not match the radio observations. Ryle and Oort were of the (also incorrect) opinion that the radio waves came from stars which were not unlike the Sun except that for some reason radio emission was unusually enhanced.

Spectral Lines of Radio Emission

In 1944, at the suggestion of Oort, his a young Dutch student, Henk van de Hulst, started to investigate whether there could be spectral lines in radio emission. The spectral lines had proved their value in optical astronomy where they are used to determine the motions of stars and galaxies as well as other properties. Radio emission with spectral lines would open a totally new window on the universe.

Van de Hulst concluded that a transition of a hydrogen atom between two of its energy levels could correspond to radiation at 21-cm wavelength, in the region of radio emission. Here an electron does not jump from one orbit to another, but there is rather a much smaller change in the state of the electron. As we learned earlier, an electron has a property called spin which one may visualize as axial rotation. Also the nucleus of a hydrogen atom, the proton, has its own spin. The electron and proton spins can be parallel or antiparallel, with the hydrogen atom in an excited state in the former case. When the excitation is released, the atom goes to its ground state, and while doing so, emits a photon which has the energy equal to the excitation energy. Since the energy is low, the corresponding frequency of radiation is low (1,420.4 MHz) and the wavelength is long, 21.1 cm to be exact (Fig. 26.4).

Hydrogen is by far the most common element in the universe, and there is no shortage of potential 21-cm radio emitters. The state of excitation may result from collisions between hydrogen atoms. The excitation is released after about 11 million years and the 21-cm photon is created. Even though the radiation process is rare, the Milky Way contains so many hydrogen atoms that put together they should create a strong signal. Indeed, the signal was observed in 1951 in observations carried out in the United States and in the Netherlands. The radiation comes from cold interstellar clouds whose existence was previously known only by indirect means.

While optical astronomy had provided the methods of mapping out the distribution of stars in the Milky Way, it became now possible to use radio astronomy to map the second major component of our galaxy, the interstellar gas. By 1958 we had radio maps of the Milky Way and a clear indication of its spiral structure. This work was done by Jan Oort, Frank Kerr (1918–2000), and Gart Westerhout. In 1951 Kerr set up the southern sky 21-cm line program and started mapping the Magellanic Clouds. This was the first detection of a radio spectral line in an external galaxy.

Hydrogen is not the only emitter of radio spectral lines. The OH molecule composed of one hydrogen atom and one oxygen atom was found in space in 1963 by its

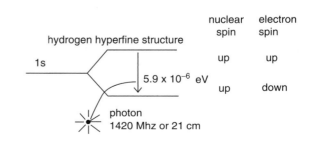

Fig. 26.4 Origin of 21-cm radiation, when a hydrogen atom goes from an excited state to its ground state

spectral line at 18 cm. The next discoveries were water and ammonia in 1968 which started the race for finding new molecules in space: in the 1970s about five new molecules were detected every year through their spectral lines and now their total number is about 130. At the same time, information on different kinds of interstellar clouds accumulated. The most significant sites of molecules in space are *molecular clouds*. They have relatively high gas densities and complex chemical reactions. A molecular cloud can be massive: it may have as much mass as 100,000 stars.

Radio Galaxies are Discovered

What are the radio sources outside our galaxy? In Cambridge University and elsewhere, especially in Australia, new radio sources were listed. In the first Cambridge catalog released in 1950, Ryle and his associates gave information on 50 radio sources. Five years later the second catalog appeared with 1936 sources, while the third catalog from 1959 contained 471 sources. Even today the brightest radio sources are named after their number in this third Cambridge catalog. For example, the bright radio source in the constellation of Cygnus is known by the name 3C405. For the southern skies, corresponding work was being undertaken in the Parkes radio observatory in Australia. The brightest sources are also named after the constellations where they are found – 3C405 is the same as Cygnus A.

The cataloging and naming of radio sources did not leave us with very many clues as to the nature of the radio sources. The problems were twofold: first, the poor accuracy of the radio position in the sky, and second, the lack of spectral lines in a typical source which could be used to obtain the redshift. One could always photograph the sky in the direction of the radio source, but the picture would contain so many different objects, near and far, that it would be often impossible to say which one of them was the radio emitter. The identification of radio sources thus became a special problem which required considerable amount of work.

The first radio sources were identified in Sydney, Australia, by John Bolton and associates. The radio source Taurus A was recognized as the Crab nebula, the remnant of the 1054 supernova. Virgo A and Centaurus A were associated with galaxies relatively close to us (M87 and NGC 5128). They were the first examples of radio galaxies, radiating strongly in radio waves. Then Cygnus A provided a surprise.

For Reber, Cygnus A was a patch in the radio map, a patch so wide that the radiation could have come from any of the thousands of objects which happen to be in this direction of the sky. The identification was not possible until 1951, when Graham Smith at Cambridge University determined the position of Cygnus A with the accuracy of one arc minute (the accuracy in Tycho's naked eye observations!). He sent the coordinates to Walter Baade, who was a staff member at Palomar observatory and had regular access to the new powerful 5-m telescope. Baade decided to slip in the photography of Cygnus A region in his next observing session. The next day he developed the photograph and began to inspect it: *"I knew that there was something unusual in the picture as soon as I cast my eyes on it. There were galaxies all over the photograph, in all more than 200, and the brightest of them was in the middle of the picture. It showed tidal perturbations, gravitational pull between two nuclei. I had never seen anything like this before. It troubled my mind so much that while driving home in the evening, I had to stop the car and think about it."*

Then it all became clear to Baade in a flash: he was witnessing a rare traffic accident, the collision of two galaxies. Baade estimated that the probability of such an accident is one in a hundred million, and here as the first human he was looking at it. Two entire stellar worlds had drifted to a collision course, and the radio emission brought us the message of the accident. Together with Lyman Spitzer, Baade published a theory which explained the radio sources in general as colliding galaxies.

Soon afterward another staff member at Palomar observatory, Rudolph Minkowski, gave a seminar on radio sources where he contemplated different theories. As was common at the time, Minkowski also believed that radio sources lie inside the Milky Way, not outside in other galaxies. He mentioned Baade's "incredible theory" only in passing. After the seminar Baade went to Minkowski and told: "I bet you a thousand dollars that in Cygnus A there is a collision of galaxies in progress." Minkowski had just bought a house, and he couldn't afford such a huge bet. Instead the men agreed to bet on a bottle of whisky, and they decided on what observational signatures would decide the bet.

A few months later Minkowski came to Baade's room and asked: "What brand?" Minkowski showed Baade the spectrum of Cygnus A which showed clearly the spectral lines which had been agreed upon as decisive evidence of a collision. They published the new results in *Astrophysical Journal* in 1954.

Who was right in the end? Baade's theory started to lose popularity after a few years when it became clear that the radio emission does not come from the colliding galaxies, but from a region outside them. But again in recent decades the collision of galaxies has become a fashionable idea. However, we now know that the radio waves originate through a much more complex route than Baade and Spitzer could fathom.

What was really significant in the spectrum of Cygnus A was its redshift 0.057. Baade and Minkowski calculated the distance to the galaxy based on the redshift; the value using today's distance scale is 800 million light years (250 Mpc). Cygnus A was at a shocking distance considering that it is the second brightest radio source in the sky. When the distance is known, it is easy to calculate that the source emits in the form of radio waves as much energy as a hundred billion stars, ten times the energy radiated by all the stars in the Cygnus A galaxy! The stars are powered by nuclear energy; what is the origin of this mysterious energy which overwhelms nuclear energy by tenfold?

Cygnus A is so bright that it would be easily seen by radio telescopes even if it were ten times further away. Its radio emission comes from two regions which are a little more than an arc minute apart; the galaxy itself is half-way between the radiating regions (Fig. 26.5). Cygnus A is an example of a *double radio source*; its two radio regions are 0.4 million light years apart from each other. The separation of the two radio regions varies from one double radio source to another. The greatness

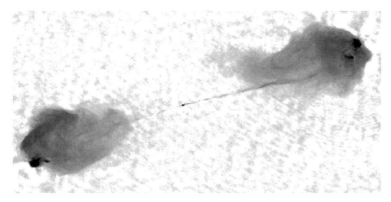

Fig. 26.5 The double radio source Cygnus A in the direction of the constellation Cygnus. The thin jets connect the active nucleus of the host galaxy with the outer radio components (VLA map courtesy of R.A. Perley)

of these phenomena is paralleled by the fact that the galaxies in their centers are among the biggest in the universe.

Discovery of Quasars

The study of the sizes of radio sources intensified in late 1950s. The specialty of the radio astronomy group in Manchester was the small-sized sources, but even their radio telescope could not resolve the structure of a few sources; they appeared point-like. One of these was 3C48. Thomas Matthews measured the accurate position of 3C48 in the sky using the Owens valley radio telescopes in California, and gave the coordinates to Allan Sandage of Palomar observatory. Sandage photographed the region of sky and found a faint star at the position of the radio source. At the end of 1960 Sandage reported his finding at the meeting of the American Astronomical Society. He concluded that it was the first real radio star in our Milky Way. He also mentioned that it could be a distant galaxy, but since its brightness was observed to fluctuate, the extragalactic explanation seemed impossible. How could a hundred billion stars vary in tune so that the whole galaxy would fade and brighten?

While Sandage and Matthews were pondering about 3C48 in America, Cyril Hazard of the Manchester group discovered a very accurate method for finding the position of a radio source and used it with his Australian colleagues. When the Moon moves in front of a radio source, the emission from the background source disappears when the rim of the Moon first blocks the beam of radio waves. The motion of the Moon in the sky is known very accurately; therefore, the time of disappearance of the radio source, as well as its reappearance a little later, tells the position of the source very accurately.

In this way the position of the radio source 3C273 was determined, and the information was sent to the Palomar observatory. Matthews found that the source coincides exactly with a star in the constellation of Virgo. The Dutch astronomer Maarten Schmidt, who was working at the Palomar observatory, made spectroscopic observations of the star and found altogether seven spectral lines in its spectrum.

Neither he nor anybody else in the observatory was able to tell which element the spectral lines belonged to. In order to progress to an answer, Schmidt started to measure the accurate wavelengths of the spectral lines using the nearest Balmer line of hydrogen as a standard of comparison. The first line turned out to be at 1.16 times greater wavelength than the nearest rest wavelength of a Balmer line, the wavelength of the second line was 1.16 times greater than the wavelength of another Balmer line, and the measurement of the third line gave again the same ratio relative to another Balmer line. Now Schmidt realized that the unknown lines were in fact Balmer lines and that the whole spectrum of the radio star had shifted 16% away from the usual wavelengths. In other words, the redshift of 3C273 is $z = 0.16$. If the redshift is taken as an indication of distance in the usual manner, 3C273 is as far as 2,400 million light years from us (a thousand times the distance of the Andromeda galaxy!).

Now it became clear why it was so tricky to interpret the spectra of radio stars. In the stars belonging to our Milky Way the spectral lines cannot shift far from their normal wavelengths! It did not occur to anybody to try the large spectral shifts which are typical of distant galaxies. In the same way the riddle of the spectrum of 3C48 was also solved: it has the redshift $z = 0.37$ and the distance of about 6,000 million light years. In spite of their great distances, 3C273 and 3C48 are easily seen through telescopes. It is simple to calculate that these "stars" are as much as a hundred times brighter than large galaxies.

New radio stars were found in quick succession. These so called quasi-stellar objects (quasars for short) appear like stars but are really equivalent to a million, million stars. In addition, their brightness often changes in a short time, for example, from one night to another. The rapidity of the change tells us about the size of the source. Light travels in one day the distance of one light-day which is about 200 astronomical units, or somewhat more than the size of the Solar System. A source which brightens significantly in one day cannot be bigger than this.[2] A quasar produces more energy from a volume of Solar System dimensions than a whole galaxy from a volume with the diameter of 100,000 light years!

Allan Sandage discovered also a large population of quasar-like objects which do not emit radio waves observably. In fact, such "radio quiet" quasars are about ten times more common than "radio loud" quasars. Today, tens of thousands of quasars have been cataloged, all too faint to be seen by naked eye and covering the sky much more densely than the visible stars. The total number exceeds millions.

---

[2] In order that the source can change its brightness in only one day, it must be possible to readjust all radiating surfaces to the new brightness level within this time. The readjustment cannot be carried out with a speed greater than the speed of light. If the readjustment happens more slowly then the quasars can be even smaller than the Solar System.

The Redshift Problem

At this point one may get a nagging feeling: Is everything in this chain of argumentation really OK? What if the quasar distances have been calculated incorrectly? Then the estimated brightnesses of quasars do not need to be so high. The distance estimates for quasars are based on the redshifts of their spectra as well as on the use of the Hubble law. Are there any other ways by which a redshift could arise in a spectrum of a quasar besides the shift caused by the expansion of the universe?

The longer is the wavelength of radiation, the lower is the frequency of oscillations. One might surmise that atomic vibrations have slowed down in quasars for some reason, making spectral lines shift to longer wavelengths. This is indeed possible, if the space-time is strongly curved in the emitting region, for example near a black hole. From the point of view of an external observer, time as well as oscillations appear to have slowed down in such a region. There have been studies to the origin of redshift in strong gravity fields, but other properties of the spectrum besides the redshift do not agree with this assumption. Later investigations concentrated on the so-called anomalous redshift mechanisms. They are not known from laboratory physics, but the hypothesis was that they could arise under the peculiar circumstances of quasars. However, at present this evidence is generally not judged

Fig. 26.6 The quasar 3C273 and its narrow jet which measures 150,000 light years in length in the image taken with the NOT telescope at La Palma. The host galaxy is seen as an elongated fuss of light around the bright nucleus. Also other galaxies are visible. The bright dot near the upper left corner is a star in the Milky Way (Courtesy of Leo Takalo and Kari Nilsson)

convincing enough in order that the established fundamental theories of physics would need to be replaced.

In both radio galaxies and quasars large quantities of energy are liberated. The similarity between the two classes of objects became even more obvious when it was realized that the radio emission in quasars may come from both sides of the quasar, in addition to the radio emission of the quasar itself. If viewed in radio waves alone, quasars and radio galaxies look very similar. In radio galaxies, their centers or galactic nuclei correspond to the quasars. This has led to the conclusion that in fact quasars are galactic nuclei. They differ from ordinary galactic nuclei by their brightness; a quasar is so bright that it outshines the surrounding stellar system. This has been confirmed directly by observing light emitted by the quasar host galaxies. One of the first host galaxies was discovered in quasar 3C273 (Fig. 26.6).

What is Behind the Huge Power of Quasars?

We may calculate the total amount of energy released during an active period of a galactic nucleus. Typically the energy corresponds to the mass of a million Suns, using Einstein's relation between mass and energy. This is a big amount, because the processes in nature usually liberate only a tiny fraction of the mass energy. For example, the Sun turns only 0.1% of its mass into radiation during its whole life, using nuclear reactions. A superstar in the galactic nucleus should have its mass equal to a billion solar masses in order that the 0.1% efficiency would give enough energy. However, calculations show that superstars like this are not long lived enough to account for the many quasars which are seen.

The potential energy of a collapsing celestial body can be an even more efficient energy source than the nuclear energy. Matter falling onto the surface of a dense celestial body brings with it plenty of energy: the matter accelerates to a high speed and the energy acquired in this process is transformed to other forms of energy when the matter crashes on the surface of the body. Some of the energy may appear as radiation. Calculations show that this process may liberate into radiation as much as 10% of the rest mass energy of the infalling matter. Then the celestial body has to be exceptionally dense, a black hole or a star near to the state of collapse into a black hole. Prior to the collapse into a black hole, the celestial body may go through a superstar stage. For a while, for a million years or so, the superstar may exist as a nuclear burning and fast rotating body. At the end, it explodes, and what is left at its center is likely to collapse into a black hole. Such processes were probably common in the early universe, inside protogalaxies, and today's big black holes in the centers of galaxies are thought to have grown by gas accretion and by merging with other black holes.

According to the current view, galactic nuclei harbor supermassive black holes, in the mass range of a few million to billions of solar masses. Such black holes have not been seen directly yet, and thus their masses are not exactly known. The best measurements of the black hole masses come from the rotation speeds of stars around them. In this way the mass of the black hole in the center of our galaxy has been estimated to be 3.7 million solar masses. In the big Virgo cluster galaxy M87 the black hole mass is a thousand times greater. The latter is typical of the masses of the supermassive black holes in quasars.

A black hole itself does not radiate, but rather the observed phenomena in quasars occur in its immediate vicinity. The black hole tends to devour gas clouds from its surroundings and swallow them inside its Schwarzschild radius. Most gas clouds do not fall directly into the black hole but remain circling around the central body for some time. An accretion disk is created where gas circles around the black hole according to Kepler's laws, and at the same time, gradually creeps closer to the center. When a portion of gas has reached the inner edge of the accretion disk, the black hole pulls it into its throat. What fraction of the gas is lost inside the throat and how much of it manages to escape is not clear at present. But it appears definitely that some gas escapes in the form of two oppositely directed jets, with high-speed outflows along the rotation axis of the accretion disk. The gas in the disk is very hot and strongly magnetic. It is thought that in general all the radiation in the quasars is associated with the accretion disk in one way or another. The origin of the energy is gravitational potential energy, part of which directly turns into radiation; some part is liberated through the processes in jets.

The size of the Schwarzschild radius in a quasar black hole is of the order of the planetary orbits in the outer Solar System. Such small separations can be resolved with current telescopes only in nearby stars; for example, a planet circling a nearby star would be just barely visible, if it were bright enough to show through the over-whelming brightness of the central star. Binary stars are often at these separations and can be imaged as two stars. However, nearby stars are at distances of a few light years while the distances to quasars are billions of light years. The Solar System would appear billions of times too small at the distance of quasars in order to be resolved with current telescopes. Therefore direct observations of centers of quasars are impossible today and in foreseeable future. The studies of supermassive black holes in quasars must necessarily use indirect methods.

Light Variations and Higher Resolutions

One way to study the inner workings of quasars is to follow their variations of brightness. As mentioned earlier, the shortest timescale of variability tells the largest possible size of the radiating region. On this basis, it was deduced soon after the discovery of quasars that their sizes are at most one light-day (200 Earth–Sun distances). Brightness variations have since then been followed in many observatories, among them the Metsähovi radio observatory of the Technical University of Helsinki and Tuorla Observatory of University of Turku (Finland). These observations have revealed even intraday variations.

The brightness variations have provided the groundwork for building a theoretical picture of the strange object OJ 287 (Fig. 26.7). The source appears to be

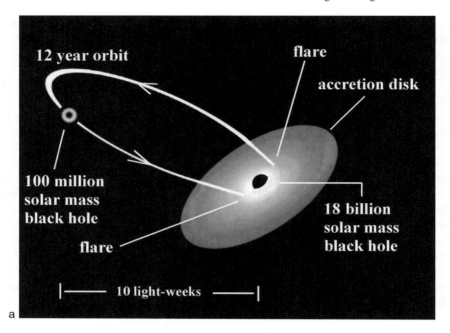

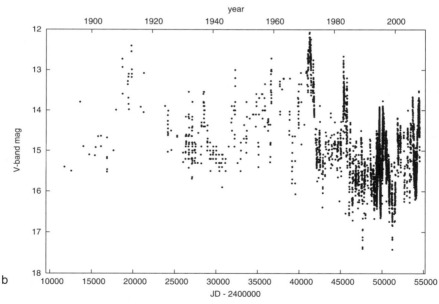

Fig. 26.7 a. Model of the highly variable quasar OJ 287 as made in Tuorla Observatory. Central black hole, accretion disk, and companion black hole are indicated. **b**. Periodic emission peaks are seen in the light curve

composed of two supermassive black holes in orbit about each other. Most of the radiation is connected with the accretion disk of the bigger black hole and regions close to it. Radio emission is born further out in the jets. The orbital revolution period of the black holes around each other is 9 years, and the masses of the black holes have calculated to be 0.1 and 18 billion solar masses.

Another way to study the structure of quasars is to improve the resolution of the telescopes. The Hubble Space Telescope achieves a resolution of about 0.1 arcsec, without the disturbing atmosphere. The new generation of large optical telescopes, such as the Very Large Telescope of the European Southern Observatory (ESO), in Chile reach similar resolutions from the ground (generally the Earth's atmosphere spreads the points of stars into dots with sizes of a second of arc or more).

The best image resolution is currently obtained in radio astronomy. As may be recalled, the initial problem of radio astronomy was the lack of resolution, and, indeed, the limiting resolution of a single dish radio telescope is still not much better than 1 arcmin; in this respect it is not much better than a human eye. But if we use many radio telescopes together and combine the signals, it is possible to achieve very good resolution, the better the further apart are the single dishes which are used. For example, if a 15-m radio telescope dish resolves two radio sources apart from the separation of 300 arcsec, two such telescopes at $300 \times 15\,\mathrm{m} = 4.5\,\mathrm{km}$ apart can achieve a 1 arcsec resolution. A telescope system like this, which in fact consists of eight radio telescope dishes, was completed at Cambridge University in 1972 by Ryle's group. It was the first radio telescope system which produced as sharp radio images as the optical telescopes could make.

Thereafter the development has been rapid. Telescopes even further from each other have been combined, so that the longest distance between the telescopes approaches the diameter of the Earth. Then the telescope separation can exceed the distance of the Cambridge line by 2,000-fold, and the limiting resolutions improve to less than 0.001 arcsec. The use of distant telescopes in a common system is called Very Long Baseline Interferometry (VLBI). During a VLBI experiment observatories in different parts of the world point their telescopes to the same radio source at the same time. Because of the increased need for high resolution, a good fraction of all radio telescopes have joined in the VLBI networks. For example, the 14-m telescope in Finland operates part of the time as a member of the European VLBI consortium. The United States operates its own VLBA system which is dedicated to interferometry full time. The VLBA antennas cover the whole country from Hawaii and Alaska to the Eastern United states and including Puerto Rico.[3]

The radio images of quasars have revealed an interesting phenomenon: quasars are made of a point-like central nucleus and of radiating clouds which escape away from the nucleus from time to time. The speed of the clouds is so high that the motion is observable as an increase of separation from the nucleus. This is very exceptional for galaxies: the transverse motions in the sky are generally not observable

---

[3] Some of the earlier interferometer systems include MERLIN of Manchester University (with a 130-km baseline) and Very Large Array of the National Radio Astronomy Observatory (NRAO) in New Mexico (a 40-km baseline).The latest project is ALMA (ESO & NRAO). In many ways it will be the superior system when it starts operation in few years time in Chile.

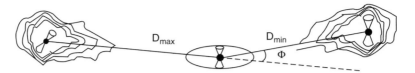

Fig. 26.8 Ejected black hole model of radio galaxies as outlined by Mauri Valtonen. Each lobe contains an ejected supermassive black hole

due to large distances. Typically galaxies move with speeds of some hundreds of kilometers per second. For a nearby galaxy this translates to about 0.001 arcsec per year. Quasars are much further away than nearby galaxies, let us say 1,000 times further than galaxies. Therefore a motion of 0.001 arcsec per year, as is observed at VLBI, corresponds to speeds of hundreds of thousands of kilometers per second. In other words, the speeds in quasar outbursts are close to the speed of light. Usually the speed seen in the sky is even 3–10 times higher than the speed of light, but such ultrahigh speeds represent an optical illusion. The true speed is close but not in excess of the speed of light.

In many cases there is a radiating band of light, called a jet, which originates at the nucleus of a quasar. It probably represents the orbital path of the matter flowing out of the nucleus. In many cases there is a continuation of the jet even to the outside of the active galaxy. Such a long jet has been found, e.g., in Cygnus A (see Fig. 26.5). The discovery of large-scale jets has led to a commonly held view that these are huge channels which transfer energy generated at the nucleus of the galaxy to the external radiation regions hundreds of thousands of light years away. So far no motion has been detected in large-scale jets, except very close to the nucleus where the motion takes place near the speed of light.

We do not know if the small-scale jets and the large-scale jets are simply connected. In fact, there is a big problem with the large-scale jets viewed as "power-lines." Considering the unavoidable energy losses in such channels, the amount of energy seen to pour out from the outer end of the jet is up to a million times greater than what is seen flowing in small-scale jets, and also a million times greater than what could be reasonably expected to come out from the surroundings of a super-massive black hole. An alternative is that large-scale jets could be trails left behind by black holes which have been thrown out of galactic nuclei. Each of the outside ra-diating regions would contain its own energy generating black hole, and no energy loss problems would arise because of the relatively short energy transfer distance (Fig. 26.8). So far there is no conclusive evidence one way or another.

Gravitational Lenses

In 1979 Dennis Walsh of Manchester University discovered two quasars which are only 6 arcsec apart in the sky (Fig. 26.9). He conveyed the news to Robert Carswell and Ray Waymann at Kitt Peak National observatory, USA, and asked them to

Fig. 26.9 The first gravitational lens Q0957+561 A and B (Credit: Bill Keel)

investigate the spectra of the quasars. They were greatly surprised to find that the two spectra were identical: they had the same spectral lines, and were equally strong and at the same redshift. It was truly amazing; up to that time quasars had been known for their individuality: every quasar spectrum is so different from the others that the spectrum is like a fingerprint. How is it possible to find identical twins, quasars Q0957+561 A and B?

The answer is that it really is only one quasar, but that it has been imaged twice because of a massive galaxy in the line of sight between us and the quasar. In this phenomenon, gravitational lensing (Chap. 25), the light from a quasar can arrive to us through different routes, bent from either side of the galaxy. In case of Q0957+561 the two routes are 6 arcsec apart in the sky. Galaxies are not designed to act as perfect lenses. They may form two, three, or more images of a single object rather than a single image. The gravitational lensing interpretation of Q0957+561 was confirmed by Alan Stockton at Hawaii when he found the galaxy which acts as a lens. Today, we know many other lenses (Fig. 26.10).

Gravitational lenses are useful tools (we saw in Chap. 25 how they reveal dark matter[4]). For example, the light in the two images of the quasar comes through different paths, so the travel times are somewhat different. One of the quasar images shows the quasar a little younger than the other image. The exact difference in travel

---

[4] In future the lenses may even reveal black holes. A lonely black hole is almost truly black, and it shows up only through its gravity. When it happens to lie in front of a distant star or a quasar, the image of the background object is magnified or split in a way which allows us to recognize the lensing action.

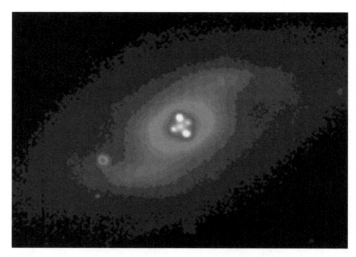

Fig. 26.10 A famous multiple image caused by gravitational lensing: the Einstein cross (discovered by John Huchra in 1984). The gravity of a galaxy splits the image of a distant quasar into four points (Photographed by A. Jaunsen and M. Jablonski at the NOT telescope, La Palma)

times can be used to calculate the overall length scale in the universe. The determinations of the Hubble constant in this way agree with the results by other methods.

Quasars and Their Relatives

In 1943 American Carl Seyfert discovered galaxies which have a bright nucleus (Fig. 26.11). Their spectra show that these nuclei are something like miniature quasars. Although the Seyfert nuclei are brighter than expected, they are fainter than the galaxies themselves, contrary to the genuine quasars. Therefore the "Seyferts" do not appear star-like, but are seen as galaxies. This intermediate class between quasars and galaxies shows that even in normal galactic nuclei there is potential for quasar-like activity. Seyfert galaxies are rather common (2% of all spiral galaxies) and are found fairly close to us; thus, they are easier to study than quasars. Similarly, radio galaxies are common; 10% of all elliptical galaxies belong to this category. When we learn more about these relatives of the quasars, we will get closer to understanding quasars. Even though the details of their functioning are still unknown, the idea that quasars are "big brothers" of Seyfert galaxies has been confirmed.

Seyfert galaxies come in three varieties. Seyfert 1's are closest to quasars, while Seyfert 3's at the other end of the classification look like ordinary galaxies and show evidence of an active nucleus only in their spectra. The Seyfert 1's are earlier type galaxies (typically Sa) than Seyfert 2's (typically Sb), which in turn are earlier than

Fig. 26.11 Seyfert Galaxy NGC 7742 photographed by Hubble Space Telescope (Credit: HST/NASA/ESA)

Seyfert 3's. "Early type" describes the fraction of stars in a bulge relative to the number of stars in the flat disk. The earlier the type, the more stars are found in the bulge. Elliptical galaxies with no disk at all are thus even "earlier" than Sa spiral galaxies.

An exciting fact is known: the mass of the central supermassive black hole is proportional to the mass of the spherical bulge. This explains at least partly the Seyfert sequence: "small-bulge" Seyfert 3's show weaker nuclear activity than the early type Seyfert 1's because Seyfert 1's have bigger black holes than Seyfert 3's. This reasoning may be extended to quasars: they have even bigger black holes than Seyfert 1's, and show therefore greater activity in their nuclei, since quasars are associated with elliptical galaxies. Radio galaxies are between Seyfert 1's and quasars in the sense that their central black holes tend to be intermediate in mass between Seyfert 1 black holes and quasar black holes.

Before the central supermassive black hole can become bright, it has to be "fed"; gas has to be channeled to its vicinity. We know at least two ways of doing it. Gravitational tides caused by a companion galaxy can perturb the disk of the galaxy and lead to a greater flow of gas into the central black hole. This could lead to the increased activity in Seyfert galaxies, as compared with "normal" nonactive galaxies. In mergers of galaxies, where a bigger galaxy swallows a smaller one, the central black holes of the two merged galaxies settle at the center of the new galaxy where they form a binary system. This binary attracts gas better than a single black hole. Indeed, in quasars one often finds signs of a past merger of two galaxies. In some cases there is also evidence of a binary nucleus.

Whatever the cause of quasar activity may be, one fact is clear: there were many more quasars in the past than today (i.e., an excess of high redshift quasars relative to low redshift ones). Similarly, bright radio galaxies were more common long ago. At redshift $z = 0.5$ the number of quasars and radio galaxies is five times greater than in our nearby universe, at $z = 1$ it is already 50 times greater, and at redshift 3 it is 1,000 times greater than locally. When the redshift is 0.5, light started from the quasar toward us 5 billion years ago, the redshift 1 represents the time 8 billion years in the past, while $z = 3$ takes us back in time by 12 billion years (assuming that the age of the universe is 14 billion years).

Earlier the distances between galaxies were shorter than today. Due to the expansion of the universe, the distance scale is inversely proportional to $1 + z$. So at the redshift 3 the average distance between galaxies was only one-quarter of what it is now. Thus tidal interactions between galaxies were stronger, and also galaxies merged more frequently than they do at the present time. These are thought to be the primary reasons for the great quasar activity at $z = 3$, and why the activity has declined since then.

When going to times earlier than what corresponds to $z = 3$, the number of quasars and radio galaxies does not increase any more. On the contrary, we see fewer and fewer quasars when we look beyond redshift 3. Why is this? In the current view, galaxies were assembled gradually from smaller pieces between redshifts $z = 30$ (corresponding to the time only 100 million years after the Big Bang) and $z = 3$. At the same time, black holes formed in centers of the protogalaxies, and they also grew in mergers between these early galaxies. It was not until two billion years post-Big Bang ($z = 3$) that there were fully grown galaxies, with big central black holes in large numbers. These could become full-scale quasars. Earlier quasars were more rare phenomena and we know only a few with redshift 6 or greater.

We saw that quasars were probably born together with their host galaxies and grew in their centers. In the next chapter we turn to the final question in this part of our book: how did the galaxies originate?

Chapter 27
Origin of Galaxies

To follow the modern view of the origin of heavenly bodies, we must go back to the ancient time where we left the story of the world periods in Chap. 24: at about 400,000 years after the Big Bang, when the universe was filled uniformly by a hydrogen–helium gas at the temperature of $3,000°C$. Today, when about 14 billion years have passed, we find that galaxies have formed and we are inside one of them, the Milky Way. Its estimated 200 billion stars and innumerable gas clouds of different sizes orbit around its center. When we look further, we see an enormous expanse of galaxies, more or less like our own stellar system. The hundreds of billions of galaxies in the sky are widely separated in space but belong to just a small number of classes being mainly elliptical and spiral, both of these made mainly of dark matter and a smaller amount of stars and gas. This suggests that the galaxies have their deep roots in just one basic process, occurring everywhere in the universe. How was the smooth and featureless state transformed into the complex system of superclusters, voids, and chains we see today?

Cosmic Eggs or Cosmic Seeds?

One may imagine two different processes of structure evolution: either some big thing may be broken up into small pieces (a popular view in old myths about a cosmic egg) or many small lumps are gathered together to make a big thing. Yakov Zeldovich (1914–1987) with his colleagues at University of Moscow considered a scenario where large structures formed first, and they subsequently fragmented into smaller pieces. The large structures would have been gas clouds more massive than clusters of galaxies. When they collapsed fastest along one axis, pancake-shaped structures would have arisen. Later the pancakes ("blinis" in Russian) fractionated into galaxies. This would explain why even today many galaxies are distributed sheet-like. However, the discovery of very distant galaxies, beyond $z = 6$ or even at a redshift as high as 10, i.e., very early in the history of the universe, contradicts the fragmentation scenario where galaxies are born much later.

P. Teerikorpi et al., *The Evolving Universe and the Origin of Life*
© Springer Science+Business Media, LLC 2009

In their correspondence in the 1590s Newton and Bentley pondered about the behavior of uniformly distributed matter in space (Chap. 23). In this connection Newton expressed a magnificent idea: "... *if the Matter was evenly disposed throughout an infinite Space it could never convene into one Mass, but some of it would convene into one Mass and some into another, so as to make an infinite Number of great masses, scattered at great Distances from one to another throughout all that infinite Space. And thus might the Sun and fixt Stars be formed, supposing the Matter were of a lucid Nature.*"

We see that Newton outlined a process of star formation. Matter uniformly spread in an infinite universe is unstable: condensations tend to form from little density seeds under the attractive force of gravitation. This is basically the same process which is today studied by astronomers attempting to understand the birth of galaxies. It is generally accepted that small units were born first and then gathered together to make larger structures. The smallest lumps that contribute to the halos of present-day galaxies were about a million solar masses. Gradually they merged to make bigger lumps, until a whole range of galactic halos were born, from dwarf spheroidals (with a few million solar masses) to giant halos (with a few trillion solar masses). The halos gathered surrounding gas and at later stages some of the gas turned into stars. Thus were the visible galaxies formed. At the same time, galaxies gathered into clusters, around cluster-sized dark matter halos. The dwarf spheroidal galaxies and globular star clusters are thought to be fossils of this early stage of galaxy evolution.

From Density Condensations to Galaxies

At the end of the Age of Radiation Domination the density of gas was low, similar to what we observe in the tenuous interstellar gas clouds today. Before it is possible to make a star out of this material, one has to compress it by ten million fold. In the initial near uniform gas, one had to be satisfied with more modest compressions, not resulting in stars but instead in early forms of galaxies and their clusters.

The reason for the condensations to develop at all is the attraction of gravity. The larger is the region, the greater is the force of gravity with respect to other forces. The main opponent to gravity is the internal pressure of gas; it depends on the temperature and density of the gas. One may calculate how big a cloud of gas one has to gather together before it starts to collapse under its own gravity; below this it does not collapse, above it does. The mass contained in such a barely collapsing cloud is called *Jeans mass*, and the size of the cloud the *Jeans length* (this critical size is *directly* proportional to the square root of the gas temperature and *inversely* proportional to the square root of the density) (Fig. 27.1).

During the Age of Radiation Domination the Jeans length was so great, as great as the observational horizon at that time, that gravitational collapse was not possible at any scale. This goes along with one's intuition that it is difficult to make structures out of radiation. Soon after the end of the radiation era, the gas pressure dropped

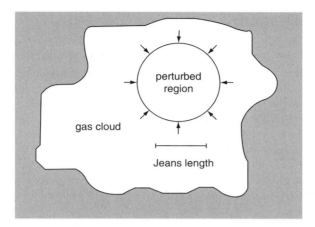

Fig. 27.1 When the perturbation in a gas cloud is smaller than the Jeans length, the perturbation does not condense. Only when the perturbed region is equal in size or greater than the Jeans length does condensation take place. In other words, the mass of the perturbed region has to be equal or greater than the Jeans mass in order that the density continues to grow. The concepts were originated by James Jeans

suddenly, as a result of which the Jeans mass was lowered to a few hundred thousand solar masses. The masses of galaxies are greater than this which means that the formation of galaxies became viable. But were there any seed condensations which could have later evolved into something so dense and big as galaxies?

How big should these initial seeds be in order that they could grow into galaxies within the available time? The condensations grow at roughly the same rate as the universe expands. For example, the perturbation may have been one part in a thousand, that is, for every thousand atoms there would have been one extra atom that we would call the initial perturbation. When the universe has expanded tenfold, there would have been ten extra atoms among thousand atoms in the same condensation. After a universal expansion by a factor of 100, there would have been 100 extra atoms per thousand atoms (a 10% perturbation). The perturbation becomes 100% after the expansion by 1,000 times, i.e., the original thousand atoms have attracted another thousand atoms to their neighborhood. At this stage the blob is so clearly separate from the surrounding gas that it collapses to some structure while the universe expands only by another factor of 2. What comes out of it depends on the mass of the condensation.

The universe has expanded by a factor of about 5,000 since Radiation Domination ended; so the aforementioned 0.1% perturbations are what we need to start the process in motion to make today's galaxies. But there is an extra twist to this. We observe small extrawarm patches in the cosmic background radiation, as discussed earlier. These patches tell us the actual level of graininess in the cosmic gas soon after the end of the radiation era. And what we measure is that the seeds were far too small, the gas was far too smooth for galaxies to have evolved from them. So why do we have galaxies at all?

We Need Dark Matter

The answer is, as we understand it today, that the decisive perturbations were not in ordinary matter but in the dark matter. The dark matter, not being influenced by radiation, could start clumping earlier even during Radiation Domination when ordinary matter had too large Jeans' length. When the patches of background radiation were born, they reflected clumping only in ordinary matter. There were much greater perturbations in the dark matter, at the 1% level or so, already at this time. Subsequently, these dark matter clumps condensed and gathered the surrounding matter in their folds. Thus the first structures to form were halos of dark matter. They became the backbone of galaxies which assembled later from numerous mergers between halos, and from matter drawn into the halos.

Where do such 1% perturbations come from? It is expected that the "seeds" existed in the dark matter relatively early on. Initially there must have been condensations like those that occur in sound waves (pressure waves), propagating from place to place. Did somebody shout out loud in the beginning? In fact, one may suggest natural processes giving rise to the sound waves. For example, they could have arisen at the end of the Age of Inflation. If the transition from inflation to normal expansion was not quite simultaneous everywhere in the universe, as is dictated by Heisenberg's uncertainty principle, then the still fast-expanding parts of universe collided with slower parts, and pressure waves were generated. These waves may have survived until the dark matter separated from radiation, at the end of the Weinberg–Salam Age (see Box 24.1), and thereafter they may have evolved into slowly contracting condensations.

One idea for the generation of seeds is the action of cosmic strings. The strings are thought to be fault lines in the space metric where the space is strongly curved. We may visualize the strings as invisible threads going through the universe which can be detected only through their gravity. The cosmic strings have never been detected but according to the Grand Unified Theories of physics, large numbers of them might have been born during the Age of Grand Unified Theories. Gradually the strings would have wound up in simple loops a few years after the Big Bang. Then closed string loops could be the points in space around which protogalactic perturbations gathered. The strings would contract and finally disappear while their energy content turns into gravitational waves and particles. Therefore it is no use looking for cosmic strings in present-day galaxies; if it is true that they started the process of galaxy condensation, then they would have disappeared well before the galaxies formed.

Formation of Large Scale Structure

The galaxies are arranged in chains and sheets, separated by almost empty voids; however, this evidence supports neither the cosmic strings nor the cosmic pancakes. We know now that the natural tendency of galaxy clustering in an expanding universe is toward these kinds of structures, when gravitation gathers small pieces together. But nobody could beforehand guess what kind of landscapes gravity can create, and these were first noticed in computer simulations by the Norwegian Sverre Aarseth at Cambridge University and his associates in 1979. Around the same time, large surveys of the distribution of galaxies were begun at Harvard by Huchra, Geller, and others, inspired, e.g., by the maps presented by Jõeveer and Einasto at the Tallinn 1977 meeting (Chap. 22). One could start detailed comparisons between the complexities of the real universe and the world built in a computer.

Today research teams around the world use high-speed computers and sophisticated computing codes to create three-dimensional maps of possible end results of gravitational clustering in an expanding universe. It has been shown that gravitation can explain the formation of galaxy structures on scales from about one million to two or three hundred million light years (for an example see Fig. 27.2).

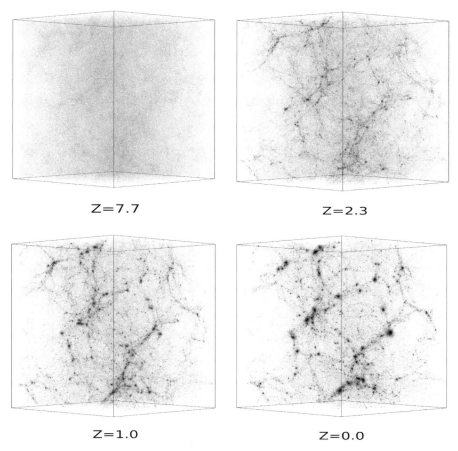

Fig. 27.2 Galaxies are distributed in space in a complex nonuniform fashion. Rather similar landscapes are revealed by computer calculations. The simulation in this series of pictures follows the evolution of 130 million dark matter particles from the rather smooth state at redshift 7.7 to the complex honeycomb-like distribution at the present time ($z = 0$). The side of the box is about 200 million light years (60 Mpc). The calculation was performed on 256 processors of a Cray XT4 Massively Parallel Processor (MPP) supercomputer at CSC, the Finnish IT center for science (courtesy of P. Nurmi, S. Niemi, J. Holopainen, P. Heinämäki)

Gravity is skillful in making complex things from simple initial states. William Saslaw (Virginia and Cambridge, England) asked what happens if originally galaxies were simply distributed like randomly thrown grains on a sheet of paper (Poisson law; Chap. 22). He has found that the way gravity pulls galaxies together, it is able to make a highly structured universe out of randomly distributed units. All it takes time. Start with galaxies randomly spaced in the universe less than a billion years from the beginning, and today you will find clustering with about those properties that we observe in the sky.

Long ago also Immanuel Kant envisioned in his *Universal Natural History* how the simple law of gravitation creates structures: "... without the aid of arbitrary notions, a well-ordered whole arises under the direction of established laws, a whole so similar to that world system which we have before our very eyes that I cannot prevent myself from holding it to be the same." Kant had in mind an overly simple hierarchy in a static universe, but we may today feel much the same seeing the agreement between the real universe and the simulated one. There are still open problems in this difficult field, but we feel confident that gravity is the main architect behind the impressive structures in the galaxy universe.

Generations of Galaxies

According to the current view, during the first two billion years, the first structures to form were small dark matter halos, intermediate in mass between large globular clusters and small dwarf galaxies of today. Each small halo had its own central supermassive star which subsequently exploded and left behind a black hole remnant. At the same time, the supernova explosions produced the first heavy elements which are necessary for forming normal stars. Galaxies from the first generation were made of dark matter primarily. Only at a later time did ordinary stars start to form. The initial evolution consisted of numerous mergers between these small galaxies and of gradual growth of giant galaxies.

The birth of the first superstars shows up in the background radiation. When the microwave radiation travels through the photon field created by the stars, the microwaves become polarized. It allows us to time the origin of the first galactic halos and of the superstars born in their centers, to about 200 million years after the Big Bang. So far this figure has still large uncertainties, but it should be fixed well by the measurements of the future Planck satellite.

The chemical composition of galaxies evolved throughout this period. Initially the composition of primordial gas was 76% hydrogen and 24% helium, with no heavier elements. The stars which formed out of this gas are thought to contain about 300 solar masses each and to live only a few million years before exploding as a supernova. Such pure hydrogen–helium gas does not exist anymore, and star formation has shifted to smaller size stars, such as the Sun. The presently important constituents of interstellar matter, e.g., carbon, nitrogen, oxygen, and heavier elements were born in subsequent processes of stellar evolution. Initially there could not be interstellar dust or planets which require heavy elements for their formation.

In galaxies, gas condenses to form stars, and at the end of their life cycles, stars return some of the reprocessed matter into interstellar medium. Some of the gas becomes tied up in small-sized, long-lived stars or in stellar remnants. This gas is removed from the circulation in the galaxy. It is replaced to some extent by gas falling from outside the galaxy, but the net effect is that generally the star formation rate in galaxies is slowed down. It affects also the appearance of galaxies: they become on average redder with time since the fraction of newly born blue stars is diminished. This evolution shows up especially in elliptical galaxies.

The Young Milky Way and Stellar Populations

We will now consider in detail the processes for a typical spiral galaxy, our own Milky Way. We think that it has followed the same route of evolution as galaxies in general. It was formed in mergers of dark matter halos which had the added mass of close to a thousand billion solar masses. The total number of original halos may have been in millions; we do not have much evidence for them except for a few leftover dwarf spheroidal galaxies that are nearby the Milky Way. The gas clouds of ordinary matter fell toward the center of the halo and fractionated into stars. Some of these stars have survived in globular star clusters, others were scattered about to form the stellar halo of the galaxy.

The stars of the first generation in galactic halos were much more massive than the stars which we see in the sky today. These stars were probably 300 times more massive than the Sun and lived only for a few million years. At the end of their lives such stars would have exploded as supernovae. The stars would therefore have produced the first elements heavier than helium and mixed them with the interstellar gas after their explosions. The center of the supernova collapsed into a black hole, heavier than 100 solar masses. The explosion also blew off any remaining gas out of the dark matter halo. Thus only one star per halo could be born, and all or most of them became black holes. For an external observer the product of this early evolution would not have looked like a galaxy at all: the dark matter was invisible, the black hole also, except for the disk of gas which possibly remained around it. Thus the universe appeared as a universe of gas clouds, with an occasional black hole swimming through it.

The evolution continued with merging of halos. It also meant that the black holes at the centers of halos were pulled to the common center of the merger where they formed binary black holes. Yet another merger, and the newly merged halo would have had a triple black hole system at its center. While the process of halo mergers went on and on, more and more black holes would be brought together. However, we remember from the three-body problem that as soon as three black holes are brought together, the system is unstable (Fig. 11.3). One black hole is thrown away by the other two, and both the single black hole and the binary (by recoil) escape from the center in a process called slingshot. Thus begins an *Age of Slingshot*, when this process dominates evolution until large halos are built up, and black hole escape becomes less frequent. The Age of Slingshot comes to end around redshift $z = 6$ when the large black holes are finally built up in the centers of full size galactic halos. The slingshot process should go on even today, but with much lower frequency since galactic mergers are now relatively rare.

What remains in the Milky Way from the Age of Slingshot? The pieces of halos which make up our Milky Way halo have been fully mixed into a single unit. Many of the black holes which escaped from the small halos left the system permanently, but a good fraction should be captured in the newly formed Milky Way. They should still be there today, circling the Milky Way center in elongated orbits, but fully invisible. To be seen, they would need to have a disk of gas around them. However, such a disk is gradually swallowed into the black hole, and the process of swallowing the disk would have been completed long ago. How many of them there are and how much of the mass of the galaxy is in these remnants of the first generation of stars is still an open question.

One way these first-generation remnants might show up is through gravitational lensing. A black hole will brighten the light of a background object such as a star in another galaxy. Since the black hole and the star are in motion relative to each other, the lensing magnification lasts only for a short time, the time during which the two are sufficiently lined up. This method has been used extensively to look for dark bodies in the Milky Way halo, and 17 lensing objects have been detected so far. Some of these have also been visually detected and found to be dim cool stars, not massive black holes. There are not enough of these low mass objects to explain the total mass of the halo.

The gas out of which the next generation of stars formed had some elements heavier than helium, even though the heavy element abundance was still low: less than 0.1% of what is found in gas clouds today. Stars could now form in their normal mass range, from below 0.5 solar mass to over 15 solar masses.[1] The lower mass ones are still around us as main sequence stars, and they are mainly found in the spherical component of the Galaxy and in globular star clusters. The heavier ones fused heavy elements in their interiors, from carbon, nitrogen, and oxygen all the way to iron and nickel. The very heaviest stars exploded as supernovae, producing even heavier elements from nickel to uranium and thus spreading more heavy elements in the interstellar medium.

As mentioned in Chap. 24, the origin of elements in stars and in their explosions was first explained by Fred Hoyle and his associates in the mid-1950s. The team consisted of William Fowler who was later rewarded by Nobel Prize for his contribution to this work, and Margaret and Geoffrey Burbidge.

---

[1] The change in the mass range of formed stars is related to cooling: even the low heavy element abundance was enough to make the cooling more efficient than in the pure hydrogen-helium gas. Collisions of H and He atoms do not produce low-energy photons that escape the star or cloud (i.e., cool) carrying energy away.

How Old is Our Milky Way?

The abundance ratios of the very heaviest isotopes can be used to study the age of our Milky Way. For example, both the uranium isotopes 235 and 238 are radioactive; their half-lives are 713 and 4,510 million years. Since the isotope 235 disintegrates faster than the isotope 238, the amount of the former isotope relative to the latter isotope decreases continuously. The current global ratio is 0.00723. Calculating backward, we find that the ratio at the time of origin of the Solar System 4.6 billion years ago was 0.31. Using this method, Rutherford concluded already in 1929 that the Milky Way must have existed billions of years before the Solar System.

What was the original 235 to 238 ratio? The Canadian astronomer Alastair Cameron and Geoffrey Burbidge and his associates were first to calculate in 1957 that stellar explosions actually produce 50% more uranium 235 than uranium 238. Therefore the original isotope ratio was 1.5, but with passing of time the ratio has become smaller in the interstellar gas. The explosions of stars have taken place throughout the history of the Milky Way. We can start from the ratio 0.31 and go back in time, considering both the raising the ratio due to supernovae, and lowering it through radioactive decay. If the supernova explosions have always taken place at the same rate as today, the isotope ratio reaches its original value about 10 billion years before the Solar System was born. On the other hand, if the supernovae were more common in the young Milky Way, as suggested by other evidence, then the original isotope ratio is reached at a shorter time. In 1980 Fowler estimated using this method as well as other isotope ratios that the production of heavy elements started 4–8 billion years before the origin of the Solar System. More recently, Roger Cayrel of the Observatoire de Paris-Meudon and associates have derived the value 8 billion years, plus or minus 3 billion years. This makes the age of our Galaxy about 12.5 billion years, conveniently smaller than the age of the universe, about 14 billion years.

The uncertainty in the derivation of the age of the Milky Way by radioactive dating is unfortunately large (especially in comparison with the accurate radioactive timing of minerals on Earth and in the solar system; Chap. 29). However, there are other methods with perhaps somewhat greater accuracy. One can use the main-sequence lifetime of low mass stars to determine the ages of globular star clusters which are probably the oldest surviving components of the Milky Way. When one plots the Hertzsprung–Russell diagram (Chap. 19) of a globular cluster, one typically finds a well-defined main sequence which ends abruptly at some point. The sequence of stars leading to red giant stars starts from the termination point of the main sequence. The lifetime of a star in the main sequence depends fundamentally on its mass. For example, the MS lifetime of a 0.8 solar mass star is 14 billion years, while a 1.1 solar mass star spends only 5.1 billion years in the main sequence before starting to evolve toward the red giant phase. The ages of globular clusters were first determined by this method by Allan Sandage in 1953. In 1970 he derived the average value of 11.5 billion years for four globular clusters, while the latest results by Lawrence Krauss and Brian Chaboyer in 2003 give the average age as 13.2 billion years, plus or minus one billion year.

Fig. 27.3 Ancient white dwarfs, about 12–13 billion years old, in the globular cluster M4. The *top panel* shows a view of the whole cluster in the constellation of Scorpius. The *box* at *bottom left* shows a small area photographed by Hubble Space Telescope. In the still smaller area (*bottom right*) very faint dwarfs are pinpointed by *circles* (credit: NASA and H. Richer (University of British Columbia); NOAO/AURA/NSF)

There is another method which uses white dwarf stars (Fig. 27.3). White dwarf stars cool down slowly, and within the age limit of the Milky Way, they must still be rather hot on their surface. They have not been able to cool much below 4,000 K. The task is then to find the coolest white dwarf star; it will be the oldest, and from its surface temperature we can calculate its age. In 2002 Brad Hansen and his associates derived the globular cluster age as 12.7 ± 0.7 billion years. In the same work it was found that the white dwarfs in the disk of the Milky Way are considerably younger which testifies to the late of formation of the disk inside the dark matter halo. All in all the evidence points to the age of the Milky Way not being far from 13 billion years, while its disk was gathered gradually billions of years thereafter.

No first-generation stars, consisting of pure hydrogen and helium, have ever been found. It probably means that all of them were massive and exploded. In recent years some stars have been discovered which contain very little heavy elements; these fit the picture of being the very first second-generation stars. More commonly one finds stars with about 1% of the heavy element abundance as compared with the Sun. Traditionally these are called population II stars, not because they came second after the first generation, but because they precede the "ordinary" stars like the Sun. This current star population (population I) represents stars removed from population II by possibly many generations. All population II stars as heavy as the Sun or heavier have already passed through their whole sequence of evolution; only the low mass stars of this population are still found shining brightly, with quite some lifetime still left in them.

The Changing Milky Way

The hazy belt in the night sky is a symbol of constancy, but when we look at it from the perspective of millions of years, our stellar system is full of change. The disk of the Milky Way consists of stars of different ages as well as of gas clouds. New stars are born continuously in the spiral arms of the disk, in bands wound up in a spiral shape. At the Sun's orbital radius in the Milky Way, the spiral is thought to be a density wave which travels in the Galactic disk of stars. Or more exactly, the wave travels around the center of the Galaxy more slowly than stars and gas clouds orbit it. For that reason, stars and gas clouds pass through the density wave once every 200 million years or so. Stars pass through the wave without experiencing major influences, but the gas clouds are compressed by the wave, and this compression is an adequate push to start the formation of new stars. It is the newly born bright stars that give the spiral arm its impressive appearance.

Because the stars and gas continue to orbit the galactic center, spiral galaxies keep producing new stars. But even in the spiral galaxies the rate of star formation weakens gradually when some of the gas is taken out of circulation when it is locked up in stars. In the end the gas is used up so thoroughly that the star formation practically stops.

Now we can summarize the history of our home galaxy. In main outline it goes as follows. A large number of smaller dark matter halos merged to make the dark halo of the Milky Way. The gas of ordinary matter fell into the "potential well" of dark matter. Already during the early stages of infall some stars formed and they remained in outer parts of the stellar halo. The heavier ones of these stars exploded and increased the heavy element content of the galactic gas. The newly enriched gas, together with the gas falling from outside, settled in a disk at the center of the dark matter halo. The disk gas made new stars, and gradually the stellar disk of the Milky Way was built up. Our Sun is one of these later formed disk stars which were made of the original light elements of the Big Bang plus a small amount of heavier elements resulting from several generations of star evolution. The level of heavy element content around such stars was high enough that the planets like the Earth could be constructed. On at least one of such planets the very complex phenomenon we call life was born.

Our present view of the gravitational origin of galaxies and their huge systems is undoubtedly impressive. It again makes one think of certain words by Kant, the early theorizer about planetary systems and the realm of nebulae. He predicted that the origin of "the whole present arrangement of the world edifice" will be sooner understood (on the basis of gravity) than the production of even a simple living being (his example was a caterpillar) will become fully clarified from mechanical reasons.

Of course, we now know much more about the universe and its structures than Kant did and understand that gravitation is a more versatile architect than perhaps suggested by the simple-looking Newton's inverse square law. We know that the fraction of ordinary matter which can make stars, planets, and life is minute in comparison with dark matter and dark energy which are important for the evolution of the expanding universe as a whole and for the formation of galaxies, those great domiciles of stars.

Nevertheless, we still agree with Kant that life is a much more complex phenomenon than even a huge cluster of galaxies with its thousands of members, each containing 1–100 billion stars and planetary systems! There are about 100 trillion ($= 10^{14}$) cells in a human body, each having about the same number of atoms. This leads us to the last part of our book where we will discuss the origin of planetary systems and questions of astrobiology.

Part IV
Life in the Universe

Chapter 28
The Nature of Life

At this time and location, about 14 billion years after the enigmatic beginning of the universe, in the outer parts of an average galaxy, in a planetary system formed 5 billion years ago around an ordinary star, we observe a special phenomenon: the surface of one of the planets is covered by a biosphere, or a complex network of organic chemistry taking place in water solution. This chemistry is driven mostly by the energy streaming from the star, and it maintains a diversity of living beings, from unicellular microbes to large plants and animals. These form complex ecological communities, with long energy-transferring interactions (food chains), which effectively circulate carbon compounds between oxidized and reduced states. In particular, photosynthesis using sunlight by the green plants and algae converts the oxidized carbon, CO_2, into reduced carbon compounds (sugars), which are used as chemical energy by other organisms. The photosynthesis binds large amounts of carbon into organic compounds (biomass), while animal respiration and decay of organic material again release CO_2 into the air. These reactions have strongly affected the carbon dioxide content of the atmosphere, and thus the climate. The photosynthesis process takes its reducing power (protons) from water molecules, H_2O, and through this mechanism it has produced the oxygen in the atmosphere of the planet.

Life and the Universe

Life is composed of the most common elements beyond hydrogen and helium, i.e., of carbon, hydrogen, oxygen, nitrogen, phosphorus, and sulfur (cf. Box 12.1). In spite of this, life differs significantly from the surrounding inanimate world. It is based on very complex chemical compounds, and is all the time performing complex biochemical reactions, which could not happen in an inanimate milieu. Thus, life induces a strong increase of order in its structures compared to a simple collection of its constituent atoms. In other words, it reduces the entropy in its system (see Box 28.1). Life seems to break the second law of thermodynamics. However, this is not the case. The order is produced by utilizing energy from the environment and is controlled by vast internal information embedded in complex molecular structures. The living system is in a strong disequilibrium with its surroundings.

Box 28.1 Entropy

It is our experience that many things left on their own gradually lose their strict order or structure, some may even disrupt into a pile of dust. In physics, the Second Law of Thermodynamics states that when a physical process goes on without interaction with the external world a quantity called *entropy* in such a closed system always increases. This contrasts with the total *energy* which is conserved in a closed system (according to the First Law of Thermodynamics).

Entropy characterizes the level of order – more entropy means more chaos. One can also say that entropy roughly means the number of separate units in a system; what is whole in the beginning, tends to break into pieces in the end, approaching the most likely state. This tendency also defines the arrow of time in real life, even though in simple mechanics the direction of time is not known to exist.

A common way to see the entropy growth in action is to consider a box filled with gas. To begin with, suppose the extremely improbable situation that at some moment of time all the myriads of molecules are found in one-half of the box, while the other half is empty vacuum. It is clear that after this moment the molecules tend toward filling the box completely and in a uniform manner. Such a situation is the most likely one and corresponds to the maximum entropy. Note that this natural tendency toward "chaos" depends on the assumption that the system (the box plus the gas) is isolated – one might well imagine external influences that could drive the system from its "most probable" state to an apparently "improbable" state. Life is such a phenomenon which superficially looking may seem to violate the law of increasing entropy. However, we should remember that life cannot exist in an isolated box, but it critically depends on a flow of energy from the environment into the living system and back out again; when one takes into account the biosphere and its wide cosmic environment (including the star giving the energy), the entropy of this whole region is indeed increasing.

Life is not just an ordered system which gets the energy and chemical nutrients from the environment. Life can maintain itself and reproduce, and with the reproduction process, it gains new properties and adapts to new environments. This adaptation potential has produced a diversity of species, and new traits have developed, such as multicellularity. The appearance of many different survival strategies has led to increasingly complex organisms and ecosystems. Development of instincts and learning capabilities in vertebrate species has improved their adaptation and survival in new ecological niches. Growth of intelligence has also produced social behavior, curiosity, and communication within the species, and even between different species.

So far, the highest level of consciousness and intelligence has been achieved in the human species. These properties are manifested in humanity's spread to different habitable locations on the Earth which has been explored and mapped. We have cataloged most other species, and basic cell and molecular biology are known by now. The exploration is driven by curiosity, the need to know what things are and how they work. Another driving motive is the desire to use natural resources, and thus humanity has greatly affected the biosphere on the Earth over the last several centuries.

We have learned much about the world, and about life as it now exists here. However, the relation between life and the universe is still unclear. How did life start and in what conditions? Is its origin a very rare event, or does it happen easily? Do the basic elements of life (C, H, O, N, P) have an innate tendency to form complex structures leading to life? Is life necessarily of "our kind," or could it be very different? If habitable places exist elsewhere, do they harbor life? If there is alien life, what is it like? How diverse or complex can life get? Could different life forms communicate with each other?

Thus far we have not yet found any life elsewhere. Based on this, we may think that life is rare in the universe – but this may just reflect our current ignorance, due to the fact that we have not been able to observe the signatures of life over large distances, even within our own Solar System or elsewhere. On the other hand, life did start on Earth fairly quickly, almost as soon as the conditions here had become suitable for it. Does this suggest that life is common throughout the universe? Again, we do not know whether those initial conditions here were very special. Thus, even if there may be habitable conditions elsewhere, we do not know whether such places ever supported the *origin* of life. Finding solutions to these questions is the aim of a new interdisciplinary field of science, *astrobiology*.

Our Changing Views of Life

Animals wander about, reproduce, and pass away. Plants grow up and flourish during the growing season. Even the tiniest creatures, invisible unicellular microbes, are able to reproduce in suitable conditions and to fill up their available growth space. Inanimate objects lack these functions. Earlier it was natural to think that inanimate things had become alive by obtaining a force of life, *vis vitae*. At death the force would exit. This theory "vitalism" was accepted widely for centuries and a lot of discoveries had to take place before it was abandoned, and we learnt to view life as a special physical and chemical process. The versatile Robert Hooke (he appeared in Chaps. 8 and 10; Fig. 28.1) saw plant cells through his microscope in 1665 and in fact used the word "cell" to describe them. Then it took a long time that study of cells really began. The Scottish botanist Robert Brown (1773–1858) noticed in 1831 a dark object inside the cell. This was the nucleus.

Matthias Schleiden (1804–1881) and Theodor Schwann (1810–1882), both from Germany, are viewed as fathers of the cell theory. Schleiden studied law and became an attorney. Only later did he pick up biology studies. In 1838 he set forward an idea that the growth and development of beings is based on the birth of cells. He envisioned that cells grew around the nucleus. The same year Schwann suggested

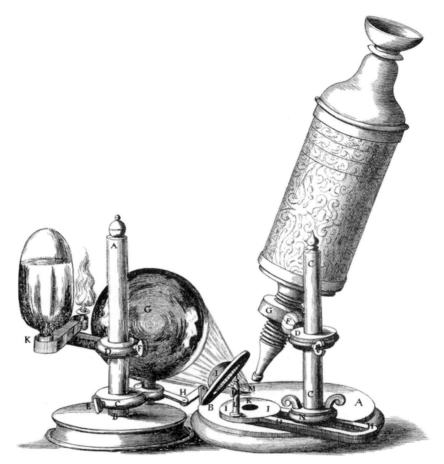

Fig. 28.1 Robert Hooke's microscope from his book *Micrographia*. He coined the word "cell"

that plant and animals have the same basic unit, the cell. Thus he tore down one border fence between the plant and animal kingdoms. Later studies showed that the essential parts of a cell are cytoplasm (a sort of a liquid), the nucleus, and a large number of small organelles.

The German Oskar Herwig described in 1876 the fertilization as a fusion of a sperm cell and an egg cell. In 1882 (the decade astronomers started to photograph nebulae) Walther Flemming presented the first photos of the division of the cell and its nucleus. Now the cell became viewed as the "atom" of life, both structurally and as an operational unit. The role of the nucleus was emphasized.

The Basic Structures and Functions of Life

Life on Earth is very diverse, ranging from giant redwoods and whales to unicellular bacteria, and from very actively replicating cells to quiescent or even dormant

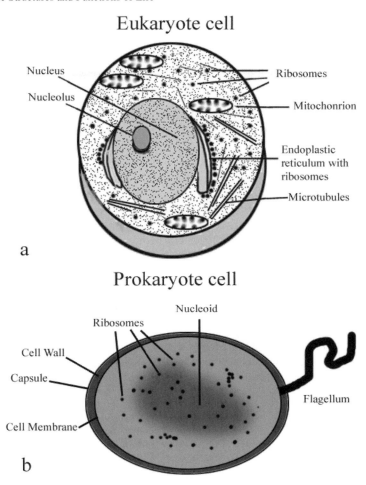

Fig. 28.2 Two kinds of cells. (**a**) Eukaryote cell (containing a nucleus); (**b**) prokaryote cell (without a nucleus)

resting stages. However, with all this diversity, life forms on Earth have basically very similar chemical structures and reactions to mediate their functions. The cellular functions of even the simplest creatures are very complex, with a multitude of chemical reactions. We will describe only those features of life common to all life forms – the hallmarks of life on Earth. The uniformity of these properties indicates they are derived from the one common origin, the Last Universal Common Ancestor (LUCA) of all life forms.

The basic unit of life is the *cell* (Fig. 28.2). It can be of different shapes, but it is usually microscopic. We can view the cell as a minute factory, where a large number of complicated actions are taking place all the time. The cell is surrounded by a semipermeable membrane with "gates" and "pumps," though which the cell gets nutrients and other molecules from the outside. The same membrane also functions in the opposite direction letting out other molecules, such as waste products.

The interior of the cells is filled with a water solution, *cytoplasm*, and with a multitude of different macromolecules (large molecules which indeed crowd the space). In the simplest cells, the volume enclosed by the membrane is one single compartment. In more advanced species the cell has a separate coordination center, the *nucleus*. The species which do not possess nucleus are *prokaryotes*, and those with a nucleus are known as *eukaryotes*. Prokaryotes are divided further into *bacteria* and *archaea*. The archaea (often thriving in extreme conditions, such as high temperatures) were not discovered to be separate from bacteria until the 1970s. The eukaryotes include many unicellular animals and plants as well as most multicellular beings (like us) composed of systems of cells. Together the eukaryotes, bacteria, and archaea form the three known *domains* of life.

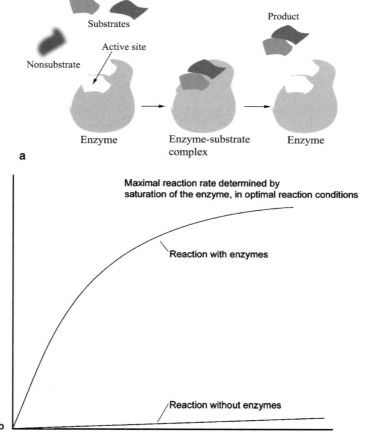

Fig. 28.3 Schematic presentation of enzymatic catalysis. (**a**) Enzyme catalysts bind to the reacting molecules (substrates) and hold them together in optimal positions so that they can react easily and efficiently to create the product. (**b**) Qualitative effect of enzymatic catalysts on biochemical reaction rates as a function of time

Chemistry of Life

Proteins are the main labor force and structural material in cells. They come in many different shapes and sizes, and each has a specific task in the cell. Some of them form essential cellular structures, such as the filamentous cytoskeletons, components of the cell walls, or the very varied "pump" and "gate" structures spanning the cell membrane. Another important function of the different proteins is to act as chemical signals, and the regulation of other protein functions. For instance, the expression of each unit of inherited information (the gene) is mediated and regulated by multiple other proteins, as is also the activation or inactivation of the gene products.

An important role of proteins is their function as enzymatic catalysts. The catalysts participate in all biochemical reactions by binding to the reacting molecules (substrates), and by holding them together in optimal positions so that they can react easily and efficiently. The presence of enzyme catalysts can enhance the reaction rates by many orders of magnitude, and make the reactions happen efficiently in mild conditions and from low concentrations of reacting materials or substrates. They also take care that the right reactions happen, and not the wrong ones. The catalysts are really essential "helpers" to make the biochemical reactions happen, and indeed, most of the cellular reactions would stop in the absence of enzymes (see Fig. 28.3). A big dilemma in regard to the very earliest life forms is how they could accomplish any of the necessary functions for their replication and survival, as they were not yet able to produce the enzymes needed for these functions.

In living cells, a multitude of proteins are needed to mediate different structural, regulatory, and catalytic functions. The human cell can produce more than 40,000 different proteins – and many of these also occur in different forms (e.g., active and inactive). But how are the proteins obtained or made?

The Discovery of Genetics and Its Chemical Basis

The Austrian monk Gregor Mendel (1822–1884) did groundbreaking work to resolve the rules and mechanisms of heredity. Over many years he grew peas in the monastery's garden and took detailed notes of the about 10,000 plants that he had grown. This way he could follow the heredity of certain properties, such as the seed color, through generations and find the patterns of the inheritance. He published his results in 1866, but they were understood only after the same phenomena had been rediscovered in the early 1900s. He adopted the concept of a heritable unit, now called the *gene*, which determines each observable heritable feature.

In early years, genes were thought to be contained in proteins which were mediating most of the cellular functions. The existence of DNA was known, but with only four alternating bases, its structure was considered too simple for coding the large amount of genetic information. This would be like a language with only four letters in its alphabet. Then, several people showed that the genetic properties were transferred with DNA, and not proteins.[1] The large coding potential of DNA was gradually resolved as its basic double helix structure was first solved by Rosalind Franklin, Maurice Wilkins, James Watson, and Francis Crick. The structure was revealed by an x-ray diffraction photograph of DNA, taken and interpreted by Franklin in 1953. Franklin herself died of cancer in 1958, at the early age of 37, well before getting proper recognition of her work, and before her colleagues got the Nobel prize based on this work in 1962.

The genetic code, formed from nucleotide triplets, was solved in Crick's laboratory in the University of Cambridge, in 1961–1966. Interestingly, the principles of the coding mechanism were predicted correctly already in 1954 by physicist George Gamow, whose important cosmology studies we described in Chap. 24. It was known that genetic information was coded in the sequence of four different nucleotides, and that these nucleotide sequences determined the sequences of 20 different amino acids in proteins. Based on this information, Gamow deduced that the genetic code had to be formed of nucleotide *triplets*.

We now know that proteins do not reproduce themselves, but instead they are produced by the instructions encoded in genetic information, written in the nucleotide sequence of the genomic DNA. In the interpretation of this message, another nucleic acid, RNA, is required.

Nowadays we often hear the term DNA, but what is it? Please inspect Fig. 28.4 while reading this paragraph. DNA and RNA are similar and closely related molecules. The NA in their names denotes "nucleic acid," which refers to the location in the cell where these both are found, the nucleus. In RNA the first letter R is pronounced "ribo" and refers to the sugar ribose, or a cyclic sugar molecule ring composed of five carbon atoms (the lower right two molecules in Fig. 28.4). In DNA, the letter D refers to deoxyribose, or a five-carbon sugar ring very similar to ribose, just with the absence of an –OH group attached to the carbon at position $2(2'$ carbon) of the ribose ring. Both types of nucleic acids are composed of nucleotides. In nucleotides, the sugar ring acts as the central molecule, which binds a nucleobase to its $1'$ carbon. As shown in Fig. 28.4 (top two rows and first molecule in the third row), the nucleobases are formed of cyclic compounds of nitrogen and carbon. In each nucleic acid four different kinds of bases are used. The bases in DNA are adenine A, guanine G, cytosine C, and thymine T. The same bases appear in RNA, except that uracil (U) replaces the thymine.

The combination of sugar and nucleobase together forms a unit called a nucleoside. To make a nucleotide, a phosphate group (lower left of Fig. 28.4) is linked to the $5'$ carbon of the sugar. As shown in the left of Fig. 28.5, the phosphate groups link adjacent nucleotides together (by a phospho-di-ester linkage) to build long nucleotide chains. The phosphate bound to the $5'$ carbon of the sugar ring is always attached to the $3'$ carbon of the previous nucleotide, meaning that the chain always

---

[1] Oswald Avery, Colin MacLeod, and MacLyn McCarty in 1944, and again Alfred Hersey and Martha Chase in 1952.

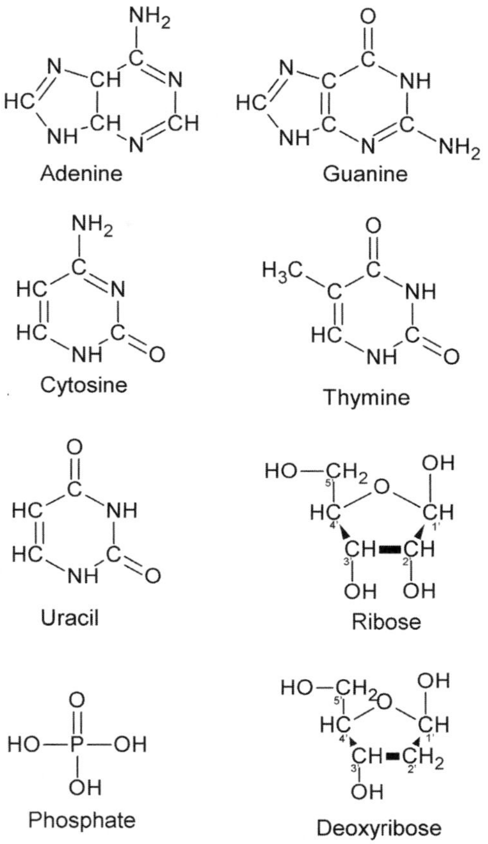

Fig. 28.4 Components of
RNA and DNA nucleotides

Adenine

Guanine

Cytosine

Thymine

Uracil

Ribose

Phosphate

Deoxyribose

grows in a unidirectional fashion, growing at the $3'$-position of the last nucleotide
as shown in Fig. 28.5, left.

In Fig. 28.5, on the right, we see the DNA strands except without the details
seen in the left side of the figure. DNA is composed of the two antiparallel copies
of the long nucleotide chains, attached to each other by the matching pairs of the
complementary nucleotides. Due to the three-dimensional structure of these base
pairs, they tend to stack on top of each other in such way that they twist the parallel
chains into a regular helical structure. So, the DNA double helix appears as a spiral
staircase, or a twisted ladder, where the two linear backbones are formed of the
long chains of sugars and phosphates, held together by the pairs of complementary
nucleotide bases. Each of the strands carries the same genetic information, but it is
readable (or active) only in one of the copies, while the other copy serves only as a
replica of the other.

The reader will note that the DNA molecule on the right has two backbone
strands which coil around one another in the famous "double helix." The bases are
shown as pins of a ladder that connect the backbones of the two strands together. To

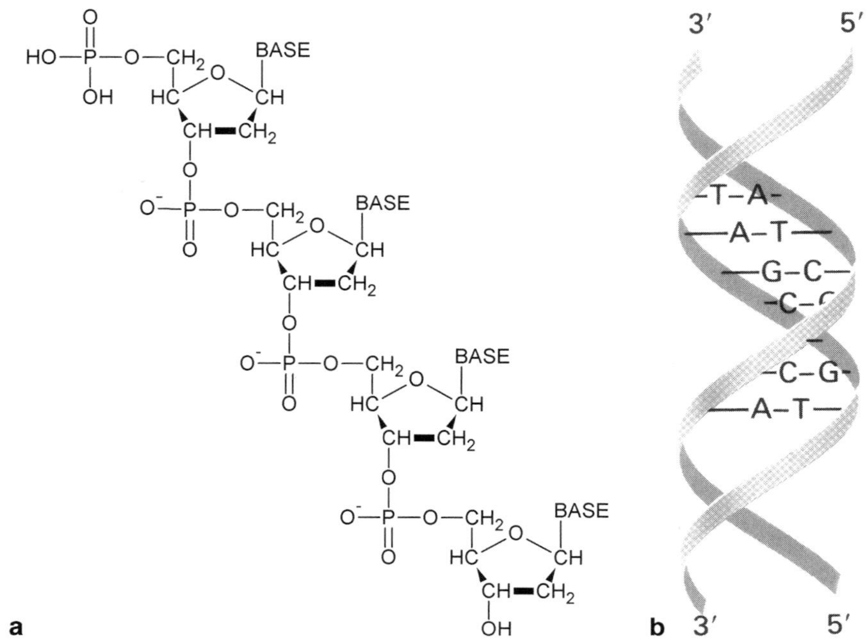

Fig. 28.5 Diagram of how the nucleotides join into a DNA strand (**a**) and how two strands connect into a double helix via complementary bases linking (**b**)

maintain the genetic information, the nucleotide sequence of the DNA is copied (or *replicated*) to a parallel strand to form the double helix. Interestingly, this other half of the helix does not run in the same direction as the original DNA strand, using exactly the same nucleotides but instead, it proceeds in the opposite direction. Each nucleotide is copied to a matching (*complementary*) nucleotide: the As link with Ts and the Gs link with Cs.

In the double helix, linked bases have complementary shapes on their ends resulting in preferable linking A to T and G to C. Then, when this replica template strand is again copied, the original informative sequence is reproduced. The double helix structure is very stable and durable, and as the replication process is very accurate, the genetic information is maintained safely in the DNA. In the replication of DNA (see Fig. 28.6), the double helix is temporarily relaxed (separated) and a new complementary copy strand is built next to each of the parent strands, now forming two daughter double helixes. Simultaneously with the DNA replication in the nucleus of a cell, the whole cell compartment divides into two, and the two daughter DNAs are directed each to its own daughter cell. Thus, each daughter cell inherits identical DNA genome. In spite of this, these cells may develop different roles and functions via specific expression of developmental genes. Such cellular differentiation is typical in multicellular organisms, where the same genetic information directs the formation of specific cell types in different organelles (like skin and internal organs).

Fig. 28.6 Schematic presentation of DNA replication

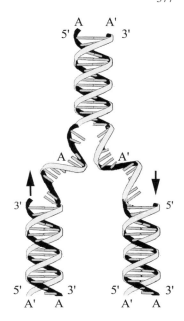

The Genetic Code and Its Expression

The genetic code is stored in the DNA nucleotide sequence in the form of consecutive *triplets of nucleotides*, which each correspond to a specific amino acid. A string of triplets, or *a gene*, specifies the string of amino acids, which must be linked together to form one protein. Using triplets of four different nucleotides, a

Table 28.1 Genetic code: Correspondence of the nucleotides triplets and the amino acids

UUU	Phenyl alanine	UCU	Serine	UAU	Tyrosine	UGU	Cysteine
UUC	Phenyl alanine	UCC	Serine	UAC	Tyrosine	UGC	Cysteine
UUA	Leucine	UCA	Serine	UAA	Stop codon	UGA	Stop codon
UUG	Leucine	UCG	Serine	UAG	Stop codon	UGG	Tryptophan
CUU	Leucine	CCU	Proline	CAU	Histidine	CGU	Arginine
CUC	Leucine	CCC	Proline	CAC	Histidine	CGC	Arginine
CUA	Leucine	CCA	Proline	CAA	Glutamine	CGA	Arginine
CUG	Leucine	CCG	Proline	CAG	Glutamine	CGG	Arginine
AUU	Isoleucine	ACU	Threonine	AAU	Asparagine	AGU	Serine
AUC	Isoleucine	ACC	Threonine	AAC	Asparagine	AGC	Arginine
AUA	Isoleucine	ACA	Threonine	AAA	Lysine	AGA	Arginine
AUG	Methionine	ACG	Threonine	AAG	Lysine	AGG	Arginine
GUU	Valine	GCU	Alanine	GAU	Aspartic acid	GGU	Glycine
GUC	Valine	GCC	Alanine	GAC	Aspartic acid	GGC	Glycine
GUA	Valine	GCA	Alanine	GAA	Glutamic acid	GGA	Glycine
GUG	Valine	GCG	Alanine	GAG	Glutamic acid	GGG	Glycine

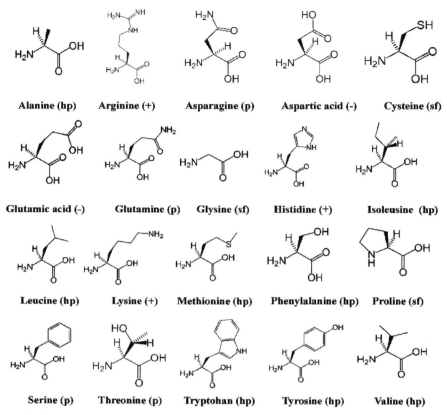

Fig. 28.7 Structures of the 20 amino acids used in biological protein synthesis. The status of electric charge of each amino acid is indicated as $(+)$ positively charged, $(-)$ negatively charged, (p) polar but noncharged, (hp) hydrophobic, (sf) special form

total of $4^3 = 64$ different amino acid codes can be formed. The different triplets and their corresponding amino acids are shown in Table 28.1. Three of the triplets (TAG, TAA, and TGA) are reserved for the identification of the end of each gene; these triplets do not code for any amino acids. ATG, or a start-triplet designates the beginning (although it also codes for methionines in the middle of the genes). A string of triplets located between the start and the stop marks is called an *open reading frame, ORF*. In protein synthesis, in most species a total of 20 different amino acids are used (although 2 additional amino acids are used by some bacteria). The formulas and the chemical properties of the 20 amino acids are shown in Fig. 28.7.

Coding of the 20 amino acids with the available 61 triplets allows the use of the more than one code for each amino acid, and indeed, two to three codes are utilized for most of them (see Table 28.1). This *redundancy* of the code means that the genetic information is not overly sensitive to minor changes. Mutations or misinterpretation can change the variable nucleotides in the triplets, but the encoded protein product still remains the same.

Fig. 28.8 Expression
machinery of the genetic
material

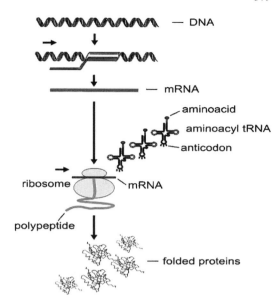

The triplet code in an open reading frame is read and converted into a corresponding string of amino acids in a process called *translation* (see Fig. 28.8). In
this process, the other nucleic acid, RNA, copies the nucleotide sequence of the
gene from the DNA in the form of a mRNA (m = "messenger") and transports it
out of the nucleus (in eukaryotes), into the cell cytoplasm, and into the translation
machinery as shown in the figure.

Again referring to Fig. 28.8, the translation machinery consists of large catalytic complexes called *ribosomes*, which are formed of two different subunits,
and each of these of either one, two, or three different RNAs and multiple different proteins (Table 28.2). The ribosomes recognize and read the code written in
the nucleotide sequence of mRNA, and combine amino acids in the corresponding sequence. The amino acids are brought into this reaction each by their specific
transfer-RNA (tRNA) molecules as shown in the middle right of the figure.

The translation process links the amino acids to each other by peptide bonds.
Refer to Fig. 28.9 while reading the following paragraph. In the peptide bond, the

Table 28.2 Number of ribosomal components in eukaryota and prokaryota

	No. of RNAs	No. of proteins
Eukaryotes		
40S subunit	1	About 35
60S subunit	3	About 50
Prokaryotes		
30S subunit	1	21
50S subunit	2	31

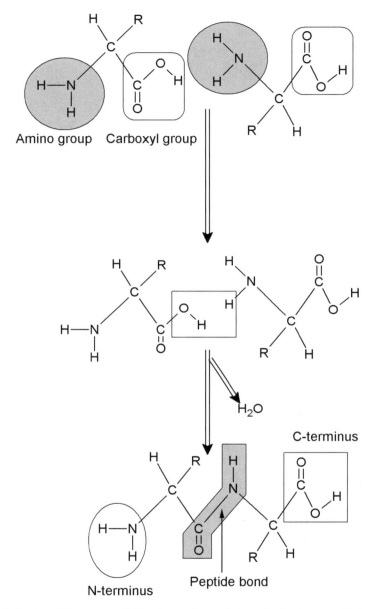

Fig. 28.9 The translation process linking two amino acids to each other by a peptide bond, with the release of water. The carboxyl and amino groups are *circled*, and the variable side chains are indicated by R

carboxyl group of the incoming amino acid is bound to the amino groups of previous one; thus, the amino acid chain grows in a linearly directed fashion, as do also the nucleic acids. The produced amino acid chain is initially called a polypeptide.

The chemical properties of the amino acids (ionic charge, polarity, or neutrality) in the polypeptide determine how they tend to interact with each other and with the water solvent surrounding them. The negatively and positively charged amino acids tend to interact with each other and bind together, the polar amino acids tend to be directed to the outside of the large molecule, and the hydrophobic amino acids tend to aggregate together and to withdraw away from the water into the interior of the molecule. These interaction forces cause the long linear molecule to fold into a very specific three-dimensional secondary structure, where each amino acid is located accurately at its own position. These correctly folded protein products may still be modified by addition of accessory molecular groups, such as sugars or phosphates in specific locations of its structure, and by binding together several protein subunits, which may be either of the same kind, or different from each other. So, in a remarkable process the final functional protein complexes are determined by the primary sequence of the genes, converted into sequences of amino acids, folded into the accurate three-dimensional structures, and modified to a final functional complex.

Genetics and the Evolution of Life

Charles Darwin (1809–1882) was the first person to clearly recognize that the new species were formed as a result of genetic variation, and that the natural selection was the driving force of evolution. He was a fresh graduate from Cambridge University when he was offered a possibility to join the multiyear expedition trip on Beagle, which should take him around the world. The highlight of the trip for Darwin was the Galapagos Islands, where he found many new species, which were still quite similar, but obviously different from one island to the next. Here the evolution of the species from a common ancestor was apparent. The material he collected from his trip was so plentiful that it took him years to study and organize it.

Darwin kept to himself the conclusions about the evolution of species he had made from his investigation. Finally, an 1858 letter from his friend Alfred Wallace (1823–1913) who was at that time in the Malayan archipelago got Darwin going. Wallace's letter contained a compact description of the theory of evolution. Darwin reported his and Wallace's findings in a joint report during a Linné society meeting in 1858. Next year Darwin published his *Origin of species*, in which he described his and Wallace's main theory: Evolution takes place though natural selection from slight variations in inherited traits. The changes taking place in successive generations show up as evolution within populations. The evolution of species is due to the refinement of blueprints of life. Even if the individuals die, the blueprints get passed on to the next generation, and natural selection takes care that the individuals with the best blueprints flourish and reproduce.

The concept of a species' development into another one had already been suggested by the Greeks Anaximander and Empedocles and later the Frenchman Jean-Baptiste Lamarck (1744–1829). But Lamarck's idea of development differed from that of Darwin: He thought that the properties acquired by an individual during its lifetime are inherited by the progeny. For instance, the giraffe's neck is long because reaching for the food from the treetops has stretched the neck of each individual a little bit, and so each offspring has inherited a neck that has been slightly longer than that of its parents. According to Darwin, the reasons for the long neck are different: the longer necks have been beneficial for the survival of the animals, and therefore slight intrinsic variations in genes for long necks have been favored or selected over the generations.

The German August Weismann (1834–1914) developed in the 1880s a theory of the transmission of hereditary properties through reproductive cells, egg cells and sperm cells. He was originally a doctor and practiced microscopic studies, but his deteriorating vision forced him to concentrate on theoretical research. His theory that the properties of each species were transferred by the reproductive cells re-enforced Darwin's evolution theory of natural selection.[2]

As Darwin adapted his theory to the origin of humans, he came into conflict with religious circles. This was probably one reason he had delayed publication for so long. Though Darwin himself did not participate in these disputes, feelings were strong between his supporters and opponents. In a debate between his supporter Thomas Huxley and Samuel Wilberforce, the bishop of Oxford, the bishop was said to have asked Huxley whether it was through his grandfather or grandmother that he claimed his descent from a monkey. Huxley replied that he was not ashamed to have a monkey for his ancestor, but he would be ashamed to be connected with a man who used great gifts to obscure the truth.

The impression conveyed by this famous anecdote may be balanced by noting that Wilberforce had earlier written a review of Darwin's *Origin of species*, where he expressed the view that scientific theories should be assessed solely on scientific grounds and one should not reject any conclusion by reason of its strangeness: "*Newton's patient philosophy taught him to find in the falling apple the law which governs the silent movements of the stars in their courses; and if Mr Darwin can with the same correctness of reasoning demonstrate to us our fungular descent, we shall dismiss our pride, and avow, with the characteristic humility of philosophy, our unsuspected cousinship with the mushrooms.*" Wilberforce did conclude, after a discussion of the facts available to him at that remote time, 1860, that in his opinion Darwin's theory is false, but for modern debaters – evolution in general and its application to humans still excite strong feelings today – he had a message: "*We have no sympathy with those who object to any facts or alleged facts in nature, or to any inference logically deduced from them, because they believe them to contradict what it appears to them is taught by Revelation. We think that all such objections savour of a timidity which is really inconsistent with a firm and well-intrusted faith.*"

---

[2] Weismann's theory had an important social aspect, too: it showed that moral properties are not inherited from parents to children, but are obtained by learning.

The Central Features of Life are Derived from the Same Origin

Recall that the essence of Darwin's theory is the occurrence of small variations in the genetic material and thus in the resulting organism. In his time the mechanism of these variations was not known. We mentioned earlier that the redundancy of the genetic code allows certain degree of variation of the nucleotide sequence without any change in the encoded protein products. Also the amino acids coded by similar sequences are structurally similar. Then, formation of the final three-dimensional protein products allows alterations in the code, so that it will still produce proteins that are not identical but similar, and can still mediate the same functions. If an amino acid in a protein is replaced by another one that has similar chemical properties (the same charge or polarity) it will behave in the polypeptide chain in similar manner as the previous one, and produce the same or similar folding structure and the same final function. As the *function* of the proteins is the most important property that needs to be conserved, the sequences can change to some extent as the genes are passed down through generations, while the three-dimensional structures and the functions of the gene products (proteins) are conserved. Based on the conserved features, a gene that is inherited from the common ancestor into two progeny lines can be still recognized as a related (or homologous) gene in these two lines, although in both lines it has accumulated certain level of mutations since the divergence from their common ancestor. The variations Darwin postulates thus arise.

The amount of mutations in the two progeny species is directly related to the time that they have evolved separately, and thus, this diversity can be used as a statistical measure of relatedness between different species. These relationships can be drawn out as phylogenetic trees, where the branches indicate the phylogenetic lineages of organisms after their separation, and the length of the branches indicates the genetic distance (or divergence) between the species. Most of the changes are either nonsignificant or deleterious, but a few actually are helpful to survival. This is the modern-day understanding of the root mechanism of evolution.

All the central features of life mentioned, i.e., the structure and composition of the genetic material, the genetic code itself and the 20 amino acids used in all the proteins, and the translation machinery are all nearly identical in all the species that live on Earth today. This indicates that these features were obtained (invented) very early in the evolution of life and apparently were present in the last common ancestor at the time when the three domains of life separated from it. The genetic relationships of all life forms can be studied by comparing the sequences of those RNAs (e.g., the 16S RNA) that are included in the translation machinery. Based on the 16S RNA sequence, a classical phylogenetic tree was first drawn by Carl Woese and his team around 1990. This tree indicated that at first the two prokaryotic lineages separated from the LUCA population, and that the eukaryotes were later derived from the fusion of these two lineages (Fig. 28.10).[3]

---

[3] However, this phylogenetic tree and the order of descent of the three domains have been later questioned, as different phylogenetic relationships between many organisms are obtained from different gene sequences. These differences may be explained by the phenomena that many genes

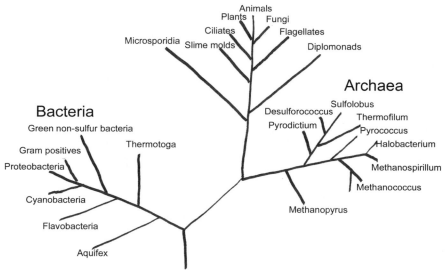

Fig. 28.10 Phylogenetic tree based on the 16S RNA sequence of different organisms and with presumed Last Universal Common Ancestor as "trunk" (adapted and modified after Woese, Kandler, & Wheelis 1990)

In addition to the mentioned genetic elements of life, another ubiquitous feature derived from the very earliest life forms is the *homochirality* of the building blocks of both the nucleic acids, and of the proteins. Chirality is a feature of molecules that have an atom (e.g. carbon) with four different bonds. As these bonds point to four different corners of a tetrahedron, the structure is nonsymmetric and can be composed in two different orientations that are mirror images of each other. Now, the bonds of the central (α) carbon in all the amino acids are nonsymmetric (Fig. 28.11), as are the bonds of the carbon number 4 in the ribose and deoxyribose sugars (see Fig. 28.4).

In all synthetically produced amino acids or sugars these molecules have the carbon bonds randomly in both orientations (and thus are not homochiral). However, all biogenic amino acids are homochiral and have the α-carbon bonds in L-configuration (L = Levo, left), and all the sugars are also homochiral and have the carbon-4 bonds in D-configuration (D = dextro, right). The origin of this specific homochirality is not understood as yet. It is clear that homochirality *per se* is a chemical obligation, as this will allow the arrangement of the monomeric subunit into nice linear polymers, while the opposite chirality would turn the structure into

---

have been exchanged between various species still after their separation to different phylogenetic lineages – and also by the fact that the evolution of different genes may not be comparable (does not happen at the same rate) over very long periods of time.

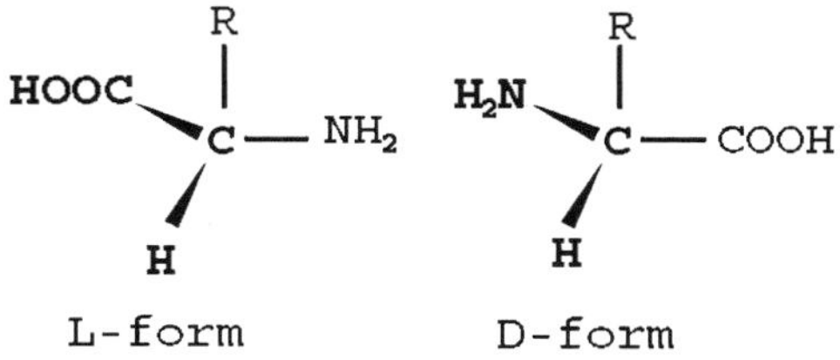

Fig. 28.11 The two possible
chiralities of the amino acids.
R denotes the variable side
chain

L-form   D-form

the opposite (kinked) direction. However, it is not understood why or how just the left-handed amino acids, and the right-handed sugars have been chosen by life. Just as little we understand, as yet, why it is just those 20 amino acids that are used in the proteins, and those four different nucleotides that are used in the DNA and RNA, respectively, instead of some of their similar analogs. So far, it seems that these may have been just random choices of molecules that were available early on in the environment, and that became established by chance, as a "frozen accident," and then fixed as "the rule." In any case, these are now some hallmark features of life on Earth. If we some time will find life elsewhere, it will be of great interest to see if that life bears the same hallmarks. These features might reveal whether that life has the same or separate origin with us.

Environmental Requirements of Life

Although life is based on cellular structure, genetic information, and its replication and evolution over time, these alone do not make life feasible. The structures and functions form a viable unit only in *environment that can sustain it*. Life needs *energy* for all its processes. The (nearly) sole ultimate energy source of life on Earth is the sunlight even for, say, an animal that consumes a plant that uses the Sun to grow. However, some bacteria and archaea are adapted to live by the energy obtained from the reducing chemistry of minerals, but these energy sources are very limited and could not possibly support the existence of a biosphere of any significant size. Life also needs *nutrients*, building material to maintain and replicate its structures. These are composed of organic carbon compounds, as well as minerals that are available in the environment, and circle between organic and inorganic compounds. Life also needs a *solvent* to dissolve and transport all chemicals. Water, the solvent of the life here on Earth, is also a major component of living organisms.

Water appears to be overwhelmingly the most suitable solvent for all biochemical reactions. The water molecule is composed of one oxygen and two hydrogen atoms, bound to each other by covalent bonds, meaning that a shared electron pair circles the oxygen and each of the hydrogen atoms (see Fig. 28.12).

The oxygen atom exerts a stronger pull on the electrons than the hydrogen, and consequently the electrons are located slightly more to the side of the oxygen than of the hydrogens. Thus, the oxygen end of the molecule is slightly negatively charged, and the hydrogens positively – the whole water molecule is an electric dipole

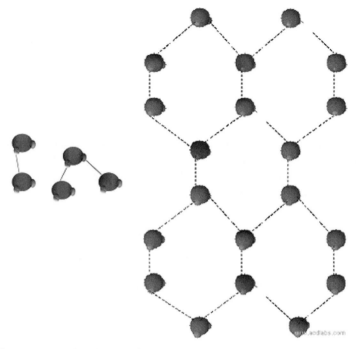

Fig. 28.12 Water molecules, (*left*); and ice, (*right*). Hydrogen bonds indicated by the lines

(a polar molecule). This property strongly affects the chemical properties of water. The electric polarity of the water molecules causes weak electric interactions, or *hydrogen bonds,* to form between different molecules (Fig. 28.12), making the water to behave as an integrated, weakly bonded network. The hydrogen bonds make the molecules attract each other, and make the liquid somewhat "sticky," or viscous. Due to this "stickiness," a relatively high temperature and a lot of heat energy are required to evaporate water into gas form. So, water stays in liquid form over a broad temperature range. The "stickiness" also resists an increase of the water temperature (increase of the thermal movement of the molecules), and thus a lot of heat energy is required to raise the temperature of water. Likewise, much heat energy is released when water cools, making it a very good temperature thermostat, both in a large environment and inside cells.

Water interacts readily with other charged molecules; this makes it a very efficient solvent for all ionic compounds made of positively and negatively charged atoms. Water also dissolves polar compounds where the positive and negative charges are still together on a molecule but separated (like water). On the other hand, it does not tend to dissolve nonpolar molecules, such as long, noncharged hydrocarbon chains. Also this feature is biologically very important, as it means that these molecules are "hydrophobic" and in water solution tend to aggregate to each other rather than with the water molecules. A very important group of molecules are the lipids, or molecules which have a polar or charged group attached on one end of the molecule, making this end hydrophilic, and a nonpolar group (e.g., hydrocarbon chain) in the other end, making this end hydrophobic. Such dual-property molecules are amphiphilic, and they assemble in water solutions to form bilayered membranes (Fig. 28.13). The hydrophilic and hydrophobic interactions also affect strongly the three-dimensional folding of all other molecules, including proteins, and help them settle into the correct functional forms.

Due to the pull of the hydrogen bonds, and by the action of surface tension and evaporation, water also behaves very nicely in the environment. By capillary action it can move against gravity, for instance into the vascular systems of plants, making it able to rise into the high canopies of tall trees. Water also moves in the capillary spaces of soil and rises spontaneously from water tables into the root zones of plants. Hydrogen bonding also affects the density of water at different temperatures in very specific ways. As the temperature cools, the hydrogen bonds become tighter and shorter, so that at the temperature of $+4°C$ water molecules are most closely pulled to each other. At this temperature water is most dense. As the temperature falls below this, the molecular configuration starts to convert toward a looser six-cornered hydrogen bonding typical of ice crystals (Fig. 28.12), and thus the volume of the water starts to expand. The lower density ice forms on the surface of water at $0°C$, and the denser $+4°C$ water is left on the bottom of the water basin. Therefore, if the waters are deep enough, or the freeze is not too severe, the $+4°C$ water can remain in liquid form under the ice cover through cold periods, which allows life to survive in deep water, protected from the freezing under the ice. This is a significant and exceptional property. For instance ammonia, which might be a somewhat suitable alternative solvent for life, is heavier in solid form than in liquid, meaning that ammonia ponds would freeze directly down to the bottom and might easily stay permanently frozen. Due to the lack of hydrogen bonding, ammonia exists in liquid form only in a quite narrow temperature range, in much lower temperature than water (between -78 and $-33°C$, at sea level). At these temperatures, all biochemical reactions would happen very slowly. In addition, ammonia is easily broken up by ultraviolet light, and its lighter component, hydrogen, escapes easily into space. Ultraviolet sunlight can break up also water, but this reaction is slower, and produces oxygen (O_2) and ozone (O_3), which block the ultraviolet radiation and thus prevent the further breakdown of water. Therefore water can exist in large quantities in the atmosphere of an earth-like planet, while ammonia cannot.

General Principles of Life

We have now learnt that the general features of life are, first, that it is cellular, i.e., confined and separated from its environment, and based on genetic information that allows the maintenance of specific chemical composition and complex structures and functions inside the cellular structures. All the cellular structures and molecules are formed of only a very limited set of chemical elements, i.e., mostly of carbon, hydrogen, oxygen, nitrogen, and phosphorus; and of tiny amounts of sulfur,

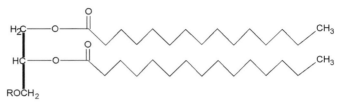

Diphytanyl glycerol diether

a   Glycerol diester

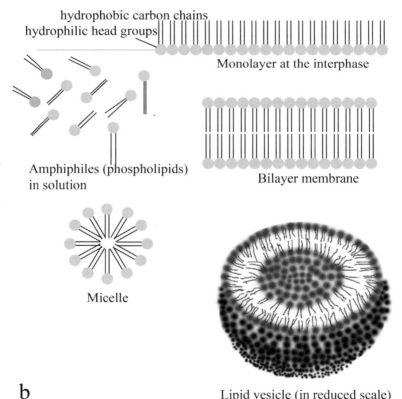

hydrophobic carbon chains
hydrophilic head groups

Monolayer at the interphase

Amphiphiles (phospholipids)
in solution

Bilayer membrane

Micelle

b   Lipid vesicle (in reduced scale)

Fig. 28.13 Lipids. (**a**) Different types of lipids: many lipids are formed of two long hydrophobic hydrocarbon chains, and of one polar side group ($ROCH_2$) linked to the glycerol backbone. Lipids with ether bonds occur in membranes of archaea, and lipids with the ester bonds between the glycerol backbone and the acid side chains occur in bacteria and eukaryotes.(**b**) Different membrane formations: amphiphilic molecules assembled into a bilayered, monolayered, and micellar formation

calcium, potassium, and some other elements. We also know that all the life we know here on Earth is structurally and functionally similar, i.e., based on the same kind of genetic material, same genetic code and its expression mechanism, and on very similar basic metabolic reactions. The likeness of all life forms indicates that they all are descended from a unique original life form, the last common ancestor. This uniformity of all the life here on Earth causes a major problem for its characterization. From this sole example of life, we cannot say whether it could be also different – or, how different could it be in other worlds. However, we can make some educated guesses.

It seems that any complex biochemistry needs to be based on the carbonaceous compounds, using water as solvent, and light from a nearby star to provide a long-term energy source. Also, presumably the principle of reproduction and natural selection (evolution) would be good driving forces for maintaining life elsewhere. These processes are driven by random changes of the genetic information, and by the selective power of the environmental conditions, which might be very different from here. Therefore, the outcome of the evolution of another biosphere in some other time and place would certainly be totally different from here. The forms and functions, and even the cellular structures and biochemistry of any truly alien life could differ a lot from the life here on Earth today.

Still, also other life might have developed some similar features that are used here, as these should be universally useful. For example, those alien life forms would have to have some means for harvesting light energy and for converting it to chemical form, and most likely some light-absorbing pigments would be used for this. It is also likely that these creatures would have means for sensing their environment, and to transmit signals between each other, be it then by means of light, chemical, or audible signals. Possibly the creatures would have evolved motility. Motility, high complexity and communication skills might have allowed invention of use of tools, and evolution of intelligence. But this may be unlikely, recalling that on Earth, life remained very simple (prokaryotic and unicellular) for most of its existence. Multicellular complex life is really recent here and could be rare in the universe.

Still Deeper into the Biochemical World

Since the discovery of the composition and the coding system of genetic material, our understanding of life has hugely increased. Powerful new methods and tools for cellular and molecular biology since 1970s have revolutionized the study of DNA, gene functions, protein structures, and of the regulation and co-ordination of the different biochemical reactions in the cells. For examples of the recent breakthroughs in molecular and cell biology, please see Box 28.2.

During the last five decades, these new methods and research approaches have revealed us the deep complexity of cell and molecular biology. Molecular interactions, and different regulatory reactions and feed-back cycles within any cell are turning out to be very multi-layered, and very well fine-tuned to react to different external and internal cues. The complexity of these intracellular molecular networks can be now analyzed with highly computerized calculations, and thus we are gradually becoming able to comprehend the biochemical world that is closed within our cells, or, the molecular basis of life.

Box 28.2 Current Status of Genetics and Molecular Biology

One example of new, hugely beneficial methods in molecular biology is the invention of the restriction endonuclease enzymes isolated from (hyperthermophilic) bacteria and archaea, which made possible the accurate cutting of DNA into specific pieces. Cloning techniques have allowed the ligation of any desired DNA fragment into different cloning vectors (plasmids or viruses) capable of independent growth and amplification in alternative host, e.g., bacteria or cultured animal cells. A polymerase chain reaction (PCR), invented by K.B. Mullis in 1983, amplifies DNA sequences very efficiently, by utilizing rapidly alternating high and low temperatures for copying the target templates, starting from primers annealing at specific sites of the target DNA.

The amplified DNAs can be easily isolated and analyzed to deduce their nucleotide sequences. Isolated sequences can be also expressed either in vitro, to produce the proteins for structural and functional studies, or in alternative host cells, in vivo, to study the actual protein functions, localizations, or interactions inside cells. Our ability to transfer new genes to beneficial organisms, such as bacteria or plants, has made it possible to accurately modify these organisms in a designed way to improve their genetic properties. This *genetic engineering* approach is already used in many applications in biotechology and is likely to become even much more applied in the future.

Efficient detection methods for the gene expression levels (e.g., using homologous nucleic acid probes) have allowed the investigation of the expression patterns of any genes of interest, and elucidation of how different genes regulate the development and differentiation of multicellular plants and animals. With the automated DNA sequencing methods one may determine huge amounts of genomic sequences, and by now, complete genomic sequences of many prokaryotic and eukaryotic organisms have been deduced.

It has been found (see the Table below) that the genome sizes vary from about 580,000 nucleotides and 470 genes in the smallest self-sustainable bacteria (*Mycoplasma genitalium*), to about 100×10^6 nucleotides and some 20,000 genes in small animals like the nematode *Caenorhabditis elegans*, $3,400 \times 10^6$ nucleotides and 32,000 genes in humans, $17,000 \times 10^6$ nucleotides and 60,000 genes in polyploid plants like wheat, and up to about $670,000 \times 10^6$ nucleotides in amoebas which have the largest known genomes of all life forms. The genetic sequences have revealed the large variation in genome sizes and complexities, and also genetic similarities between closely related species. So, the difference between humans and chimpanzees appears to be only 1% on the DNA sequence level.

Table: Genome sizes and coding contents of different types of organisms

Species	Domain	Genome size in millions	No. of genes	Percentage of coding
Mycoplasma genitalium	Bacteria	0.58	470	100
Escherichia coli	Bacteria	4.8	4,288	100
Methanococcus jannaschii	Archea	1.7	1,738	100
Saccharomyces cerevisiae	Eukaryota	12	6,144	70
Caenorhabditis elegans	Eukaryota	100	18,266	25
Arabidopsis thaliana	Eukaryota	100	25,498	31
Homo sapiens	Eukaryota	3,400	32,000	28 Total
Triticum aestivum	Eukaryota	17,000	60,000	
Amoeba dubia	Eukaryota	670,000		

The availability of the sequence data for very large numbers of genes (stored in huge data banks) also allows systematic investigation of the expression patterns of different RNAs or gene products inside cells. This again makes it possible to study their molecular interactions and regulatory relationships. In *systems biology* these aims are approached in a high-powered, automated, and computerized manner. Very large collections (arrays) of molecular probes are used in miniaturized configurations (DNA microarrays embedded in *microchips*), to test simultaneously the amounts of many thousands of cellular RNA species. Then, real-time changes in the RNA expression patters can be analyzed under different external conditions (e.g., stresses or growth factors) to elucidate how these factors affect the gene expression. Likewise, the total cellular protein or metabolite profiles can be analyzed under different conditions to see how the cells respond to these conditions.

Chapter 29
The Origin of Earth and its Moon

We have narrated how *Homo sapiens* has step by step discovered the vastness of the universe by inventing methods to measure distances and properties of celestial bodies. Along with deep space, we have deep time. The huge distances revealed are hard to imagine. Similarly painful for common sense are the huge lengths of time that one has to accept in order to understand the origin of the Earth (and our galaxy, of course). Anything much shorter than a tenth of a second is difficult to comprehend and anything much longer than the age of our grandparents goes easily beyond our normal thinking. We have to use various indirect methods to get to grips with very long times, millions or billions of years.

Historic Estimates of the Age of the Earth

A famous determination for the age of the Earth was made in 1654 by the Irish bishop and scholar James Ussher using the Bible. He followed time backward from the birth of Christ, and concluded that the universe and the Earth had been created 4004 BC. Such a *biblical age determination* of the Earth (in use long before Ussher) was considered valid up to the nineteenth century, until new evidence of relevant time scales began to emerge from geology, paleonthology, astronomy, and physics.

George-Louis Leclerc (1707–1788), the count of Buffon, a Frenchman, challenged Ussher's results in 1779. He argued that the fossil records known at the time could be formed only if the Earth had an age of at least 75,000 years. This radical suggestion was the first *geologic age determination* of the Earth. This was in reasonably good agreement with Isaac Newton's estimate. In his 1687 *Principia* he had suggested that the Earth could have an age of 50,000 years. He based his estimation on the cooling of an iron sphere scaled linearly to the size of the Earth. The count of Buffon made similar experiments with spheres of different sizes.

Soon, a Scottish geologist James Hutton (Fig. 29.1) introduced a novel concept. He suggested that past events could be understood by studying the present ones. Processes, such as the accumulation of sand on the beach or outpour of lava and

P. Teerikorpi et al., *The Evolving Universe and the Origin of Life*
© Springer Science+Business Media, LLC 2009

Fig. 29.1 James Hutton (1726–1797) founded modern geology: slow processes have formed geologic features

ashes from volcanic eruptions, have remained the same over the times, so they can be used to study geologic layers and rock types. Hutton's ideas were published in 1788 in the *Theory of the Earth*. He also argued that geologic layers formed over long periods of time, in contradiction to the prevailing catastrophe theory, which suggested that geologic features were formed, say, by the biblical Great Flood.

The Scotsman Charles Lyell (1797–1875) studied in Oxford. His book *Principles of Geology* (in three volumes in 1830–1833) became so highly regarded, that it caused the idea of catastrophes to recede. He emphasized that the same physical laws that are valid now have prevailed throughout the past, and that geologic processes have always proceeded in the same way and the same pace as they do at present. We know now that this is not exactly so, as the processes have varied in intensity over the past.

Next, influenced by Lyell's work Charles Darwin introduced a new aspect into the age discussion by considering the evolution from the simplest organisms to humans. He estimated that the geologic processes should have taken 300 million years, a time he considered also sufficient for the evolution of life. John Joly, an Irish geochemist, came up in 1899 with a roughly similar result, 90 million years for the age of the Earth, by calculating how long would it take for the salinity of oceans to accumulate from salts carried by all rivers combined. He assumed, somewhat incorrectly, that the yearly transport of salts remained the same and that no oceanic salt is removed. Thus, at the start of the twentieth century, it appeared that the geologic age of the Earth was 100 million years or perhaps a little bit more.

The third way of estimating the age of the Earth is *physical age determination*. In 1862, William Thomson, also known as Lord Kelvin, revised the study of Count of Buffon by calculating how long would it take for a $1,000°C$ sphere to cool to $15°C$.

Kelvin used Joseph Fourier's theory of thermal conduction. (In fact, Fourier came up with similar results, but did not dare to publish anything as radical at the time!) Kelvin obtained a value of 98 million years for the age of the Earth. Considering the possible errors in his estimate he said that the age was somewhere between 20 million and 400 million years. If nothing were heating up the Earth's interior, this estimate would have been reasonable.

The age of the Sun and the age of the Earth should be comparable since they presumably formed together. Being different kinds of bodies, their age estimates are based on quite different methods. As a first step one can calculate that if the Sun was made of carbon and burned in free oxygen, it would take only 10,000 years for the Sun to burn up at the rate it is shining at present. It is now obvious that the Sun's energy cannot be chemical in origin.

Lord Kelvin considered the possibility that the solar energy we see today is produced from the heat released by the material that fell in to make up the Sun plus the much smaller present-day in-fall of meteorites (the kinetic energy is transferred into heat). He estimated that it is likely that the Sun could not have been shining for 100 million years, and most definitely not 500 million years. He also suggested that the Sun would fade in a million years or so. This upper bound for the age of the Sun was quite grim. These numbers provided also a natural upper limit for the age of the Earth and agreed with Kelvin's separate estimates for the age of the Earth.

Kelvin noted that his 100-million year estimate for the Sun was in contradiction with Darwin's 300 million years for the Earth. Darwin gave way and reckoned that it was possible that Kelvin could be correct. Soon, however, the conflict between the estimates grew worse as it turned out that Kelvin had overestimated the ages of the Earth and the Sun. His new calculations resulted in an age of 20 million years for both of them. This was assuming that no internal energy sources existed and that the Earth had cooled from a molten state, the highest possible original temperature. The time to cool to the present state thus provided a maximum age estimate.

Conflict of Cooling Ages with Sedimentation Ages and Its Resolution by Radioactivity

The geologic sequence was originally defined by fossil records. A certain kind of mix of fossils in a soil layer defined a geologic era. The boundaries between different mixes were quite sharp. Often one could measure separate deposits in a way analogous to tree rings. The length of each era was then measured from sediment deposit thickness. By matching these layers and their boundaries one could build a sequence of geologic eras. One can measure the current deposition rate of sediments in centimeter per year in the ocean floor. One can add up the thicknesses of all the known geologic eras. This divided by the thickness deposited per year is an estimate of the age of the Earth, the *sedimentation age*.

Contemporary with Kelvin's estimates, Edward Poulton, a professor of zoology at Oxford, estimated that at the present sedimentation rate, the time elapsed since the Cambrian time was about 400 million years. Geologist John Goodchild ended up with an amazing figure of 700 million years for the start of the Cambrian period. The current estimate is 542 million years. Evidently the Earth and the Sun had to be older than this. A conflict thus appeared between the age estimates of tens of millions years made by the physicists and some biologists vs. the hundreds of millions argued for by geologists.

All the age estimates were profoundly revised with the advent of *radioactive age determination*. Ernest Rutherford (who discovered the atomic nucleus; Chap. 16) and Frederick Soddy measured in 1902 how much heat is produced by radioactive radiation or how much energy is liberated by one gram of radioactive material. The results were mind boggling. One gram of radium produced over a thousand times more energy than one gram of carbon turned into chemical energy. This idea was on the right track, but it led to ages conflicting with the views of Lord Kelvin. In 1904 Rutherford gave a talk at the Royal Institute. He saw Kelvin in the audience: "*I realized that I was in trouble at the last part of my speech dealing with the age of the earth. . . . To my relief, Kelvin fell fast asleep, but as I came to the important point, I saw the old bird sit up, open an eye and cock a baleful glance at me! Then a sudden inspiration came, and I said Lord Kelvin had limited the age of the earth, provided no new source was discovered. That prophetic utterance refers to what we are considering tonight, radium! Behold, the old boy beamed upon me.*" The young physicist argued that the Sun could have been present for well over 20 million years, perhaps even billion years, and that the Earth would not meet its demise in a short time because of Kelvin's dimming Sun. The newspapers of the following day declared that "doomsday has been postponed."

Radioactivity turned out to lead us even deeper in time. A story tells that Rutherford was walking at the university with a stone in his hand. He bumped into a geologist and asked him "Adams, how old do you think the Earth is?" Adams argued that based on various methods it is about 100 million years old. Rutherford replied to his astonishment, "The age of this stone is 700 million years." He had just derived the age by using the decay rate of uranium. In 1907 Bertram Boltwood measured ages of various rock samples to range from 400 to 2,200 million years. Later the timing of geologic strata has improved and has now errors less than 1 million years. We know also that the Solar system has an age of 4.567 ± 0.001 billion years and that the Earth is nearly as old. See Table 29.1 for decay rates of several radioactive isotopes and Box 29.1 for examples of how radioactive dating is done.

Table 29.1 Isotopes in common use in dating minerals

Original isotope	New isotope	Half-life (years)	Range of use	Comments
Rubidium 87	Strontium 87	47×10^9	>100 Myr	Good for granite
Thorium 232	Lead 208	13.9×10^9	>200 Myr	
Uranium 238	Lead 206	4.5×10^9	>100 Myr	Widely used
Potassium 40	Argon 40	1.2×10^9	>0.1 Myr	Widely used
Uranium 235	Lead 207	0.7×10^9	100 Myr	Widely used
Samarium 147	Neodymium 143	106×10^9	1,000 Myr	Rocks and meteorites
Carbon 14	Nitrogen 14	5,370	In archeology	

Box 29.1 Examples of Radioactive Dating

As an example of radioactive dating, consider the potassium isotope which weighs 40 units (^{40}K). This isotope disintegrates to 89% calcium and 11% argon, both of which have atomic weight 40. Out of the two elements, argon is an inactive gas which can be trapped inside a rock. However, if the rock is melted, it diffuses out. The ratio of argon and potassium isotopes 40 tells us how long the rock sample has been in the solid phase. One could say that the solidification of the rock starts a clock ticking; in the beginning there is no argon 40 in the rock at all. When time goes on, the fraction of the argon isotope increases so that after 1,200 million years there is already 11% of argon in relation to the potassium isotope. This is the half-life of potassium 40. By measuring the ratio of argon 40 and potassium 40 in the sample, the age of the rock is determined.

Here is another interesting example. Consider a mineral crystal zircon of the element zirconium (Zr). The zircon crystal is very stable. It survives weathering and even partial melting and metamorphosis in Earth's mantle. The most common form of zircon is $ZrSiO_4$. Small impurities occur in the crystal formation, where the zirconium atom is replaced in less than a percent of cases by a uranium ($USiO_4$) or a thorium ($ThSiO_4$) atom. This is possible because of the similar sizes of U and Th nuclei to those of Zr. Lead (Pb) on the other hand cannot enter the crystal, because it is simply too big, so when these crystals form they are lead free, but contain small amounts of U and Th in addition to the much more abundant Zr. As time passes the U and Th atoms experience radioactive decay. The ^{235}U atoms decay into ^{207}Pb with a half-life of 703.8 million years, and ^{238}U atoms decay into ^{206}Pb with a half-life of 4,468 million years. We can now calculate the age of the zircon crystals simply by counting how many ^{235}U atoms have changed into ^{207}Pb and ^{238}U atoms into ^{206}Pb. If half of the ^{238}U have changed into ^{206}Pb then the crystal has an age of one half-life or in this case 4,468 million years. The age can be calculated also for other uranium-bearing minerals. In practice, the age determination is rather done by constructing isochrones, lines of same age, by plotting the measured ratio of ^{206}Pb$/^{238}$U against the ratio ^{207}Pb$/^{235}$U to make sure no outside contamination of U or Pb is having a contribution.

Discovery of Tectonic Plate Motions

It is difficult for us to comprehend that the "fixed" stars in the sky change their relative positions over thousands of years. Similarly, it is difficult to realize that the solid rock under your feet also moves and changes shape. The movement of both the stars in the sky, and the mountains and the continents is counterintuitive because they are both used as references against other things that change much faster, such as

Fig. 29.2 Antonio Snider-Pellegrini (1802–1885) drew the first map showing the motion of continents

planetary motions, tides, or weather cycles. The huge ages from radioactive dating have opened up a new view of processes in the Earth in which a human lifetime is less than the blink of an eye.

The nice fit of the East coast of South America and the West coast of Africa bewildered geographers when the first relatively accurate maps of these shorelines were published. The usual explanation early on was that the Atlantic Ocean was a wide canyon and that the continent in between had sunk. In 1858 the French geographer Antonio Snider-Pellegrini suggested that South America and Africa had been ripped apart to the present distance. His brave theory was set strictly into a biblical context. On the fifth day of creation all the continents were one big landmass. On the sixth day of creation a long fracture appeared. Volcanic gases erupted and drove the continents apart. At the same time the Earth shrank and the seas overran the continents creating a gigantic flood (Fig. 29.2).

For a geologist this explanation did not make any sense. Half a century later when the motion of continents was brought up again, the poor reputation of Snider caused a negative reaction for the new theory. In 1910, the American F.E. Taylor proposed that the formation of mountain chains was due to movements of continents. He reasoned that the Himalayas had folded and risen to heights because the Eurasian continent was moving South and colliding with Indian subcontinent.

Alfred Wegener (1880–1930) is usually viewed as the father of tectonic motions. The German scientist was struck by the similarity of fossils on both sides of the Atlantic and suggested in 1912 that 200 million years ago during the Mesozoic era all the continents formed a single supercontinent *Pangaea* surrounded by an ocean Panthalassa (see Figs. 29.3 and 29.4 while reading the following). The supercontinent broke into parts and the present continents

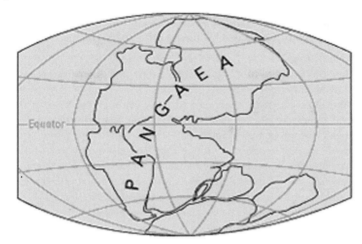

Fig. 29.3 The single supercontinent Pangaea 225 million years ago, surrounded by an ocean Pan-thalassa. The fitting of coast lines and the continental shelf edge is shown. The reader will easily recognize North America, South America, Eurasia, and Africa. Try to find out Antarctica, India, and Australia (credit: United States Geological Survey)

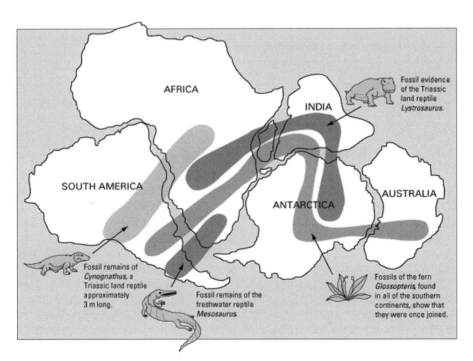

Fig. 29.4 Part of the single supercontinent Pangaea. The map shows the fossil evidence from the present-day continents (credit: United States Geological Survey)

began to drift apart. Wegener published his views in 1915 in his book *The origin of Continents and Oceans*. He argued his theory with several points:

1. Two kinds of crust exist on Earth. The lighter one forms the continents and the heavier one the ocean bottoms. It was the continents that can float on the ocean floor and even slide on it if lateral forces exist.
2. The coastal lines of South America and Africa fit well together. If the shallow continental selves are also considered, the fit improves. In that case also North America and Europe fit well (Fig. 29.3).
3. When this fit is made, one may notice that the geologic features also align so well, that it appears that the continents were at some point torn apart.
4. Fossils on each side of the Atlantic Ocean fit each other at respective locations. This suggests strongly that a land connection existed. See Fig. 29.4.
5. The geologic features suggest that continents have also moved across different climatic zones, one can find coal layers at high latitudes, although they were formed near the equator. Similarly close to the equator one can find signs of polar icecaps.

Wegener was not able to tell what was driving the movement of the continents. This slowed down the acceptance of his theory. Even in 1948 a renowned American geologist declared that "The theory of continental movements is fiction. It is an exciting idea that has let the imagination astray." Soon after this the idea gained wide support with the detection of two additional items:

6. Magnetic patterns in both bedrocks and seafloor (paleomagnetism).
7. The topography (surface features) of the seafloor.

We return soon to the modern view of the motion of the continents and its physical significance after a description of how the Solar System and the Earth were formed.

Origin of the Earth as Part of the Solar System, a Modern View

Our Sun, the planets, and also the smaller Solar System bodies were formed from an extended interstellar gas cloud. This part of the cloud was cold (about$-260°C$). It was dense in astronomical terms (about a million atoms per cm^3), opaque, and completely black in appearance. Such clouds still exist in space today. The cloud was made mostly of hydrogen and helium gas with about 1% of various heavier elements. Some matter was in the form of ices and mineral dust such as olivine crystals, graphite, very tiny diamonds, and other minerals. The cloud collapsed on itself and in doing so it flattened and sped up like a ballerina in a pirouette because of the conservation of angular momentum (Chap. 15). Eventually a *protoplanetary disk* formed with an ever-increasing temperature and density toward the center. This is where the Sun would form. The central plane of the disk also became dense, and here planets were going to form.

   Closer to the protosun, the protoplanetary disk was hotter than further out. At the time when most of the dust had sedimented onto the central plane of the disk, the temperature at the distance where Earth was going to form had risen to about 700°C. When heated matter in the protoplanetary disk starts to cool, various minerals start to crystallize. Depending on the initial conditions such as temperature and cooling rate, different kinds of crystals form. These are still found in meteorites. The most pristine matter in the Solar System is found in *carbonaceous chondrites*. These meteorites contain two types of very old material: the carbon chondrules, which are small, black, and usually spherical; and calcium-aluminum inclusions which are paler and somewhat larger than the chondrules (up to 1 mm).

Ages can be measured from isotopes of different elements. Meteorites are aged from the ratios of U and Pb isotopes, Al and Mg isotopes, and Rb and Sr isotopes. Al/Mg isotopes provide a relative age telling about the history of the solar nebula. Based on the decay of the short-lived ^{27}Al isotope, this analysis suggests, for example, that the different chondrules in a meteorite were formed at the same time (within 1 million years). Rb/Sr isotopes also measure relative timings of the young protoplanetary nebula, but these are used more to study relative ages of individual meteorites. The U/Pb isotopes give absolute ages, because the amounts of the parent and daughter isotopes can be measured directly. By measuring several chondrule ages from a meteorite one can tell the time of formation of that individual meteorite. The absolute age obtained for carbon chondrules in pristine meteorites is 4.567 ± 0.001 billion years – the best value of the age of the Solar System now available.

The time scales involved in planetary formation are relatively short: in the following keep in mind that 50 million years is only 1% of the age of the Solar System. As we will see the formation of a planetary system was nearly instantaneous – at least in astronomical and geologic terms.

The details are still studied, but the formation of the planets is thought to proceed along the following lines: The dust particles start to stick together in the central plane of the protoplanetary accretion disk. These conglomerates grow in a few ten thousand years into big loose dust piles, may be of the size of 1 km. As they orbit inside the protoplanetary disk, they collect more dust and hit other masses of about their own size. Gravity sets in and starts to make them more compact, sort of taking out the "loose space." Once the planetesimals reach a size of about 800 km in diameter, the self-gravity of the body is now so strong that the body becomes spherical in shape. This transition is not abrupt, but happens as the body grows. Also around this time the planet has grown large enough so that it starts to collect dust and gas from its surroundings by gravity thus growing more rapidly.

At a distance of about Jupiter's present orbit, the temperature in the protoplanetary disk is low enough for ices to remain frozen. This means that there is more solid material to build a planet. As Jupiter kept on growing it reached a mass size of about 30 times Earth. A new process now sets in. The forming planet is now so massive that its gravity can hold even the lightest of the elements, hydrogen, and helium. With its growing mass it scoops up all the available matter in the vicinity of its orbit. Dust, ice, rocks, and gases all increase its mass, until it has swept clean the neighborhood of its orbit. It takes only about 30 million years from the sedimentation to achieve this point. The same process takes place at a slightly more leisurely pace with the formation of the three other giant planets Saturn, Uranus, and Neptune.

Debris of this planet formation process is left around as dust, asteroids, comets, and 1,000-km-sized Kuiper belt objects (or "plutoids") – beyond the orbit of Neptune.

As the planets grow and form, the temperature at the center of the protosun keeps on rising. Once it reaches about 4 million degrees, nuclear reactions begin and the Sun is born. The absolute timing of this is difficult. Possibly this takes place within a few million years of the accretion disk sedimentation time. The Sun shines intensely particularly in ultraviolet as a T Tauri star, and the solar wind, made of fast-moving particles streaming from the Sun sweeps clean the remaining gas from the Solar System.

The Early Earth and the Origin of the Moon

As Earth grew, encounters with other candidate earth rivals became scarcer, but a few times it collided very violently with bodies of quite respectable size. The Earth had practically formed after only 50 million years after its raw material (dust) had settled to the central plane of the protoplanetary disk. At this stage, about 95% of Earth's mass had accumulated, but several big impacts were still to occur. One of them was the impact which we think formed the Earth's Moon.

Prior to the Apollo flights several competing theories were given to explain the origin of the Moon. As early as 1909, Thomas J. J. See, an American astronomer with a checkered and controversial carrier, suggested that the Moon was a body captured by the Earth. Another theory was put forward in 1878 by George Darwin, the son of Charles Darwin. He suggested that the Moon was slung off the molten Earth because of its rapid rotation. Reverend Osmond Fisher elaborated in 1892 that the scars of this event showed up as the large oceans, the Pacific and the Atlantic. However, F.R. Moulton and H. Jeffreys later showed with a detailed calculation that this was not a physically viable explanation (although it persisted in text books until the 1960s).[1]

The Apollo flights in the early 1970s changed the ideas of the Moon's origin, since the astronauts brought back samples. Surprisingly, the rocks were similar to terrestrial basalt (in the darker lunar "seas" or maria) and anorthosite (in the lighter colored highlands). Anorthosite is a rock type principally made of the feldspar mineral plagioclase, the most common mineral found in the Earth's crust. Most of the other minerals in Lunar rocks are also similar to what we find in the crust of the Earth. The mean Lunar density of $3.3\,\mathrm{g\ cm^{-3}}$, turned out to be the same as in the oceanic crust of the Earth indicating that the Moon cannot have a large dense iron/nickel core like the Earth has. This is also reflected in the lack of a magnetic field on the Moon. At best, the iron core can be about one-fourth of the Moon's mass in contrast to the Earth's iron core which makes one-half of its total mass.

---

[1] A third theory was that both bodies were formed at the same time in situ from the protoplanetary disk. This was originally suggested in 1943–1946 by the late Academician Otto Y. Schmidt, and followed up by V.S. Safronov and is presently studied, e.g., by Dr. E. Ruskol, at the Schmidt Institute of Terrestrial Physics, in Moscow.

Fig. 29.5 Schematic sketch of the giant impact hypothesis of the origin of Earth's Moon. The idea was first published in 1975 by William Hartmann and Donald Davis, and independently studied by Alastair Cameron (1925–2005) and William Ward

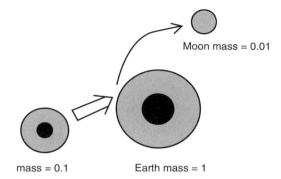

Moon mass = 0.01

mass = 0.1   Earth mass = 1

These facts suggest that the formation of the Moon was not independent of Earth's formation.

A new theory based on these observations was that our Moon was formed in a near-grazing impact between the proto-Earth and a ten times less massive, Mars-sized body, as sketched in Fig. 29.5. The impact was so violent that both bodies melted, the cores eventually fused together, and part of the impact debris fell back to the Earth and part of the ejected debris ended up on orbits around the Earth and formed the Moon. Quite obviously, this impact also evaporated any atmosphere that might have formed before it. Moon samples provide us with a timing for this impact, 4.527 ± 0.010 billion years before the present, or about 40 million years after Earth had began to accrete.

Also other strong impacts took place. The early atmosphere was likely rather thick, but it was removed repeatedly and finally by the "Last atmosphere stripping impact." In doing so it most likely melted quite a bit of the Earth also. The time of this impact can be inferred from the measured isotopic ratios of noble gases such as the $^{129}Xe/^{130}Xe$ ratio (^{129}Xe is formed by the decay of ^{129}I, while ^{130}Xe is stable). The impact took place 4.45 billion years ago, or 120 million years after dust settled to the central plane of the protoplanetary disk to form the Earth and 80 million years after the Moon forming impact.

The previous impacts could have been due to objects that were on orbits similar to Earth's orbit. However, during the next 800 million years or so, a large number of comets and asteroids entered the inner Solar System. Many of these hit the Earth and the Moon. The intense bombardment was caused by minor bodies, which had ventured too close to giant planets. The giant planets changed the orbits of these small bodies, with about half of them being ejected from the Solar System and half being ejected to its inner parts. This clearing action caused mainly by Jupiter intensified the early bombardment and at the same time reduced the later attacks, making the forthcoming evolution of Earth and its life somewhat easier.

Evolution of Earth and the Relevant Timescales

It is difficult to make an accurate picture of the early Earth. The main reasons are that the Earth is geologically active, erosion takes place, and the crust of the Earth

Table 29.2 Geologic times (in millions of years)

Hadean Eon	4,567–3,800
Archean Eon	3,800–2,500
Eoarchean Era	3,800–3,600
Paleoarchean Era	3,600–3,200
Mesoarchean Era	3,200–2,800
Neoarchean Era	2,800–2,500
Proterozoic Eon	2,500–542
Paleoproterozoic Era	2,500–1,600
Mesoproterozoic Era	1,600–1,000
Neoproterozoic Era	1,000–542
Phanerozoic Eon	542-present
Paleozoic Era	542–251
Cambrian Period	542–488
Ordovician Period	488–444
Silurian Period	444–416
Devonian Period	416–359
Carboniferous Period	359–299
Permian Period	299–251
Mesozoic Era	251–65
Triassic Period	251–200
Jurassic Period	200–146
Cretaceous Period	146–65
Cenozoic Era	65-present
Paleogene Period	65–23
Neogene Period	23–0
Present	

is essentially recycled. The radioactive dating techniques described earlier have improved recently to an astounding accuracy so as to reveal very early processes in the evolution of the Earth. At present, the timing of the geologic eras is good to within 1 million years or better for the whole age of the Earth. Table 29.2 shows the geologic timescales (note that two or more Periods comprise a geologic Era and two or more Eras form an Eon).

Recall our description of radioactive dating using very durable zircon crystals (Box 29.1). The oldest terrestrial zircon crystals from 4.2 billion years ago have been found in the Jack Hills region in Western Australia, but these are in metamorphosed rocks. The zircons are from a time *before* these rocks were partially melted and reprocessed. The oldest bedrock, 3.9 billion years in age, is found in the Isua region in Western Greenland. South Africa and Western Australia have bedrocks with an age dated at 3.5 billion years. Most parts of the continents are significantly younger than a billion years, while the bottoms of the deep ocean, the oceanic floor is nowhere more than about 250 million years of age. All in all this means that if one wishes to study the Earth as it was in its youth then the number of sites for obtaining geologic samples is small. As mentioned the timing of important events in the past, based on isotopic measurements, provides us with a solid framework to build on. We can combine information from meteorites, astronomical nebulae, dynamical studies of the Earth, and isotopic studies of terrestrial minerals to obtain the following picture.

Starting from the time of settling in the protoplanetary disk, about 4.567 billion years ago, the body forming the Earth started to grow rapidly. The material was already hot, perhaps 750°C. The Earth began to segregate when it reached a radius of about 1,000 km. Iron and heavier elements sank to the core and silicon and lighter elements occupied parts closer to the surface. The Earth accreted also gases, part of these formed a protoatmosphere, possibly consisting of H_2, H_2O, CO_2, CO, and N_2. Part of this atmosphere escaped immediately into space and parts of it were lost in big impacts. A good fraction of volatile gases was pulled into the Earth, for example in the form of crystal water in hydrated minerals. The energy liberated in the formation process was kept at bay from being released rapidly by water vapor, a very effective greenhouse gas. Thus the surface of the early Earth was hot, maybe 1,700°C, and fully molten. The Earth was covered by a magma ocean, essentially liquid rock.

The Earth cooled down by thermal infrared radiation. At some point when the temperature fell below about 550°C, maybe after 200 million years, the liquid magma ocean started to solidify. A thin solid crust formed. The crust was punctured repeatedly by asteroids and comets. The Earth cooled. When the temperature fell to about 250°C, a long rain began. The first ocean formed, covering the whole globe. At the same time, the greenhouse effect and the total pressure of the atmosphere were reduced significantly as water was removed from the atmosphere into the ocean. The change in the Earth's conditions was quite dramatic. Air pressure was reduced maybe to one-tenth of its original value; the temperature fell significantly. The Earth was left mainly with a nitrogen (N_2) and carbon dioxide (CO_2) atmosphere. This is our understanding at present.

Plate Motions

The Earth cooled further, volcanoes started forming, some of which reached above the surface of the water. As hot basaltic lava poured out of volcanoes, it got mixed with water and created minerals containing crystal water, e.g., serpentine. The rock formations grew larger. At some point they became too heavy to be supported by the thin crust and sank into the mantle. They melted partially at relatively low pressures and temperatures, and as a result, lighter rocks than the basaltic floor separated from the basalt. These lower density rocks thus "floated" on the ocean, eventually becoming the continents. The first signs of both oceans and possible continents come from the Jack Hills zircons dated at about 4.2 billion years ago, only 400 million years after the time of formation of our planet.

The erosion rate from these early continents is thought to have been high, maybe one million times higher than at present because of the high temperatures and the high CO_2 partial pressure. The erosion of silicate rocks turns out to be an effective means of removing carbon dioxide from the atmosphere and burying it into the mantle. Calcium was released from Ca-silicates in the Earth's crust by erosion and transported to the sea. Due to high CO_2 partial pressure in the sea, carbonates precipitated and subsequently the CO_2 content of the atmosphere decreased. In subduction zones (similar to the present Pacific coast of North and South America), these sediments were pulled down into the mantle. Carbon dioxide was later released back into the atmosphere by the volcanic activity. The continents grew larger and at some point plate tectonic movements began. The trigger for tectonics could have been the rising hot convective cells of the mantle, or the tidal forces of the nearby Moon and the breaking of the crust.

Plate tectonic movements have been solidly established and measured in recent years. Here cm per year is an appropriate unit of speed. The Atlantic Ocean is spreading at rates from 2.2 cm per year (North Atlantic) to 3.5 cm per year (South Atlantic). In the SE Pacific Ocean, the Nazca plate is receding from the Pacific plate at a record speed of 15 cm per year and forming new basaltic sea bottom! Besides the spreading of the ocean bottom, other kinds of movements occur. In California, the Pacific plate is slipping against the North American plate at a rate of 5 cm per year. In Indonesia, the Australian plate is subducting under the Eurasian plate (about 6 cm per year). The lighter continents basically float on the sea bed. The implication of this is that all sea bed is much younger than the continents.

Plates also collide. These regions are seismically active. If two continental landmasses collide, they often get folded into high mountains (Himalayas, Alps). Such areas are usually not associated with high volcanic activity. Mountains form also when a seafloor gets subducted under a continental plate. Here the former seabed is partially melted in the mantle, and gets an upward buoyancy, and eventually forms a chain of volcanoes as in the Lesser Antilles in the Caribbean, or in the ring of fire surrounding the Pacific and including the Andes, Sierra Nevada, and the Cascades.

Structure of the Earth

We cannot "look" at the interior of the Earth, but fortunately there is a natural tool to study its structure: a seismic wave. It is caused by earthquakes and registered by seismic stations at a large number of locations. A cross section of the Earth reveals three main parts: the crust, the mantle, and the core. See Fig. 29.6 while reading what follows.

The crust (0–40 km) is the outermost part of the solid Earth. Above the crust one can find the hydrosphere, biosphere, and the atmosphere. The crust is made mostly of silicates and is enriched in the elements Si, Al, K, and Na. It is thin under the oceans, only about 10 km, but below continents its thickness can reach 40–50 km. The oceanic crust is slightly denser than the continental crust because it is enriched in Fe- and Mg-silicates.

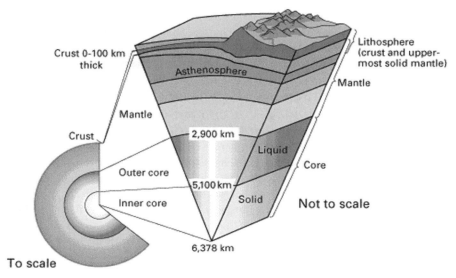

Fig. 29.6 Interior structure of the Earth (credit: United States Geological Survey)

The mantle (40–2,890 km) is divided into the upper mantle and the lower mantle. In the mantle rocky material and minerals move by convection, cooler parts sink down and hotter parts rise up.

For tectonics the most crucial part of the mantle is the partially melted, and nearly solid upper about 100 km of the mantle. This part combined with the crust is called the lithosphere. The lithosphere can be up to 200-km thick under old continents and a few tens of kilometers thick under oceans. The lithosphere is "floating" on a more "fluid" but not liquid layer (the asthenosphere).

The core (2,890–6,370 km) is formed mostly of iron and nickel. The outer core at a depth of 2,890–5,150 km is liquid, and the inner core, which has a radius of about 1,220 km, is solid because of the very high pressure. It is an interesting coincidence that the temperature in the core of the Earth is 5,500K, or about the same as the photosphere or the visible "surface" of the Sun.

The convection in the liquid outer core is quite strong, maybe a million times stronger than in the mantle. Because the matter of the outer core is highly conductive, we have an interesting environment, where rapidly moving and rotating electric fields interact with each other. This interaction causes the magnetic field to form. The source of energy for the convection is the heat from the radioactive decays in the inner core.

Earth is quite unique among terrestrial planets in the sense that it has a relatively strong magnetic field. Mercury, Venus, Mars, and the Moon all lack a strong magnetic field. Mercury, Mars, and Moon are so small that they lack a solid inner core. Venus is somewhat of a problem and is not understood well. It has a mass close to Earth's mass and thus should be similar internally. It should have a magnetic field like Earth's, but it does not. Venus rotates very slowly, once every 243 Earth days, which may suppress the field generating processes in the core, or Venus may lack an inner solid core because heat loss from its core could be less than in Earth's core.

Climate, Atmosphere and, the Greenhouse Effect

The climate of the planet is determined by its radiation balance, and the greenhouse effect formed by its atmosphere (this effect was first discussed by Joseph Fourier in the 1820s). The temperature of the Earth can be calculated by combining the energy received from the Sun and from internal terrestrial sources and by subtracting from it the energy loss radiated by the surface of the Earth. Let us assume at this point that there is no atmosphere. For the present Earth we obtain from this calculation a temperature of $-20°C$. It appears that our planet should really be an ice world! This is indeed what would happen if we had no atmosphere or if the atmosphere was pure nitrogen gas (with or without oxygen). The temperature calculated here is about $35°C$ below the present average temperature of about $+15°C$.

Earlier in the Earth's history, say 2.5 billion years ago, the Sun was fainter. The energy received from the Sun was lower by about 10%, meaning that temperature should have been even lower, about $-28°C$. Yet from geologic records we know that the Earth was for most of the time free of ice. The Earth would have needed some sort of a blanket to bring up the temperature to at least above the freezing point of water. There was indeed such a gaseous blanket, which was several kilometers thick and made of greenhouse gases water, carbon dioxide, and methane. The exact amounts of each gas varied with time.

The heat-trapping effect of the greenhouse gases has been essential for life in both early times and in the present Earth. The greenhouse effect arises in the following way. Radiation from the Sun heats the Earth. The Earth warms up and radiates itself, but now mainly in the longer wavelengths of the infrared (some of the solar IR radiation is also reflected from the surface of the Earth). Part of the outbound energy in the IR is absorbed by greenhouse gases, which re-emit it later to all directions. Some of it escapes into space and a part of it causes the lower atmosphere to warm up. This warming up of the lower atmosphere is the greenhouse effect. At present, the effect on Earth is about $35°C$, quite substantial indeed! Half of this is produced by water vapor (not clouds!), about one-fifth by carbon dioxide, and the remainder mostly by methane and ozone and several other minor greenhouse gas constituents. It should also be pointed out that molecular oxygen, molecular nitrogen, and hydrogen are not greenhouse gases as they do not absorb infrared radiation.

Our globe has been in a kind of a balance with this. During episodes of reduced greenhouse gases the Earth has slipped into a state of glaciation from which it has recovered by volcanic activity. Should it go the other way, one could face a threat of a runaway greenhouse heating effect.

Only a couple of degree change in either direction in the greenhouse effect could have serious consequences. A significant reduction, for example, in the water vapor content of the atmosphere would cause a rapid temperature drop. Similarly, a sudden melting of methane calthrates in the deep seas would cause an increase in the methane levels and thus a larger greenhouse effect. Nowadays much concern is given to the forecasts of the global temperature increase due to rising CO_2 concentrations. The average CO_2 has increased by about 30 ppm (ppm = parts per million) in 20 years and is now at a level of about 390 ppm. This increase of CO_2 has to be taken very seriously. Using a rough calculation we can estimate the effect of this increase to the global temperature. Assuming one-fifth of the global greenhouse is due to CO_2, we can estimate that its contribution is about $35°C/5 \approx 7°C$. In 20 years the relative increase of CO_2 has been about 8%. Multiplying these two numbers together we get a value that the increase of carbon dioxide in just 20 years has increased the global greenhouse by about 0.6°! This gives us an idea of the severity of the problem. A 3° increase in the global temperature would cause dramatic effects to all of Nature including the human population because of changing fresh water resources, rising sea level, and adaptability of crops.

Our calculation is simplified and does not take into account the added modifications of real Nature such as a large number of feedback mechanisms, changes in cloudiness, water drop size, aerosol content, seawater spray interaction with atmosphere, and feedbacks between various atmospheric constituents, occurring on different timescales and on different length scales, not to mention some special interactions such as the ones between the atmosphere, Antarctic and Greenland ice shelves, and the ocean surface and deep currents. Also the slightly variable Sun's luminosity has to be taken into account. Many of these are already implemented in computer simulations of climate to the limits of computer capacity, but still our rough estimate gives a value close to the measured one. The models for predicting climatic changes are complicated and at present can at best provide only good guesses, but the climate forecasts generally tend to agree with a temperature increase of about 1.5–4.5°C in the next century.

Chapter 30
Emergence and Evolution of Life

There were two main historical views of the origin of life. In one, life appeared at some time in the distant past. In particular, in Jewish and Christian tradition the Genesis was interpreted as revealing that God had created all living beings as they appear now. Another view was held by the ancient Greeks, who thought that life can form at any time spontaneously and directly from inanimate materials. Following this tradition, as late as a few hundred years ago it was believed that worms generate from mud and mice appear from dirt. Spontaneous generation was accepted by such men as Newton, Descartes, and by William Harvey (1578–1657), who discovered the circulation of blood.

However, some expressed doubts: Francesco Redi (1626–1697) showed that maggots do not appear in old meat if it is protected from flies, and Lazaro Spallanzani (1729–1799) showed that microbes do not grow in boiled, sealed bullion. Finally, the French chemist Louis Pasteur investigated thoroughly different microbes, and showed that many of them spread through air and caused various phenomena, such as infections of wounds. The careful sterilization experiments of Pasteur finally ended the concept of everyday spontaneous generation of life from inanimate substances (Fig. 30.1).

Chemicals and Structures of Life

But, even if life does not easily start from inanimate matter, it was reasoned (as a variant of the first historical view mentioned earlier) that life must have had at least one initial origin from inanimate substances in the distant past. Modern origin-of-life theories were pioneered by the Russian Nobel laureate Alexander Oparin (1894–1980) in his 1938 book *Origin of Life*. He stated that living structures could not possibly have been able to instruct their own synthesis, or the synthesis of their building blocks, but that they must have been assembled by spontaneous chemical reactions, from pre-existing compounds. English chemist J.B.S. Haldane (1892–1964) expressed similar views.

P. Teerikorpi et al., *The Evolving Universe and the Origin of Life*
© Springer Science+Business Media, LLC 2009

Fig. 30.1 Louis Pasteur (1822–1895) showed that life does not arise spontaneously in a short time span

We do not fully understand how life originated on the Earth. However, we know what are the main functional components of life. Life is based on genetic information, encoded in the nucleotide sequence of DNA, and interpreted via RNA copies to produce proteins, mediators of biochemical functions in the cells. DNA is critical for living beings, as it contains instructions of what proteins are needed in the cell, and how they are made. It also transfers essential information from one generation to the next. However, on its own, DNA cannot utilize its information content. It is like a computer's hard drive which needs to be read by software (RNA) to bring the information to the executable program for display on the screen. However, the proteins, the expressed information, are also needed for the replication of DNA, as well as in each of the steps of the gene transcription and translation. Thus, genetic information is part of a cyclic sequence, where the products of the process are needed for its own maintenance and function. This creates a classic "chicken-and-egg" dilemma. How could the whole system get started, when the products are needed for making the information, the information is needed for making the products, and neither can be made without the other?

RNA World

Clearly, one part of the present-day cycle must have occurred initially and others added later. Although RNA in today's life serves mainly as the transporter of genetic information from DNA to proteins as described earlier, it can also function as a catalyst in many different reactions. Its catalyzing potential is not as good as protein catalysts', but in spite of this, specific RNA molecules perform some of the most

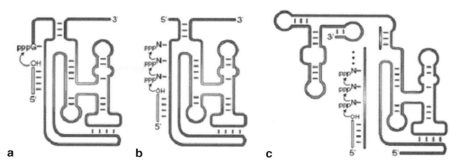

Fig. 30.2 Replicase-ribozyme produced in vitro via multiple rounds of selection for replication activity (Reprinted by permission from Macmillan Publishers Ltd; G.F. Joyce: The Antiquity of RNA-based evolution. Nature 418:214. copyright (2002))

central functions of life (e.g., RNAs inside the ribosome complex join the amino acids to each other in the translation process). It is believed that present-day life process from DNA to RNA, and then to protein, was preceded by a simpler life form, based only on RNAs and proteins. This hypothetical era, when the genetic code would have been encoded in RNA sequences, is called the *RNA-protein world.*

Going back still farther, the earliest protein synthesis machinery had to exist before the first proteins were synthesized (or, before protein synthesis was "invented"). Since the central components of today's machinery are still RNA, we can assume that it was initially composed only of RNA molecules. Thus, complex RNA molecules had to exist and replicate on their own prior the invention of protein synthesis. Thus, the first function in the origin of primitive life supposedly was replication of RNA molecules. This hypothetical era is termed the *RNA world.*

The catalytic property of any given RNA strand requires that the strand be folded into a secondary structure which can interact with the raw materials in specific ways. The folding of the strand is determined by the nucleotide sequence of the strand via base pairing of nucleotides within the strand, and by formation of multilayered interacting stems and loops. This indicates that the catalytic strands need specific sequence information, although they do not code for any proteins. Today's catalytically active RNA strands are called RNA enzymes (*Ribozymes*), and they are able to mediate many different types of chemical reactions. For example, ribozymes can be produced, which catalyze the replication of short RNA strands, either by extension of the molecule itself, or by catalyzing the extension of a separate strand (Fig. 30.2). It is thought that some kind of ribozymes may have mediated the first replication functions. This started molecular evolution which then proceeded toward more complex life.

Conditions on the Early Earth

Not much is known about the conditions prevailing on Earth during the first billion years, except that after some millions of years after the Earth's formation the very hot initial temperatures had cooled down to allow the precipitation of liquid water. Some geologists even argue that due to the dimness of the young Sun, the temperatures may have cooled down over the first half billion of years close to, or below the freezing point of water. The conditions on the young Earth were most likely very varied, being repeatedly heated up by still-frequent meteorite impacts. Local hot spots and very reactive chemical environments occurred in the vicinity of volcanoes and geothermal outlets. Conditions also varied due to the strong winds and tides created by the Moon, which was in a very close orbit at that time. Also the diurnal variation was fast, as the Earth rotated around every 5 h. No sedimentary (water deposited) rocks or geological records remain intact from this early Hadean era, as they all have been melted and resolidified through later tectonic processes.

The oldest sedimentary rocks that have remained fairly intact are the 3.9-billion-year-old rock formation in Isua, on the west coast of Greenland. These rocks contain small carbon inclusions, possible remains of living organisms. Over time, this carbon has been converted to graphite, so there are no cell structures or biochemical compounds to be found any more. However, the biological origin of the carbon is suggested by its enrichment in the lighter isotope ^{12}C, in relation to the heavier ^{13}C, as compared to the constant ratio of these isotopes existing in the carbon dioxide in air. (Carbon appears as two stable isotopes ^{12}C and the rarer ^{13}C which makes up about 1% of natural carbon on Earth.) Biological processes strongly favor the use of the lighter carbon isotope, and therefore the enrichment of ^{12}C indicates biogenic origin of the compounds.

It is now known that also some hydrothermal nonlife processes can fractionate carbon isotopes, casting some doubt on the biological origin of the carbon in these very old rocks. However, the sedimentary structure of the Isua rock indicates that it formed over millions of years in the bottom of deep water, not near the hydrothermal sources. This calm sedimentary origin indicates that the carbon particles originated from overlaying water, probably from photosynthetic plankton that lived in seawater. Even clearer remains of early life, containing fossilized structures of unicellular micro-organisms and some chemical compounds derived from membrane lipids, are detected in the next oldest sedimentary rocks at Pilbara, Australia, and Barberton, South Africa. These fossil records indicate that life was established in significant amounts on Earth at least 3.5 billion years ago. If life was abundant already at 3.9 billion years ago, then it had to start during or right after the era of the "heavy bombardment," or the massive impacts assumed to have happened during the later part of the formation of the Solar System, about 4 billion years ago.

Prebiotic Synthesis of the Building Blocks of Life

To understand the origin of life, we should ask how could the initial RNA polymers have been assembled, how did they obtain the genetic code and the potential to synthesize proteins, and where did the ribonucleotides and amino acids come from? As outlined by Alexander Oparin, the assembly of the initial polymers had to be by means of spontaneous, progressive chemical reactions, starting from simple precursors, and gradually leading to the assembly of more complex molecules. All the initial building blocks and structures of life had to be first assembled in natural physical conditions, without any aid from biological catalysts (complex, information-containing molecules).

Nucleotides and amino acids were needed as building blocks for the assembly of the initial polymers, the RNA genomes, and the proteins. The building blocks had to be formed spontaneously from their organic precursors, small molecules. The most important atoms in these precursors are carbon, hydrogen, nitrogen, oxygen, phosphorus, and sulfur. We have discussed how hydrogen and chemically inert helium were born in the Big Bang. The others formed in stars which then dispersed them into interstellar clouds to subsequently form into later generations of stars with planets. It is known that these relatively abundant elements can react in suitable energizing conditions to make small, *reduced compounds*, such as hydrogen cyanide (HCN), ammonia (NH_3), methane (CH_4), and formaldehyde (CHOH). Reduction means that electrons, usually with hydrogen atoms, are added to the central element, and at the same time plenty of energy is stored into the compounds, making them suitable precursors for further chemical reactions.

Experimental study of prebiotic synthesis of organic compounds started in 1952–1955, when Harold Urey (1893–1981) and his student Stanley Miller tested how the elements of life (C, H, N, O, P, S) can be converted into biomolecules in the simulated early atmosphere of Earth. It was assumed that the atmosphere of the giant gas planet Saturn represented the pristine gas composition of the Solar System, and that the early atmosphere of Earth would have been similar to this, i.e., composed of water, methane, ammonia, and hydrogen. So, reactions of these gases in different mixtures with other gases were tested in laboratory: the gases were closed in a glass vial, over water, and electric sparks were used as an energy source to simulate lightning in the early atmosphere (Fig. 30.3).

Surprisingly, a large variety of organic acids, including several different amino acids, appeared in these conditions by running the reactions for just a few days. The yields of the products depended on the composition of the gas mixture. Efficient production of organics required reduced gases (methane *or* molecular hydrogen). If oxidized carbon CO_2 was used as the carbon source, or molecular oxygen was admitted in the reactions, no organic compounds were produced.

The outcome of the Miller–Urey experiment was really exciting. It clearly proved that synthesis of organic compounds can happen fairly easily from inorganic precursors. However, the hypothesis about the early Earth atmosphere appears to have been wrong. There is now evidence that the "first" atmosphere of hydrogen rich gases was stripped away by the heavy bombardment or the intense early solar wind. The "second" atmosphere may have come from volcanic gases, and from volatiles brought by comets.[1] It was composed mainly of CO_2, N_2, and H_2O, with some CO and H_2. As mentioned, these neutral gases did not produce any biogenic reactions in the experiment.

---

[1] The isotope mix of the noble gases in today's air matches that produced from the decay of radioactive elements in the Earth's crust and not the observed isotope composition of interstellar clouds from which the Earth would have been formed.

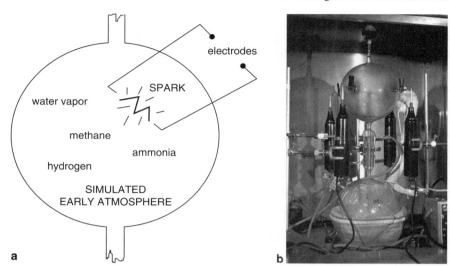

Fig. 30.3 (**a**) By passing electrical sparks through mixtures of water vapor, methane, ammonia, and other gases Miller and Urey produced amino acids in 1953. (**b**) The experiment has been repeated many times; the photo shows the equipment at NASA-Ames Research Center (credit: NASA)

However, the early production of amino acids does not seem problematic, since their synthesis may have occurred in various natural places where small, reduced compounds can react in energizing conditions. Such conditions have occurred, e.g., in geothermal areas under the sea floor, where the sea water filtrated deep into the hot crust and dissolved minerals, carbon, and sulfur. Hydrolysis of water would have produced adequate reduction power (H_2), and the high temperature and pressure would have driven the reduction of the compounds. The reduced compounds were then released with the hot water into the sea bottom at specific outlet sites (similar to the hydrothermal vents now occurring e.g. at the midocean ridge in Atlantic). Sulfides reacted with iron and nickel (and other metal) ions, prevalent in the early seawater, and formed sulfide precipitates, which accumulated in porous structures similar to the *black smokers* seen in such places today. The metal sulfides are active in catalyzing different chemical reactions. These geothermal vents could have served as efficient hatcheries of small organic compounds, including amino acids (Fig. 30.4).

The necessary material for the origin of life may have arrived to Earth also from outer space. Many small organic compounds have been detected in meteorites, in interstellar clouds, and associated with cosmic dust particles, where their synthesis appears to be driven by the abundant UV radiation of massive stars. In addition to small organics, up to seventy different amino acids, different organic acids and sugars have been detected in carbonaceous chondrite meteorites. It is quite possible that extraterrestrial prebiotic chemistry has made a significant addition to the inventory of organic compounds on Earth.

Fig. 30.4 A hot "black smoker" undersea geothermal vent in the Atlantic Ocean (credit: National Oceanic and Atmospheric Administration)

Also, prebiotic synthesis in the early atmosphere has to be reconsidered. Feng Tian and others at University of Colorado, Boulder, published calculations in 2005 suggesting that molecular hydrogen may not have escaped as fast as has been thought. Instead, the early atmosphere may have contained up to 40% of H_2, making it conducive for the synthesis of organic compounds. This new suggestion, although not very well verifiable, just shows how poorly we know the conditions of the early Earth, and how differences in the early environment may affect the possible routes of prebiotic chemistry.

Possible prebiotic synthesis routes for different nucleotides have been much studied, e.g., by the groups of Juan Oro (1923–2004) of the University of Houston and Leslie Orgel (1927–2007) of the University of California, San Diego. These pathways need to proceed through several different steps (1) the synthesis of the nucleobases, (2) the synthesis of ribose sugar in a ring form (formed with the 5′ carbon in right-handed (D) orientation, as described in Chap. 28), (3) the covalent binding (in β-orientation) of the bases into the 1′ carbon of the ribose ring, and finally, (4) phosphorylation of the ribose 5′ carbon. In contrast to the synthesis of the amino acids, the synthesis of nucleotides in prebiotic conditions through all these steps is very difficult and is not yet completely understood.

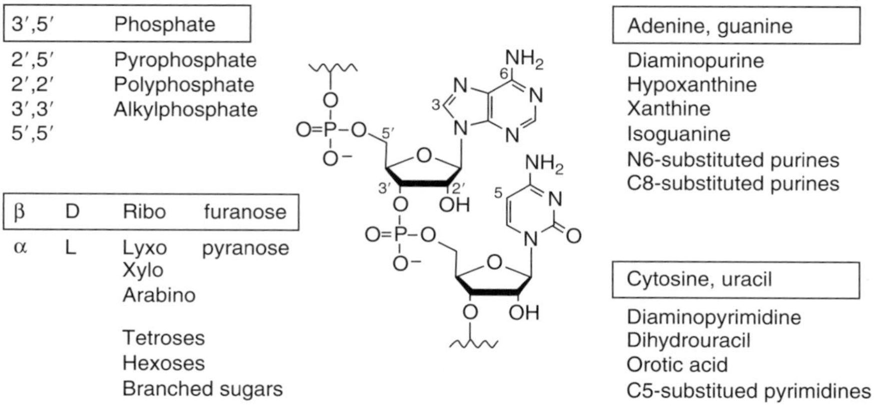

3′,5′	Phosphate

2′,5′	Pyrophosphate
2′,2′	Polyphosphate
3′,3′	Alkylphosphate
5′,5′	

β	D	Ribo	furanose
α	L	Lyxo	pyranose
		Xylo	
		Arabino	

Tetroses
Hexoses
Branched sugars

Adenine, guanine

Diaminopurine
Hypoxanthine
Xanthine
Isoguanine
N6-substituted purines
C8-substituted purines

Cytosine, uracil

Diaminopyrimidine
Dihydrouracil
Orotic acid
C5-substitued pyrimidines

Fig. 30.5 Polymers formed by phosphodiester linkages (containing phosphorus and oxygen) between the 5′ and 3′ carbons of the β-D-nucleotide. The nucleosides are formed from the adenine, guanine, cytosine, and uracil bases and of four-carbon cyclic form of ribose, in D-orientation (Reprinted by permission from Macmillan Publishers Ltd; G.F. Joyce: The Antiquity of RNA-based evolution. Nature 418:214. copyright (2002))

The Riddle of Prebiotic Assembly of Polymers

It is likely that many, or rather, a huge number of RNA strands, with adequate length and variation had to be present to provide even one polymer that had the potential to copy itself, and later, to copy also other RNAs. Thus, efficient spontaneous formation of RNA polymers was required to get the functional RNA world started. The prebiotic polymerization of nucleotides is difficult to explain by known RNA chemistry, because it requires energy and does not happen easily. In optimized laboratory conditions, polymers of about 40–50 nucleotides can be produced. In these experiments, the nucleotides are polymerized in water solution, in the presence of clay minerals. The fine-layered clays, composed of positively charged mineral grains, bind the negatively charged nucleotides and place them in suitable positions to promote their reactions with each other. Further on, the presence of the clays significantly stabilizes the ready-made RNA polymers, which otherwise would be very easily degraded by hydrolysis. These water/clay conditions have been much studied, e.g., by James Ferris at New York Center for Studies on the Origins of Life (Rensselaer Polytechnic Institute).[2]

Next we describe problems in the polymerization. The reader should refer to Fig. 30.5, which shows the form of nucleoside subunits and their phosphodiester linkages in the RNA polymers. The figure also refers to alternative building blocks,

---

[2] David Deamer's team (University of California) has found other environments promoting RNA polymerization, though to a lesser extent than clays. In cold (−18°C) ice solutions, the water remaining between the ice crystals concentrates the precursors, and the low temperatures help to slow down the reacting components, thus allowing the formation of linkages between nucleotides. In such conditions polymers of up to 16 nucleotides have been obtained in the course of a few days.

which cannot be used in the RNA polymers. Thus, RNA nucleosides are formed of the adenine, guanine, cytosine, and uracil bases, linked to ribose sugar, as already described (Figs. 28.4 and 28.5). *The adjacent ribose sugars have to be bound to one another via phosphodiester linkages between the 5′ carbon of one ribose to the 3′ carbon of the previous one.* The phosphodiester bond is formed via a phosphate moiety, containing phosphorus, P and oxygen, O. For this, the nucleosides first have to bind a phosphate group (Fig. 28.4), or to be phosphorylated at their 5′ carbon, to convert them into nucleotides. In the early Earth this was a problem because soluble phosphates were hardly available. Possibly some phosphate dissolved from an inorganic calcium phosphate mineral (hydroxylapatite), although this is only minimally soluble in water. It is also possible that phosphates were obtained from volcanically produced linear polyphosphates, or from their breakdown products. Even if these sources could provide the adequate dissolved phosphates, the phosphorylation of nucleosides proceeds only with great difficulty and in laboratory conditions can be completed only in presence of urea, ammonium chloride, and heat. Further on, polymerization of the nucleotides also requires that they are activated by some high-energy bond (for example, by binding of a nucleobase analog, or an amino acid) at the 5′ position, to provide energy for the binding reaction between the nucleotides.

A further difficulty in the polymerization of the ribonucleotides is that in a mixture of monomers, many different reactions can take place. To make a functional polymer, the phosphodiester linkages must form exactly between the 3′ and 5′ carbons of the adjacent nucleotides. However, the ribose ring has reactive –OH groups in carbons at positions 5′, 3′, and 2′. In prebiotic conditions all these groups can react with each other, and also cyclic compounds can be formed between the 2′ and 3′ OH-groups. Furthermore, the phosphate molecules could have formed different polyphosphate linkages between different carbons. All these varying bonds would have produced dead-end products for further polymerization.

As described by Gerald Joyce (The Scripps Research Institute, La Jolla), a leading student of prebiotic RNA chemistry, the lack of specificity has indeed been a major problem of prebiotic reactions. The spontaneous reactions starting from hydrogen cyanide, or from cyanoacetylene, cyanate, and urea can lead to a number of different nucleobase analogs. But of all the analogs, only adenine and guanine purines, and cytosine and uracil pyrimidines were eventually used by nature for formation of the functional nucleosides. In the composition of the nucleosides in prebiotic conditions, the existing bases could have been connected to the ribose components, just as well, both in α- and in β-configuration, and the furanose (four-carbon) ring of ribose could have formed just as well in L and D isoforms (left- and right-handed; described in Chap. 28). Ribose sugar could also have formed a five-carbon (pyranose) ring by binding of the 5′ and 1′ carbons. Prebiotic polymerization reactions between all different nucleotide analogs and isoforms would have also led to a wide variety of different phosphate linkages between different carbon atoms of the ribose. Altogether, these reactions would have easily used different purine and pyrimidine variants, bound with different derivatives of different cyclic sugars, formed both in L- and D-configurations. These very random nucleoside analogs could then have been phosphorylated at different carbon positions, and then again, the randomly phosphorylated nucleotide analogs could have been connected to each other in a number of different ways as shown with light lettering in Fig. 30.5. None of these alternatives would have produced functional RNA polymers.

Only the correctly formed and polymerized nucleotides would have been functional templates for replication via complementary base pairing. We do not understand how life, in the absence of any selective enzyme reactions, choose to use exactly these nucleotide components and their specific isoforms, or how it could control formation of the phosphodiester bonds to occur only between the 5′ and 3′ carbons of the nucleotides.

A further problem in the accumulation of long RNA polymers is their inherent instability. RNA polymers are very easily broken into parts by hydrolysis, and their functional sequence could have been easily lost via multiple copying mistakes or mutations. Considering all these chemical obstacles, it seems that the whole reaction cascade for the formation of functional polynucleotides (including the synthesis of the nucleoside bases and ribose, assembly of nucleosides, their phosphorylation and activation, and finally, the polymerization and stabilization of the polymers) has been very difficult in the prebiotic conditions. These processes seem so unlikely that it has been proposed that some other information storing and transfer mechanisms preceded the RNA world and then "guided" the formation (or provided catalysts for) the RNA-based world. But it is not easy to explain how the transfer from a more primitive genetic system into RNA could have happened.[3]

Production of the Genetic Code

Although RNA molecules can mediate several kinds of chemical reactions, we now know that protein catalysts are clearly superior in their versatility and efficiency. Thus, invention of the genetically encoded protein synthesis gave a huge advantage to the developing life. It has made possible the appearance and evolution of the DNA-based genomes, of complex cellular structures, and the biochemistry now typical of life. The invention of protein synthesis was so crucial for the evolution of life that Anthony Poole and associates at Stockholm University (Sweden) have named this phase as the *Break-through organism* or *Riborgis eigensis*. However, the initiation of the genetically encoded protein synthesis must have been an "accidental" – or completely unexpected – turn of the RNA-based chemical evolution. Indeed, there cannot possibly be any anticipation of evolutionary "inventions" before they first happen. As we know, the protein synthesis requires a machinery, composed of the catalytic ribosome complexes (to read the genetic code and form the peptide bonds), of the tRNAs and amino acids, and of catalysts that bind the amino acids to the tRNAs (as discussed in Chap. 28). The core components of this complex machinery

---

[3] The other alternative is that we just have not found yet those conditions, chemical pathways and selection factors that have made possible the prebiotic chemistry and evolution. New promising pathways for the prebiotic synthesis of nucleotides, directly from the formamide, are currently being tested in the laboratories of R. Saladino and E. DiMauro at Universities of Tuscia and Rome, Italy

are formed by RNA molecules. As both the central catalytic functions involved in the translation process (aminoacylation of the tRNAs and peptide bond formation) can be mediated by RNA enzymes, it is conceivable that a primitive form of this machinery was produced by the RNA world. However, such machineries do not exist or appear "by accident"; they must have developed (by gradual evolution) for some earlier function. As the protein synthesis did not exist as yet, the primitive translation machinery had evolved *from something else*. It has been proposed by David Penny's group at Massey University (New Zealand) that the original function of this molecular machinery was the replication of the RNAs and that all the components of the later translation machinery (protoribosomes, proto-tRNAs, and amino acids) were already involved in this function. Orchestrated early function of these components could have then facilitated the conversion of their pre-existing interactions toward the development of the translation process.

Penny's team hypothesizes that the early replication machinery may have been based on the cutting and ligation activities of the ribozyme enzymes (or, of the early ribosomes). This hypothesis is supported by the fact that these activities are still very common for the ribozymes, while the direct polymerization of individual nucleotides is not. It is possible that the ribosomes recognized the target RNA sequence as triplets of nucleotides and copied it into a new strand by ligating together complementary triplets brought in by the tRNA molecules (see Fig. 30.6). The amino acids

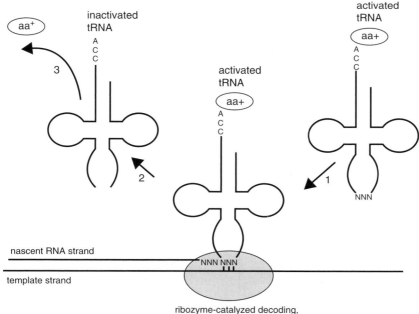

Fig. 30.6 Hypothetical transreplication by activated RNA. The template RNA strand is at the *bottom* horizontally. The replicated RNA strand is shown horizontally on the *left* (adapted from Poole et al. 1998, The path from the RNA world. J. Mol. Evol. 46:1)

could have been bound to the tRNAs either to charge them with energy, or, to provide proper folding to the molecules. This replication process brought the amino acids close to each other, and thus made possible the formation of peptide bonds between them. Possibly the early replication process was overlapping with early translation process for some time. The evolution of the genetically encoded proteins eventually produced proteins which could catalyze replication, this function was taken over by the protein enzymes, and the ribosomes and tRNAs evolved into a pure translation machinery. Both functions were strongly favored by natural selection.

The Final Step: Formation of Cellular Life

Gradual invention of the RNA polymers, the genetic code, and the translation machinery for protein synthesis provided most of the core components that were required for the self-sustainable life. However, for the whole system to function and to evolve, all these parts need to interact and support each other. They need to be bound together. At some point a surrounding membrane, or a cell structure, was acquired to confine the genomes, translation apparatus, and the different gene products into one package. Only now the different prebiotic molecules could form a functional entity that could interact with its environment and evolve via natural selection.

The origin of cell membranes is unclear. On one hand, it might have been a fairly easy process: different lipid molecules (fatty acids or other long hydrocarbon chains) spontaneously aggregate with each other in aqueous solutions to form micelles or membranes, and may form vesicles (Fig. 30.7). Such spontaneous vesicles

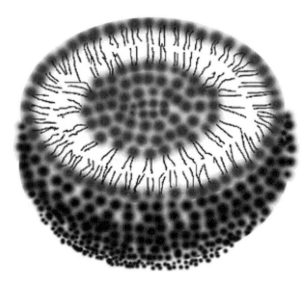

Fig. 30.7 Different lipid molecules (fatty acids or other long hydrocarbon chains) spontaneously aggregate to each other in aqueous solutions to form micelles or membranes, and may form vesicles

may enclose different molecules, at random, from the surrounding solution, and in this way also the functional RNA polymers could have been enclosed inside protocellular structures. In recent years the behavior of such spontaneous membrane vesicles has been much studied. It has been shown that, e.g., vesicles formed of oleic acid molecules (chains of 18 carbons) are semipermeable in such a way that they allow transpassing of small molecules, such as single nucleotides or amino acids, but do not allow passage of larger polymers of these subunits. This kind of selective transpassing might have allowed the uptake of the building blocks from the environment, while retaining the polymerized bioproducts inside the vesicles. These self-assembled membranes and vesicles can also expand and grow by integrating more of lipid molecules from the environment, and they can even spontaneously divide into small new vesicles! Thus, the formation of vesicle-like protocells might have proceeded spontaneously in an environment that provided abundant lipids. However, the source of these lipids is still rather unclear: formation of long carbon chains requires a lot of chemical (reduction) energy and does not happen easily.

Evolution of the Biosphere

After its first appearance, cellular life evolved and produced a fairly self-sustainable existence. This life contained all the hallmarks now typical for life. It possessed the genetic code, protein translation machinery, and produced enzymes for its energy production and for synthesizing its nucleotides and amino acid-building blocks. At some point this cellular community diverged to separate phylogenetic lineages, which produced the three domains of life, the Bacteria, the Archaea, and the Eukaryota. The parent community that gave rise to these domains is the Last Universal Common Ancestor, as discussed earlier (Fig. 28.9). It is not clear at what point this division happened: for instance, it is hazy whether LUCA was based on DNA or RNA genomes. Neither is the branching in the three domains very clear. In any case, after the division, life has diversified and adapted to new conditions, and eventually, has occupied all habitable environments on the Earth.

The basic requirements for the survival of even the most primitive life are a suitable energy source, nutrients for the composition biomolecules, liquid water as a solvent for all the chemicals, and conditions which do not destroy the biomolecules. Presumably the simple earliest life forms were not able to perform complex energy conversion reactions. Thus, energy should have been easily usable in direct chemical conversions, in the form of energy-rich (reducing) molecules, such as molecular hydrogen, methane, ammonia, and hydrogen sulfide. Also some minerals (like iron and sulfur compounds that can be easily oxidized) provide a good source for reduction energy (hydrogen atoms or electrons). Necessary oxidized nutrients, such as phosphates, nitrates, and sulfates, would have been needed, as well as soluble metal ions (Fe, Ni, Cu, Mg) for use as cofactors in different enzymes. Such small reduced compounds and soluble metal ions can be produced prebiotically in few places, such as hydrothermal and other volcanic systems. It seems likely that early life got started in the vicinity of such energy sources. Then, life learned to recycle the energy and nutrients, by using the existing biomass, or the ready-made energy-rich biomolecules, in different microbial food chains. This allowed life to diversify into new species, with new life strategies. Then life could spread further out from the limited volcanic or lithotrophic ("rock-eating") food sources and develop new metabolic pathways. The ultimate escape from the rock-bound life was the invention of photosynthesis – the ability to convert the light energy into chemical form.

Photosynthesis, untapping a new unlimited energy source, allowed life to spread out from the rocky places into illuminated seawaters. In the sedimentary rocks of Barberton, South Africa, and Pilbara, Australia, dating back up to 3.5 billion years ago, microbial fossils occur in different environments and types of sediments (Fig. 30.8). The surrounding sediments indicate that they have been formed either in porous rocks in hydrothermal environments, or in deep or shallow water sediments. The carbon and sulfur isotope fractionation data of the samples support the notion that the organisms have been using lithotrophic or photosynthetic carbon fixation, respectively. These microfossils also appear varied in size and shape, so life had already diversified to multiple different species.

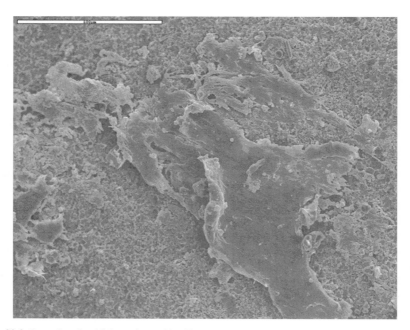

Fig. 30.8 Part of a microbial mat formed by filamentous, likely anaerobic, photosynthetic microorganisms on the surface of a 3.33-billion-year-old beach (Barberton greenstone belt, South Africa). The microbial mat was first partly calcified and then silicified. The inserted bar is 100-μm long (courtesy of Frances Westall)

The new life forms rapidly adapted to environments providing suitable nutrient and energy sources on the young Earth. Examples were hot springs (over 85°C) and their surroundings, lukewarm ponds, cold waters on the polar areas, and water under ice, shallow waters and terrestrial tidal areas and pools, surroundings of terrestrial volcanoes, and deep subsurface rocks and aquifers. From the perspective of our aerobic, temperate world we would consider many of these places hostile. Indeed, the typical conditions were very different from the present world. Without any molecular oxygen in the atmosphere, all life was anaerobic. Also, no ozone was protecting surface areas and shallow waters from strong UV radiation, and life had to use some means of shielding to avoid massive damage. The high temperatures, or high salinity, alkalinity, or acidity of geothermal systems required special modification of cellular membranes or macromolecules to prevent their degradation. However, it seems that microbial life forms were adapted to these environments from early on, and some species have thriven in them ever since. We call these species *extremophiles*, though they may much more resemble the original life forms than do the temperate aerobic species.

The early species may have been quite diverse and adapted to different environments, stretching over the whole planet. Still, for the first 2.5 billion of years of life history, all life forms remained simple and unicellular, leaving only very little of identifiable remains in the fossil record. Precambrian stromatolite fossils have been identified, thought to be built by colonies of cyanobacteria. Living structures similar to these fossils are still found in some shallow waters and may represent possible survivors of the early life forms which spread into new shallow habitats to eventually modify the global environment. (Fig. 30.9).

Fig. 30.9 Modern stromatolites at Lake Thetis, Australia. Corresponding fossilized stromatolites over 3 billion years old have been identified (courtesy of Ruth Ellison, Glass Zebra Photography)

Effects of Life on the Atmosphere and Climate

After the oceans had formed the first real atmosphere of the Earth consisted most likely of CO_2, N_2, H_2O, CO, and H_2. No oxygen was present. The details are not well known, but the present understanding is that the air pressure was about 10–20 times the present value and was dominated by carbon dioxide CO_2. The amount of carbon dioxide started to reduce once the erosion of silicate rocks set in. Calcium was released from Ca silicates of the Earth's crust by erosion and transported to the sea. CO_2 dissolved in sea water reacted with calcium to form carbonates, which precipitated to the sea bottoms. In this way, the carbon dioxide content of the atmosphere gradually decreased.

After the intense bombardment ended, the newly evolving biosphere started to influence the atmosphere. Many of the early life forms were using inorganically or organically produced hydrogen, and used it to reduce CO_2 to methane, CH_4. It appears that methane, produced by single-cell Archean microbes, became an important greenhouse gas three billion years ago. In absence of oxygen, methane accumulated in the atmosphere, apparently up to rather high concentrations (100–1,000 ppm; cf. modern air with less than 2 ppm of CH_4 and 390 ppm of CO_2). As a very strong greenhouse gas, methane induced significant global warming, and the temperatures rose to 75–80°C, despite the Sun being fainter than at present. High level of methane could also create a hazy smog, which protected the Earth surface from the UV rays to some extent.

The different microbial species were able to reduce or oxidize carbon, sulfur, and nitrogen compounds. Thus the biosphere could cycle these elements between their organic and inorganic forms, and change the composition of the atmosphere. A revolution happened in the biological energy production, when the cyanobacteria evolved a photosystem containing an adequately strong oxidizing complex that could use water as the electron donor for the photosynthetic reaction. This reaction strips two electrons from a water molecule, and transfers them to build up the photosynthetic reduction power (see Box 30.1), and at the same time releases oxygen as the waste product. Around 2.2 billion years ago, the first signs of significant atmospheric oxygen appear in the geologic records. Even small amounts (1–2%) of oxygen could effectively remove the strong greenhouse gas, the methane. The first appearance of oxygen in the air apparently led to a severe global ice age, the Snowball Earth. A second series of ice ages took place 800–600 million years ago, just before the Cambrian period, at the time when oxygen rose to the present level of about 21%.

These global ice ages lasted for long periods, and together with the drastic change of the atmosphere, led to strong reduction of all living species. This can be seen by a strong increase of the biopreferred ^{12}C in the inorganic carbon pool, as recorded in the $^{12}C/^{13}C$ ratio in carbonate sediments of that time. Over these times, existence of life was pushed to mere survival under the ice.

During both total glaciations, the tectonics worked under the icy covers, and the volcanic eruptions eventually returned sufficient amounts of carbon dioxide to the atmosphere to restore adequate greenhouse effect. The global temperatures

Box 30.1 Processes, Global Impact, and Signature of Photosynthesis

The first form of photosynthesis was, by necessity, the simplest one: it could have been similar to the photosynthesis still used by the halobacteria, where a membrane-bound pigment protein (bacteriorhodopsin) reacts to light, and via its conformation change, pumps protons across the membrane, thus creating an energy-rich proton gradient.

With time, more effective pigment molecules (chlorophylls and bacterio-chlorophylls) were invented, which could efficiently harvest the energy from the photons into their electron excitation stage and use this excitation energy to take electrons from suitable donors. In many photosynthetic bacteria this reaction is anaerobic, using H_2, H_2S, S, or organic matter as electron donors. At some point, the cyanobacteria developed a photosynthetic reaction system which was energetic enough to oxidize water.

In this process the chlorophyll excitation energy is used to take two electrons from the water molecule. These electrons are shifted to the electron acceptor, while two protons are released into the medium, and molecular oxygen is released as a waste product. These energy-rich compounds are then used in a separate reaction to convert CO_2 into sugar molecules

$$6CO_2 + 12H_2O => C_6H_{12}O_6 + 6H_2O + 6O_2$$

The carbon reduction is done by the ribulose-1,5-bisphosphate carboxylase oxygenase enzyme, or more simply *rubisco*. The rubisco enzyme strongly favors the use of the lighter ^{12}C isotope, at the expense of the heavier ^{13}C carbon. As the rubisco-mediated carbon fixation is very efficient, it has allowed massive binding of the carbon in photosynthetic organisms. Through different food chains, these serve as the primary (and nearly as the sole) energy source for other living beings on Earth. Accumulation of the ^{12}C isotope in the biomass has caused its depletion in the atmosphere, and from inorganic carbon deposits on a global scale. Particularly, the most efficient carbon-fixing system, mediated by rubisco I, typically produces ^{12}C depletion of -28 to -30‰, and strong ^{13}C isotope enrichment in the carbonate sediments is called the *rubisco signature*.

rose above freezing, waters opened and were exposed to sunlight, and allowed the new strong proliferation of the cyanobacteria. The oxygenic atmosphere was initially harmful, or even toxic to the organisms adapted to anaerobic conditions, and led to a major change in the global microbial populations. It provided a new possibility for oxidative metabolism, or a more effective way to utilize energy bound in the organic compounds. This was used by many new bacterial species, and also by a new, more complex organism, the eukaryotes. These had an aerobic bacterium adapted as an intracellular organelle, now known as mitochondrion, which allowed efficient oxidation of the organic compounds inside the cells. This new metabolism gave a strong benefit to the biosphere in the oceans. Also, the aerobic atmosphere allowed the formation of a significant amount of ozone in the upper atmosphere, blocking the strong UV radiation, and allowing the spread of the biosphere to shallow waters, and eventually, to dry land.

The time of the first appearance of the oxygenic microbes has been much debated. There are claims for traces of oxygenic photosynthesis already during the earliest known (or putative) biogenic fossils, based on the oxidized mineral formations, e.g., the banded iron formations, occurring in the Isua rock and the Pilbara deposits in Australia. As the fossilized microbial structures do not contain any identification markers for the ancient species, one has to rely on preserved biochemical evidence. The oldest proof for oxygenic environments can be read from some organic deposits strongly enriched in ^{12}C isotope. This indicates that the carbon is fixed by rubisco I, functioning only in the oxic atmosphere (the *rubisco signature,* Box 30.1). Such deposits, at Tumbiana, Australia, have been formed at 2.7 billion years ago, and some older ones at Steep Rock, Canada, at about 3 billion years ago. The oxic conditions give indirect evidence for cyanobacteria, but clearer evidence can be read from the fossilized molecular markers (the membrane lipids bitumes and hopanes; particularly 2-methyl-bacteriohopanepolyol), typically synthesized only by cyanobacteria in aerobic conditions. Likewise, the lipids produced by eukaryotes in oxic conditions (sterols) have been detected in the Pilbara formation, giving the latest definite date for the appearance of eukaryotes to about 2.8 billion years ago. Also, it appears that the oxygen produced at the early stages accumulated only locally in aqueous environments and became soon bound in oxidized minerals. Therefore it started to accumulate in the atmosphere only some 500 million years later, at about 2.2 billion years ago when the oxidizing sinks had all been filled.

Over the 2 billion years following the first appearance of cyanobacteria, abundant organic deposits accumulated in the oceans. It is not known how diversified the species were during this time, as the organisms did not make hard body structures, and very few fossilized remains have been preserved in the sediments. However, some remains of multicellular algae have been found dating back to 1.2 billion years, clearly showing that multicellularity existed by this time. The oldest remains of the first soft-bodied animals (radially symmetric fossilized impressions) date back to about 580 million years, or to the time just preceding the end of the proterozoic eon, or right after the global glaciations of the cryogenian period. Cyanobacteria, with their oxygenic photosynthesis, also became adapted as the symbiotic organelles forming the chloroplasts of eukaryotic algae, and later, of the higher plants. As the higher plants some hundred million years later colonized the dry land, this allowed photosynthetic carbon fixation on the continents, providing the energy source for large and complex food chains.

The oxygenic atmosphere, and the more efficient metabolism associated with aerobic respiration, allowed the appearance and diversification of multicellular organisms. Multicellularity allowed the differentiation of body parts and their adaptation to different beneficial tasks. This provided new possibilities for the creatures for new energy sources and nutrients. The multicellular algae and plants could grow their roots into the soil where water and soluble nutrients were available, and extend their photosynthesizing leaves toward sun light. Animals were able to search for their food and find new food sources. The sexual reproduction strongly boosted evolutionary potential, allowing the repeated recombination of genetic materials in each generation.

Catastrophes Affecting the Evolution of the Biosphere

The major atmospheric change, the rise of oxygen, discussed earlier was a result of increased efficiency in the use of solar energy by cyanobacteria, and thus was caused by life. The appearance of oxygen – itself a very toxic gas to the early anaerobic world – caused significant stress to the cyanobacteria themselves and also to other organisms. It was both a catastrophe and an opportunity.

The biosphere has encountered also other kinds of catastrophes. Global ice ages, during which the whole Earth or most of it was covered by a thick ice layer, could have been devastating to surface life. From geologic records we know that the first life forms or the spore-forming plants colonized the continents only about 450 million years ago, so no surface life was exposed to the earlier ice ages. All life was at that time either close to the seashore, below the sea surface or in subsurface rocks in the crust. Still, during the Precambrian ice age when photosynthesizing algae had evolved, the thick ice caused a potential source of a catastrophe. Fortunately, the ice may have not fully covered the tropical regions. Also, light can penetrate through ice up to three meters thick to support life, and some life can also thrive in water pockets inside of ice, as seen now in certain ice lakes in Antarctica. If the Earth for some reason should fall now into a global ice age, the continental surface life could have similar possibility for survival as is now seen in the Antarctica.

Earthquakes and tsunamis have become regular news. In recent decades, the highest count of human casualties, near 300,000, was from the December 26, 2004, tsunami caused by one of the largest seafloor earthquakes in recent history. Earthquakes in populated areas can leave millions of people homeless. However, these tragic events do not cause widespread devastation to nature.

Cataclysmic geologic events come in different sizes. Single small volcanoes can cause serious local destruction. A large volcano eruption has a global effect. The 1833 eruption of Krakatoa ejected $25\,km^3$ matter in the form of lava and ashes ($10\,km^3$ dense rock equivalent). Much of the ashes ended up in the upper atmosphere up to $80\,km$ causing a significant drop in global temperatures for several years. Basaltic magma from the hot mantle may reach the surface also through crustal cracks. In the Lakagígar region in Iceland such a fissure has occurred several times.

The last one was in 1783–1784 when $15\,km^3$ of lava surfaced causing a small basalt flood. Large amounts of poisonous gases were released. One-third of the Iceland's population died of famine, and about three-quarter of the livestock died because of fluorine poisoning. In Europe, tens of thousands of people died because of thick sulfurous haze. North America experienced in 1784 the coldest and the longest winter in recorded history.

Even larger eruptions take place, but fortunately they occur less often. Lake Toba in Indonesia is an example of a caldera of a supermassive volcanic eruption. This eruption, about 73,000 years ago, may have been the largest one in the last several million years. The equivalent dense rock volume of the ash layer from this event is $800\,km^3$, about 20 times larger than the largest historic volcano eruption of Tambora in 1815. As a result, all of South-East Asia was covered by meters of volcanic ash. It created a global temperature drop of about $3°$ for several years, a real volcanic winter. The anthropologist Stanley Ambrose (University of Illinois) has suggested that the ancient eruption caused an evolutionary bottleneck in human evolution during which the human population was reduced seriously.

As volcanic eruptions occur on different size scales, so do fissure eruptions. Tens of "Large Igneous Provinces" (LIPs) have been identified on continents and ocean bottoms. The formation of LIPs is somewhat of a mystery, but they may have been caused by massive mantle plumes. What is known is that the LIPs are formed geologically on a short time of a couple of million of years, and that the lava pours out of the ground in massive volumes. The basaltic flood causing the Deccan traps in present-day India 60–70 million years ago involved a volume of $500,000\,km^3$ of lava and covered a surface area of about the size of France. If we compare these numbers to the historic fissure in Iceland (see earlier), we can only imagine how much destruction such a basaltic flood could cause. In this respect, the timing of the massive Siberian trap is of special interest as it is very close to the Permian/Triassic global extinction about 252 million years ago.

When the Earth was young asteroids and comets were useful in importing essential building blocks, such as water, silicates, carbon, and nitrogen for the atmosphere. After the first billion years they can be considered rather as being a possible hazard for the established life. There is good evidence that a 10-km asteroid hit the Earth 65 million years ago in what is now Yucatan, Mexico, and that it resulted in the demise of the dinosaurs as well as many other forms of life.

At present, the threat from asteroids and comets is from two kinds of sources. The main belt of asteroids between Mars and Jupiter consists of tens of thousands of dangerous size asteroids and countless smaller ones (Fig. 30.10). Fortunately these objects are on relatively stable orbits that practically never bring them to the central parts of the Solar System. Over the course of 4.6 billion years some objects have experienced an orbital evolution that has brought them into Earth crossing orbits. To find these potentially risky Near Earth Objects (NEOs), several sky patrols have been set up since the mid-1980s. A potentially hazardous object is defined as one that approaches the Earth to within 0.05 AU, or 7.5 million km (about 600 Earth diameters) and has a size of about 200 m or larger. In August 2008 the number of such objects was 1,400. The total number of known NEOs was about 5,500,

Fig. 30.10 Impact crater in Arizona was formed about 50,000 years ago by a nickel–iron meteorite 50 m across. The explosion was equivalent to 150 Hiroshima atom bombs (credit: U S Geological Survey)

of these 750 larger than 1 km. The probability that any of the known NEOs will hit the Earth in the next 100 years is less than one in 10,000. As techniques and surveys improve it is expected that almost all large NEOs will be known, say by 2020. Repeated observation of the known NEOs will enable astronomers calculate their orbits accurately. If an impact is foreseen early enough, then there should be plenty of time to divert the asteroid. To do this the orbit of the asteroid should be changed by at least an Earth radius. To be on the safe side, we can estimate that a NEO should be diverted about 40 years in advance so that it will miss the Earth by at least one Earth radius. This far in advance only a small "kick" of about 1 cm s^{-1} would be required.

The threat from comets is different. Comets entering from the outer parts of the Solar System will not give a long warning time. They are typically detected when they have approached the Sun to within Jupiter's distance. If an impact on Earth is calculated, then the time left is only about 5 years. Only swift action and a large velocity "kick" could prevent the impact. We can estimate the likelihood of such a randomly entering comet hitting the Earth.[4] With ten comets entering the inner Solar System per year, on average it would take about 50 million years before one comet would have struck the Earth. As this is just a statistical estimate, the next impact can be 5 years from now or 50 million from now, or any time in between. Furthermore, the impact speed would be an order of magnitude higher than that of a NEO. This has direct implication to the resulting destruction as the energy released in the impact is proportional to the speed squared or about 100 times greater.

---

[4] Numerically it is about the same as the ratio of Earth's cross section and the effective surface area of a sphere with a radius of the Earth's orbit: P (collision/comet) $= (R_E/1\,AU)^2 = (6.4 \times 10^6\,m/1.5 \times 10^{11}\,m)^2 = 1.8 \times 10^{-9}$

A large, say, 500-km comet or asteroid would have very serious effects to life on Earth. Its kinetic energy would be large enough to vaporize all the oceans and melt the crust of the Earth to a depth of several hundreds of meters. Should this happen today, it would cause a nearly complete destruction of life in Earth. Fortunately such objects are a tiny minority among comets or asteroids today compared to the impacts in the early history of the Earth.

Benefits of Catastrophes

There are always two sides in a coin. So is the case with catastrophic events. Very rapid swings in environmental conditions may cause local extinctions, while larger events may have a serious impact on continental or even a global scale. Over 95% of living species may perish. For the species that survive, the new conditions will create a unique new chance for rapid evolution as species adapt and re-establish themselves to the newly forming ecosystems. We have seen this happen in the history of the Earth. The well-known case of the rise of mammals after the extinction of dinosaurs after the asteroid impact is not unique. After the global ice ages 600–800 million years ago a massive radiation of life forms, the Cambrian explosion, took place. The largest of the extinctions, the Permian/Triassic extinction 252 million years ago, set the conditions for the diversification of land plants, reptiles, bivalves, and crabs as well as dinosaurs. This extinction coincides in time with the formation of the Siberian trap, the largest volcanic event known on Earth, and possibly also with a massive asteroid impact found recently under the Antarctic ice shelf.

If a serious catastrophic event would be experienced tomorrow by the Earth's biosphere, it would no doubt be able to recover, for example by starting from relatively highly developed and adaptive species such as roaches and rats! Even for the case of an extremely massive asteroid like that mentioned earlier, a new start could still be made from the highly diversified domains of bacteria and archaea.

Chapter 31
Life and our Solar System

The complex and wonderful phenomenon of life has been found so far only on Earth. Signs for life have been and are being searched for in other bodies of our Solar System as well as in other planetary systems. If we consider where life or prebiotic chemistry could take place, then there are a number of interesting targets in our neighborhood. Even bodies which cannot support life now are worthy of attention because they may tell us where things can go wrong for life.

An Overview of Unlikely and Likely Suspects for Life (And Why)

When planets formed, they, as defined by the International Astronomical Union (IAU), swept clean their orbital neighborhoods (see Box 31.1). The four inner Earth-like planets in our Solar System (Mercury, Venus, Earth, and Mars) formed from rock and iron/nickel solids in the hot inner part of the solar nebula close to the young Sun. In the inner four, the denser iron subsequently settled to the center to form an iron/nickel core with a less dense rocky mantle. An atmosphere was formed by volatile gases delivered to the young Earth by comets and asteroids and cycled through volcanic activity as described in more detail in Chap. 29.

Whether or not this atmosphere is retained depends on the gravity of the planet along with its nearness to the Sun. If the gravity is low, then the thermal motions of a significant fraction of the atmospheric molecules will be larger than the escape velocity so that they will leave the top of the atmosphere never to return. Over time, the atmosphere can thus be lost. Although both the Earth and its Moon are the same distance from the Sun, the Moon with a much lower mass has lost its atmosphere. Mercury, also lighter than Earth, is practically without an atmosphere because its closer distance to the Sun resulted in high thermal molecular speeds and the subsequent loss of atmosphere. The Earth is not immune from such thermal loss either. Lower mass atoms and molecules, such as helium and hydrogen, move rapidly, even at our distance from the Sun, and thus cannot be held by the Earth.

P. Teerikorpi et al., *The Evolving Universe and the Origin of Life*
© Springer Science+Business Media, LLC 2009

Box 31.1 The definition of a planet

The International Astronomical Union (IAU) is an organization which has about 10,000 professional astronomers as members. In 2006 the IAU held its General Assembly in Prague (Czech Republic) and made a decision about the new definition of "planet" in the Solar System. Traditionally it was regarded that the Solar System has nine planets in addition to thousands of smaller bodies such as asteroids and comets. However, when the ninth planet's (Pluto's) mass was determined using its moon Charon (which was discovered in 1978), Pluto was found to be quite small, smaller than our Moon and about 20 times less massive than Mercury. Later on other small objects were discovered in the outer parts of the Solar System, at distances as far as Pluto or even farther away, some of these comparable with Pluto in their size and orbit. Should these be called planets, too? What is a planet? The problem was debated within the astronomical community for years, and finally the 2006 the IAU General Assembly voted in favor of the following definition which included the categories "planet," "dwarf planet," and "Small Solar System Body":

1. A planet is a celestial body that (a) is in orbit around the Sun, (b) has sufficient mass for its self-gravity to overcome rigid body forces so that it assumes a hydrostatic (nearly round) shape, and (c) has cleared the neighborhood around its orbit.
2. A "dwarf planet" is a celestial body that (a) is in orbit around the Sun, (b) has sufficient mass for its self-gravity to overcome rigid body forces so that it assumes a hydrostatic equilibrium (nearly round) shape, (c) has not cleared the neighborhood around its orbit, and (d) is not a satellite.
3. All other objects orbiting the Sun shall be referred to collectively as "Small Solar System Bodies."

According to this definition there are now eight planets in the Solar System. For example, Pluto is a dwarf planet, and a great majority of asteroids are Small Solar System Bodies.

Hydrogen atoms are essential to life, and to retain them on Earth, they are bound in much more massive molecules such as water (Fig. 31.1).

For our Moon and Mercury, this lack of an atmosphere combined with slow rotation results in high daytime and low night time surface temperatures. An atmosphere tends to moderate temperatures, retaining heat like a blanket at night and reflecting more sunlight than bare rock would in the daytime. We know the importance of liquid water as the solvent for all known life forms. A planet with no atmospheric pressure is particularly unsuitable for life in that any surface liquid water will boil away – liquid water can exist then only deep underground or under a frozen ice cover. Thus a planet's mass and distance from its star play both a role in it retaining

Fig. 31.1 The inner planets Mercury, Venus, Earth, and Mars in size comparison. These are called terrestrial as they all have a solid, rocky crust (credit: NASA)

an atmosphere and remaining suited for life. The more massive inner planets Venus, Mars, and Earth have atmospheres.

Another mass-related factor is the internal activity of a planet. We have discussed how volcanic vents or energy sources may have served as early habitats on Earth. A low mass object has a larger surface area per unit mass than a higher mass planet. Both the Moon and Mercury have lost most of their internal heat so that volcanic activity is virtually null on these bodies. Their inert surfaces preserve a valuable record of the impacts that played an important role in the story of life on Earth (Fig. 31.2).

Beyond the orbit of Mars and even crossing the inner Solar System are small asteroids and comets. The largest asteroids are several hundred kilometers in diameter and are even large enough to have pulled themselves into a round shape. They, thus, according to the IAU, are accorded the status of dwarf planets. These and smaller asteroids along with comets cannot permanently hold an atmosphere of any consequence. Also they are thought to be without any internal volcanic activity.

Although they are too small for permanent atmospheres, small asteroids or comets are of considerable interest in regard to the origin, evolution, and future for life. We have discussed the effects of impacts on the early atmosphere and extinction of life on Earth. Also recall that some primitive unmodified meteorites have been found to contain building blocks for biomolecules. Far from the Sun and water, ammonia and methane (compounds of hydrogen with oxygen, nitrogen, or carbon) were solid icy particles in the Solar Nebula and condensed into small bodies. Although it is controversial, some astrobiologists have suggested that prebiotic chemistry may occur in fluid regions in the cores of comets or they may be transport vehicles for bacterial spores spreading life from world to world. These objects deserve a detailed discussion.

The four outer planets Jupiter, Saturn, Uranus, and Neptune are much larger and more massive than the inner planets. Because of their greater distance from the Sun and the higher escape velocities, they all have extensive atmospheres consisting of hydrogen, helium, methane, and other relatively light gases. Jupiter and Saturn

Fig. 31.2 Mercury's cratered surface with signs of old geological activity imaged by the MES-SENGER probe in January 2008. The large double-ringed crater is about 200 km in diameter (credit: NASA/Johns Hopkins University Applied Physics Laboratory/Carnegie Institution of Washington)

are thus considered gas giants. Uranus and Neptune could be called icy giants. All outer planets are encircled by rings and a large number of satellites. Several of the satellites will be of high interest as we discuss the possibility for hideouts for life. A small satellite of a giant planet like Jupiter can be close enough to have its interior heated by tidal effects of the planet (Table 31.1).

Table 31.1 Physical properties of the planets

	Mercury	Venus	Earth	Mars	Jupiter	Saturn	Uranus	Neptune
Distance (AU)	0.39	0.72	1.00	1.52	5.20	9.53	19.2	30.1
Radius (km)	2,439	6,052	6,378	3,397	71,398	60,000	26,320	24,300
Mass ($M_\oplus$)	0.06	0.82	1.00	0.11	317.9	95.2	14.6	17.2
Density ($g\,cm^{-3}$)	5.4	5.2	5.5	3.9	1.3	0.7	1.1	1.7
Year (years)	0.24	0.62	1.00	1.88	11.86	29.42	84.36	165.5
Siderial day	58.6 d	243.0 d	23 h 56 m	24 h 37 m	9 h 56 m	10 h 33 m	17 h 14 m	16 h 07 m
Solar day[a]	176 d	116.8 d	24 h 00 m	24 h 39 m	9 h 56 m	10 h 33 m	17 h 14 m	16 h 07 m
Satellites	0	0	1	2	63	59	27	13

[a]Solar day is the average time from a sunrise to the next sunrise (note the long solar days of Mercury and Venus; e.g., in Mercury it takes 88 our days from sunrise to sunset!)

Mars, a Likely Suspect

Mars has always fascinated the human mind. It is red and quite prominent when seen at its closest when the quicker Earth passes Mars in its orbit (at this time Mars is opposite the Sun in the sky). "At opposition," even through a small telescope, it appears large enough so that details can be seen on its surface. Mars orbits the Sun at a distance of about 1.5 AU, and one round takes nearly two years. Its orbit is tilted by about 25°, meaning that it experiences a seasonal cycle similar to our own. It spins around its axis in about 25 h, so in that respect it also resembles our Earth.

In a typical opposition the distance of Mars from Earth is about 60 million km. At this distance one second of arc corresponds to 300 km. In good conditions this is approximately the limit of resolution from ground-based telescope observation. In most situations the resolution is poorer, and surface details become blurred. At these resolutions it was quite obvious that the polar caps of Mars grew during winter and receded during summer. Also during the Martian spring, a dark area appeared at the edge of the melting icecap. This proceeded toward the equator. Some observers suggested the possibility that plant life was emerging after the cold winter. As water flowed toward the equator larger areas became covered with plants.

If the air above the observing site is unusually still, then for fleeting moments, conditions may occur when the resolution can improve substantially. A telescope with lens of 22 cm could give a resolution of 0.63 seconds of arc, or 200 km on the surface of Mars (about 1/34 of its size). This was the resolution Giovanni Schiapparelli (1835–1910), the director of the Milan Observatory, could obtain for brief periods. During these times, he memorized the details in Mars and made quick drawings. Besides the large structures, such as ice caps, he saw sharp narrow lines, which he called, canale, or channels, which meant only that they were narrow streaks between two end points. An error occurred when this was translated into the English "canal," which had the implication that they were built structure. On Earth this would be by humans and on Mars by the Martian civilization!

Percival Lowell (1855–1916), a businessman, diplomat, and writer on Eastern cultures, became intensely interested in astronomy. He took the idea of Martian canals seriously and even founded his own observatory, in Flagstaff, Arizona, especially to study Mars and other planets. The enthusiastic Lowell mapped the dark and pale areas of Mars, and even the network of canals. He suggested that this was an irrigation system constructed by Martians to divert water from the polar areas to other parts of the dry planet. From the patterns he even deduced the location of the capital city. A peaceful civilization capable of huge coordinated efforts was imagined.[1]

For more down-to-earth astronomers, the idea of canals was really set aside via a combination of careful observation and theoretical calculation. In 1909, a Greek astronomer Eugene Antoniadi observed Mars with the Paris observatory's 83-cm

---

[1] Sometimes the enthusiasm to pave the way in one direction may lead to unexpected avenues in other directions. As we told before, such important phenomena as the rotation of galaxies and the cosmological redshift were discovered in Lowell's observatory, and also the (dwarf) planet Pluto.

telescope under exceptional conditions. He observed such a multitude of small details that he was astonished, but did not see any signs of canals. On the theoretical side, Alfred Wallace (the co-originator of the theory of evolution, Chap. 28) did a calculation of the surface temperature of Mars using the intensity of sunlight on it as determined by its distance from the Sun, Mars' rotation, and the assumption that the surface would warm up until its thermal emission was in balance with the energy being received from the Sun. Wallace obtained a depressing answer that at its warmest Mars was as cold as Siberia. Canals there would freeze solid and a civilization building irrigation systems would be impossible.

Now we know that on the equator in day time the Martian surface temperature can rise above freezing, up to $+15°C$ in summer time, but otherwise the planet's surface is quite cold. Besides Mars' distance from the Sun resulting in sunlight which is less than half as bright there as on Earth, daily extremes are enhanced because Mars' thin atmosphere does not trap heat well via the greenhouse effect during the day nor retain it at night. Temperatures in Siberia sound quite comfortable compared to Martian winter nights when the thermometer plummets to $-120°C$! The average temperature of Mars is about $-60°C$, but daily swings are large.

Despite its coldness, Mars is thought to be a likely place for humans to settle some day. Newcomers, despite the light cycle similarities, will experience many differences to Earth. The radius of Mars, 3,400 km, is slightly over half of Earth's, and its density, $3.9\,gcm^{-3}$, is about 70% of Earth's. Together these two imply that the Mars settlers would have only 39% their weight on Earth. Without protective gear the air outside the Martian base would cause them to gasp for their breath, as the air pressure is about 1% of Earth's. Also it is composed 95% of carbon dioxide, which is not useful for breathing. Without protective gear, a human would suffocate in a couple of minutes. Unprotected Mars settlers would also experience serious sunburns due to the high UV flux.

The Martian canals and the presumed planet Vulcan are interesting examples in showing how human imagination – in itself important for science! – can drive us to the wrong direction, and also how careful scientific observations can correct for these mistakes. Urbain Le Verrier discovered in 1859 that Mercury's elliptical orbit had an anomaly in its motion which was not explained by the gravitational effects of any of the known planets. He suggested that the culprit is a planet inside Mercury's orbit. This was reinforced by a French amateur astronomer E. M. Lescarbault who saw a planet-like object moving rapidly close to the edge of the Sun. In the coming decades, similar objects were observed several times in front of the Sun. Some detections were also claimed during solar eclipses, but nothing was seen during the eclipses of 1901–1908. Albert Einstein explained the anomaly in Mercury's orbit with the theory of relativity in 1915, thus laying Vulcan to rest.

The idea of Martian canals was picked up by writers. The Martians attacked the Earthlings in an 1889 story *The War of the Worlds* by H.G. Wells. Forty years later it was broadcast in the radio, and massive panic was caused among the listeners who took it seriously. Some science fictions have portrayed Martians as friendly, among them the 1939 novel *Out of the Silent Planet* by C.S.Lewis. In fact, canals did not totally die until the first space probes gave us clear views of Mars unaffected by distance and the Earth's atmosphere.

Missions to Mars

The Soviet Mars missions started with Mars 1 launched in 1960–1962 which never made it to Mars. The first American mission Mariner 3, launched in 1964, had the same deplorable fate. The thoughts of a plant-covered planet with canals were finally completely shattered when the first 22 black and white photographs from the Mariner 4 were sent back on July 15, 1964, by radio transmission. The first large sets of images were obtained in 1969 by the flyby space craft Mariner 6 & 7 missions: a total of 198 photographs covered about 20% of Mars' surface. A couple of years later Mariner 9, the first spacecraft to orbit another planet, succeeded in transmitting 7,329 images to the Earth and mapping 80% of the planet. These pictures, essential in planning for the next major missions, also told a rather grim story. Mars appeared to be a sandy desert with no signs of life. It appeared to be as dry as the Sahara or the Atacama desert. These pictures also revealed river beds, craters, huge extinct volcanoes, and large canyons, such as the 4,000-kilometer-long Valles Marineris.

The constituents of the Martian atmosphere remained a mystery until the Martian Viking Missions. In 1909 at the height of the canal hype William Campbell at Lick observatory made spectroscopic observations of Mars and detected no water, in conflict with the hypothesis of water-laden canals. The moisture from the surface waters should have been seen in the atmosphere. Initially, it was thought that Mars might have a fairly substantial atmosphere. At McDonald Observatory in Texas, Gerard Kuiper (1905–1973) was able to detect carbon dioxide, the first constituent of the Martian atmosphere. The presence of a relatively thick atmosphere was suggested by the size difference between the larger UV images which showed the planet's solid surface plus its atmosphere and the smaller near-infrared images which just showed the solid surface disk. In the 1950s the best guess of the Martian atmosphere was made by Gérard de Vaucouleurs as 98% nitrogen, 1% argon, 0.25% carbon dioxide, and less than 0.1% oxygen.

The first accurate measurements of the Martian atmosphere were made by the following two space probes, Viking 1 and Viking 2 which landed on Mars in 1975. During their several years of operation they revolutionized our knowledge of the planet. The Martian atmosphere is now known to have 95% carbon dioxide, about 3% argon, and only 2% nitrogen. Oxygen is found in very limited quantities (0.15%) and water is even scarcer (0.03%). The average atmospheric pressure on the surface is only 8 mbar, or about 1/120 of the air pressure on Earth. The Martian atmosphere is indeed thin with no possibility of maintaining liquid water on its surface. During the Martian year the atmospheric pressure changes substantially as a good part of the carbon dioxide condenses on the winter pole into carbon dioxide hoarfrost. The variations in pressure are about $\pm12\%$, with an asymmetry between Northern and

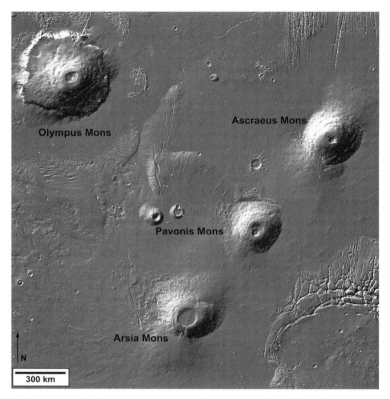

Fig. 31.3 This image of volcanoes on Mars was derived from Mars Orbiter Laser Altimeter data (onboard the Mars Global Surveyor). Olympus Mons is the biggest known mountain in the Solar System (credit: NASA)

the Southern winters. It is interesting to note that on the top of 25-km Olympus Mons the atmospheric pressure is less than 1 mbar (Fig. 31.3).

The Viking Landers Searching for Life

The Viking landers were the first and so far the only missions that have searched for life on Mars. Each of the two spacecrafts was programmed to run three biological tests. In addition, a gas chromatograph searched for chemical compounds in the upper surface layers of Mars and measured the atmospheric composition near the surface. Though not a proper biological test, it would have detected if local concentrations of oxygen, ozone, methane, formaldehyde, or gases related to life were present. It found some water but failed to detect any organic gases.

The three other tests were tailored to search for life. They were performed in closed chambers. The *gas exchange experiment* measured the production and the consumption of CO_2, N_2, CH_4, H_2, and O_2. The initial Martian atmosphere was removed from the sealed container by flushing it by helium, and then a mixture of He, Kr, and CO_2 was introduced. Nutrient solutions were added with neon as a diagnostic gas. The gas contents were regularly measured. It probed whether something was breathing in the sample. The *labeled release experiment* also measured whether something was breathing. Here the nutrient solutions contained radioactive carbon ^{14}C. The experiment monitored continuously for possible releases of ^{14}C by organisms living from the nutrient.

The *pyrolytic release experiment* searched for ^{14}C bound photosynthetically or chemically to CO or CO_2. After incubating the sample for five days, it was heated to $120°C$ to remove unreacted CO and CO_2. The sample was then heated to $650°C$, and organic products were collected in a vapor trap. Finally the remaining matter was combusted to CO_2 and any evolved radioactive gas was measured.

At the time of the Viking experiments, in 1975, one of the three main domains of life had not been yet identified. In 1977 Carl Woese, a professor in microbiology, and George Fox, a post-doctoral scientist in biological engineering, defined a new domain, *Archaea*. This domain has turned out to contain almost exclusively extremophilic organisms. These are microbes that survive or even flourish in conditions we would consider extremely harsh for eukaryotic organisms, such as humans and plants. These conditions include extreme drought, high salinity, high (UV) radiation levels, temperature extremes, high acidity/alkalinity. Presently it is considered quite possible that the life that could now be present in Mars is archaea or bacteria. Due to the abundance of different metabolic systems in these organisms, the appropriateness of the Viking tests for the general detection of life can be questioned. It would detect life in most cases on Earth, but it is not clear it would do so in Mars. It is also possible that life was killed by exhaust fumes or water before the experiments were performed. Also, microbial life is very unlikely to survive on top of the dust, which was where Viking collected its samples. UV flux there would be detrimental. A slightly deeper dig, from one or two meters, could have been better. Those areas are already shielded from the intense UV radiation and having a slightly higher pressure may contain pockets of liquid water.

After the Viking missions, there was a long pause of over twenty years before new space probes were sent to the red planet. But then an armada of missions followed which have much increased our knowledge of Mars. The Pathfinder landed on Mars on July 04, 1998. This roaming vehicle studied Mars for two months. Quite soon after it Mars Global Surveyor was launched and arrived in its Martian orbit in 1999. It operated until contact was lost in 2006. Mars 2001 Odyssey arrived at a Martian orbit in 2001. In June 2003 the European Mars Express was launched and started to circle Mars on December 2003. It carried a lander *Beagle 2*, which failed on landing. The orbiter, however, has turned out to be successful. Two Mars Exploration Rovers, *Spirit* and *Opportunity*, have been roaming around Mars since early 2004. The Mars Reconnaissance Orbiter has been orbiting Mars since 2006. Most recently in 2008, the Phoenix lander searched for water and conditions suitable for life in the Martian soil. None of the missions since Viking has really been directly searching for life.

The Martian terrain has now been mapped in good detail. The paler Northern Hemisphere is rather flat low land with only a few craters, the darker Southern hemisphere is higher land with a high density of craters. The difference in the colors comes from different colors of the dust covering the regions. One of the most striking Earth-based telescopic features on Mars is the dark Syrthis Major, a "peninsula" of darker, slightly more cratered terrain extending into the Northern hemisphere. In satellites it does not look very special. Conversely, the large Martian volcanoes and the deep canyons are not very striking from the Earth. A large 200-km impact crater Hellas on the Southern hemisphere is quite prominent both with ground-based telescopes and from satellite observations.

The large Martian volcanoes tell about an important difference compared to Earth. Our volcanoes that form from rising mantle plumes tend to form volcano chains, as the crust on top of the plume moves as part of the tectonic plate. One such beautiful example is the chain of the Hawaii islands. In Mars, the few volcanoes that are present are much larger and more massive than Earth volcanoes. This means that the Martian crust does not have active tectonic plate movements, so lava from volcanic activity just accumulates in one place to build a gigantic volcano. We cannot say if Mars has volcanic activity today. Valles Marineris (the Valleys of Mariner) on the Tharsis plateau is a striking example of former geologic activity (see the color supplement). It is 200-km wide, 4,500-km long, and up to 11-km deep. This canyon would stretch from "coast to coast" across the United States.[2]

Possibilities for Life on Mars and Signs of Water

An interesting modern picture is emerging about the possibility of life on Mars. This is due to all the data we have accumulated from the recent and ongoing missions and also from studies of extremophilic microbes on Earth. One of the most critical requirements for life as we know it is liquid water. All terrestrial life is cellular and the solvent in all cells is water. Of course cells contain other important molecules, but water is omnipresent. When viruses become crystallized in their dormant phase then water is not used as a solvent, but on the other hand they are not then "alive" any more.

Mars is now a bone-dry planet. The amount of water in the atmosphere is low; converting to an equivalent amount of precipitable water it amounts to less than 0.1 mm. The Martian atmosphere has an average pressure, 8 mbar, and can fall down to 5 mbar in summer or in winter, which is below the triple point of water (6.1 mbar, 0.01°C). This means that if liquid water were somehow released on the Martian surface it would either boil or freeze very quickly.

It appears that the past on Mars has been quite different. Some scientists argue that the red planet has always been very dry. There are however many signs that indicate that water has been present in larger amounts, and even in liquid form.

---

[2] Valles Marineris is not a valley formed by water erosion, but is rather a rift valley, similar to the East African rift valley, which includes e.g. the large lakes in East Africa and the Dead Sea. A rift valley forms when two land masses move apart and a small slice of land in between them sinks.

Mars has a weaker gravity than Earth does. The light gases of the atmosphere have escaped into the space at a pace, which corresponds to a thinning of the Martian atmosphere by a factor of 10 in 1 billion years. If we calculate this backward in time we may note that some 2–3 billion years ago Mars had an atmospheric pressure comparable to the present Earth's atmosphere. It is also likely that the atmosphere contained gases in different proportions than presently, possibly it was even more hospitable to life. With more water and carbon dioxide in the atmosphere it had a stronger greenhouse effect and thus higher temperatures.

In studying Viking Orbiter images in the early 1980s scientists came up with the idea of a large Martian Ocean. This explained why the Northern low lands were so flat and void of craters. Two long shorelines were found, each thousands of kilometers long. It appeared that at some time a third of Mars had been covered by an ocean 2-km deep. The last waters of this ocean would have evaporated or frozen one or two billion years ago. Such a large body of water would provide a cradle for life. But there seemed to be one conflict in this picture: why was the shoreline of this one single ocean not at a constant elevation? This could be due to a wandering Martian axis of rotation, which as suggested by computer simulations, would happen on timescales of hundreds of millions of years.

Iceland has awesome examples of catastrophic floods. Many of these created canyons, giving birth to the formation of rivers. Some are now dry wide riverbeds, many of them are relatively deep. There are signs of catastrophic floods in Mars. The liquid could in principle be something else, like lava, but liquid water seems most likely because of the shape and the erosion patterns. There are two types of known structures on Mars that could be due to water flows, large *outflow channels* and smaller *valley networks*. The outflow channels occur on young terrain on the Northern hemisphere and the valley networks are local features, usually on older terrain of the Southern hemisphere.

The outflow channels can be up to 2,000-km long and 100-km wide. They start from what is called chaotic terrain, and show collapsed edges, streamlines, eroded craters, and water flow marks, just as a dry massive riverbed. These are thought to have formed from catastrophic floods of large underground water reservoirs. They end up in what appears as large lakes or oceans. Examples of these kinds of sites are Tiu Vallis and Ares Vallis.

The valley networks appear sometimes as dendritic drainage systems of smaller rivers. In other cases they look like single river-like structures with only a few tributaries. Such an example is the Nirgal Vallis. These are old and were not created by catastrophic floods. If water flowed in these valley networks, the amounts were only something what one could expect from rain or glacier meltwater or ground water. Possibly there was a river, although no clear riverbed is usually found. It is also possible that the ground water caused the soil on top of it to collapse. Many of these valleys terminate abruptly (Fig. 31.4).

Even on smaller scales, on the edges of some small craters or slopes there are still signs of water. These were first found by the high-resolution images taken by Mars Global Surveyor. In appearance they look like small gullies found in terrestrial hills and mountains in various deserts and semideserts. Some of these can be found in

Fig. 31.4 Nirgal Vallis as photographed by Mars Global Surveyor (credit: NASA/JPL/Malin Space Science Systems)

the Newton Basin craters, and on the edge of Nirgal Vallis and Dao Vallis. Tens of thousands of gullies have been found, and their lengths are typically from a hundred meters to a few kilometers. Observations from the Mars Global Surveyor show that the gullies are active at present. Changes were seen in a crater between January and May 2000.

Mars Global Surveyor pictures have revealed layered formations. If they are sedimentary, it means that they would have formed in water. The mineral gray hematite found in Meridiani Planum and at least partially identified as the "blueberries" found by the Opportunity Rover is also a sign of previous presence of ground or surface water. On Earth carbonates form naturally in the combination of erosion and sedimentation processes and end up forming white carbonate rock formation. But no highly concentrated carbonate formations have been found on Mars. There is a way out of this apparent conflict: if the early oceans were of high acidity because of high atmospheric CO_2, then the sedimentation could have taken place into sulfur and magnesium-rich sulfates, which have been found in high concentrations for example by the Spirit Rover.

It seems like there has been plenty of water on Mars, but where is it now? Is there water presently in Mars? The answer to this question is a clear "yes." We mentioned the recent discovery of changing gullies suggesting liquid underground water aquifers. The polarcaps are formed of water ice, as confirmed after a profile measurement of the polar icecaps. The north polarcap is about 3-km thick and about half the area of the Greenland ice sheet. The south cap is about a bit larger and up to 3.8-km thick. The combined amount of water in the polar ice (3–4 million km^3) is sufficient to cover Mars under 20 m of water. This is still only about 20% of what is needed to explain the catastrophic flooding recorded in several outflow channels and estimated by other means.

The glaciers of the water ice caps do not alter much with the seasons, because temperatures and air pressure stay low. The seasonal growing icecaps seen from earth are formed by a thin carbon dioxide snow, only a couple of centimeters thick. This was witnessed and photographed by Viking 2.

What appears to be a large frozen lake has been identified from the Mars Express data in the region of southern Elysium Planitia, near the Martian equator. The lake measures about $800 \times 900\,km$ and is probably several tens of meters deep. This area appears as a flat region with broken ice plates at the edges. If it is indeed a frozen lake, then it might preserve life from the time the liquid lake froze.

The possibility of large underground water reservoirs has been speculated since the hydrogen mapping done by the gamma ray spectrometer on the Mars Odyssey orbiter in 2002. The experiment detected an abundance of hydrogen. This hydrogen is thought to be in the form of water – either in ice, liquid, or crystal water, in the top few meters of the surface. Finally, the Phoenix lander confirmed in August 2008 the presence of water ice just a few centimeters below the surface.

One could also search for life by looking at some gases that have a short lifetime in the atmosphere and thus need to be produced continuously to sustain any detectable quantity. Oxygen and ozone would be such, but the levels have been very low since the time of the Viking missions. Methane is destroyed in the Martian atmosphere by sunlight on a time scale of 300 years, so if methane is found it must be produced practically continuously. Methane can be released from geologic activity. Life can also produce methane, as happens on Earth in bogs and swamps and also in the guts of cattle. This biotic methane is produced exclusively by microbes in the Archaea domain called methanogens. Scientists using a special spectrometer on board of the Mars Express reported the detection of methane in localized areas of Mars at levels of ten parts per billion. It appears that at present geologic activity is not sufficient to explain the amount of methane detected leaving a definite possibility that life is producing it. More recent observations have also suggested that water in the atmosphere is more common in that same area, which may point to a common source. These results are in agreement with life, but do not prove that it exists.

More amazing was a claim of formaldehyde in the Martian atmosphere. Formaldehyde has a life time of only 7.5 h in Martian atmosphere, so the formaldehyde must be created during the same day! It can be formed from methane in principle, so the detection itself was not very surprising. The formaldehyde reported in 2005 was weak and cospatial with the previous methane discovery, but the quantities were about 130 parts per billion. From the amount of methane observed one would expect formaldehyde to be present in quantities much lower than methane.[3]

---

[3] Vittorio Formisano (Italian Institute of Physics and Interplanetary Space) offers different scenarios for explaining the formaldehyde, such as surface chemistry caused by solar radiation, chemistry from hydro- or geothermal activity, or life itself. Worth mentioning is that it may be difficult to explain the origin of methane needed to form the formaldehyde.

Histories of Life on Mars

Despite the harsh conditions now, it appears that there had been times on Mars when there was an ocean or large lakes, a thicker atmosphere, more volcanic activity, and a stronger greenhouse with more water, methane and carbon dioxide in the atmosphere, causing warmer conditions. In such conditions life could have begun to develop on Mars. In the last billion years conditions became harsher and the atmosphere thinner, freezing the remaining water. Where did the life escape if it was still present at that time?

There are two options. Some Martians may have evolved into very sturdy microbes that could stand in conditions very near to the surface (a common bacterium *Bacillus subtilis* can survive only 20 min on the Martian surface, so the microbes could be something similar to *Deinococcus radiodurans*). The more likely option is that the life sought shelter underground or in ice. Although liquid underground water has not yet been directly seen or sampled in Mars, it seems likely that somewhere in the moist underground is the best place for life to thrive. There will be no sunlight available, so the primary producers have to be chemotrophic – these kinds of terrestrian life forms are known in the domains of bacteria and archaea. The secondary producers and predators could then thrive on these organisms. If life sought shelter in brine water pockets of exposed water ice, then phototrophic primary producers could be possible, but hardly anything more complex.

Seven holes in the Martian soil were recently imaged by the Mars Odyssey Thermal Emission Imaging System (Fig. 31.5). These are circular, 100–250 m in diameter. The walls or the bottom are not seen. Their darkness and the altitude of the Sun

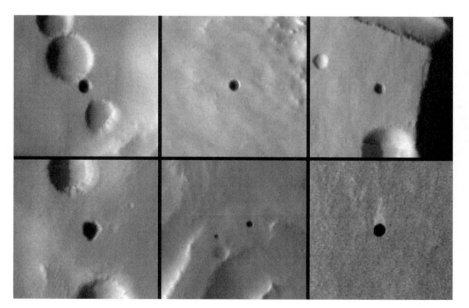

Fig. 31.5 The holes on Mars discovered by the Mars Odyssey spacecraft are located on the flanks of Arsia Mons, one of the volcanos in Fig. 31.3 (credit: NASA/JPL-Caltech/ASU/USGS)

suggest some of the caves are deeper than 80 m. They can be much deeper. Such caves, with their interesting variation of conditions, could be promising habitats for life. If these are open caves then the air pressure will not differ much from the surface pressure. If they are entrances to deeper underground systems, then higher pressures might occur deeper down and the gas constituents may change. Water could be seeping from the walls lower down. The light conditions vary along the cave, and there could be a region with a suitable amount of scattered light with reduced UV for photosynthetic organisms to live alongside with chemotrophic life providing a possibility for a full ecosystem. On Earth, sites where gradients in conditions occur tend to be locations where life thrives. These black caves could be such places in Mars.

Venus – Hot and Dry

Venus, the third brightest object in the sky, has been identified from Babylonian scriptures. It was known to Maya and to other Mesoamerican Indians. Their complex calendar system may have been influenced by Venus. The solar year cycle of the seasons was represented by a 365-day Haab "year." Then there was is another 260-day Tzolkin "year." After a "calendar round" of 52 Haabs, these two calendars were synchronized. It is clear from Mayan scriptures that Venus, associated with war, was important.[4] Especially critical was the first appearance of Venus in the morning sky after it had been "away" for a week or two passing between the Earth and the Sun in its orbit. The puzzling 260 days may be related to the time Venus is visible as an "evening star" or a "morning star". The Mesoamericans may have known that it was the same star in both cases. Among the Mediterranean cultures, this realization was ascribed to Pythagoras.

When Galileo viewed Venus with a telescope, he saw that it had the same phases as the Moon as a result of orbiting around the Sun, but with no surface features. Even with modern telescopes Venus shows no clear features, because of the thick cloud cover. The atmosphere of Venus, or its optical effects, was seen by the Russian Mikhail Lomonosov (1711–1765) when he observed the 1761 transit of Venus over the Sun. He concluded that "Venus is surrounded by a distinguished air atmosphere similar (or even possibly larger) than that is poured over our Earth" (see Fig. 9.3). In 1932 Walter Adams and Theodore Dunham using new red sensitive Kodak plates in a spectroscope detected the main constituent of the Venusian atmosphere to be carbon dioxide with a lack of oxygen and water. Its closeness to the Sun along with the thick clouds and mistaken ideas that it was somehow younger than the Earth led many to picture Venus as a hot jungle planet perhaps populated by dinosaurs. Those who correctly interpreted the spectrum realized that Venus would have a strong

---

[4] Every 584 days, its synodic period, Venus repeats a given configuration in the sky, say, appearing at a maximum angle of 47° from the Sun high in the evening sky. Five Venus synodic years equals 8 Haabs or about 2,920 days. This would ease the prediction of events of Venus.

greenhouse effect, but no water. A hot dry desert is closer to reality than the hot jungle.

Long wavelength thermal radio waves from the hot surface can make it through the thick atmosphere. The surface temperature was thus derived by C.H. Mayer and colleagues in 1958. Two measurements of Venus's brightness temperature at 3.15 cm gave values around $320°C(620 \pm 110\,K, 560 \pm 73\,K)$. Such high values met some skepticism, but have since been even raised by space probes.

The cloud cover made it difficult to unveil the rotation of the surface of Venus. In the astronomy text books of the 1950s several possible rotation times were given, 225 days (synchronous with the orbital period), or 37 days, or slightly less than 24 h (analogy to Earth and Mars). In 1962 Richard Goldstein and R.L. Carpenter from the Jet Propulsion Laboratory measured a retrograde rotation for Venus with a long period of about 240 days. This was done by bouncing radar waves off Venus. The edge coming toward us has a Doppler shift to shorter wavelengths that go away to a longer wavelength. The difference in the wavelengths gives the speed of the rotating equator which can be divided into the circumference around Venus to give the rotation period.[5]

Space Missions to Venus

The years 1961–1962 saw a rush of probes to Venus. The first Venera missions, a couple of Sputniks, and the US Mariner 1 ended up in failure. The first mission to send data from Venus was Mariner 2. It was sent in 1962 on a flyby orbit and passed Venus within a distance of 35,000 km. This probe confirmed the slow retrograde rotation, high surface temperature and pressure, the carbon dioxide atmosphere, and the continuous cloud cover at 60-km height. The race was intense with eleven following Soviet missions ending with failures. In June 1967 Soviets launched Venera 4, which was able to send data until it reached a height of 25 km. This was the first probe to measure properties within the Venusian atmosphere. Two days later Mariner 5, a flyby mission, was launched.

In 1969, the twin missions Venera 5 and 6 succeeded in measuring the atmospheric properties in detail. The next year Venera 7 was the first space probe to send back data from the surface of another planet. The rest of the Venera series were successful. Venera 8 confirmed the high surface temperature and atmospheric pressure measured by Venera 7. Venera 9 and 10 in 1975 measured various properties of the atmosphere. They also sent back the first TV pictures from Venusian soil. The next Soviet missions, Venera 11 and 12 then 13 and 14 detected lightning and

---

[5] To map the surface of Venus, again long wavelength radar waves can penetrate the clouds. The technique uses two dishes, the rotation Doppler effect plus the time delays from near and far parts of the planet's disk. The first radio echo maps were made in 1962 and 1964 by R.L. Carpenter. He identified several areas with different radio delays The first "high-resolution maps" of Venus were obtained in 1972 also by radar. They had a resolution of about 20 km. The telescope used was the 300-m Arecibo telescope and was run by D.B. Campbell and R.B. Dyce from Arecibo and Gordon H. Pettengill from MIT.

thunder and measured surface minerals. The last two Veneras, 15 and 16, launched in 1983 were orbiters which mapped the surface with radioaltimeters. Meanwhile, US had sent two Pioneer Venus probes, which repeated or predated many Soviet measurements. Two Russian missions to Halley's Comet, Vega 1 and 2, also carried balloon landers to study Venus's atmosphere. In 1989, the US launched the Magellan mission, which made a very detailed map of 84% of the planet. In 2005 the European Space Agency launched Venus Express, the first spaceprobe to perform an overall study of the Venusian atmosphere and its interaction with the solar wind. At its closest, Venus Express goes down to an altitude of 250 km and at its furthest, it is 66,000 km away from the planet.

After this rush of missions, a short summary is appropriate. Venus is nearly the size of the Earth (radius: 6,052 km). The distance from the Sun is 0.72 of that of the Earth. Considering only the closer distance, we would expect the planet to have a temperature 18% higher than Earth's. Without the greenhouse effect, this would be a comfortable 33°C. The temperature measured by several probes is on average 464°C, higher than the melting point of lead (328°C), and it does not vary significantly from site to site. The average pressure on the surface is 92 bar, 90 times that of the Earth and over 10,000 times more than on Mars. Because the equator is nearly in the orbital plane, there are no seasons. The thick atmosphere smoothes the temperature differences between the long day and night times, which last for about quarter of an orbit each.

Venus's surface lacks significant impact craters and appears quite young without tectonic plate boundaries or mountain chains. There is tectonic activity in the form of several volcanoes. It is likely that the curst is significantly thinner than on Earth, and it is possible that parts of the crust melt episodically. Near the surface, the atmosphere consists of 97% carbon dioxide and 3% nitrogen. Water has been found in tiny amounts of 20 parts per million. Oxygen has been detected only in the upper atmosphere, where it is thought to be formed by photodissociation of CO_2.

Could there be any life on Venus? In the 1950s, some still viewed it as a moist and cloud-covered world, a source of inspiration for science fiction. Observations from the ground and particularly when those from space probes became available, ideas of the Goddess of Love's planet being a cozy place for life dwindled to near extinction. Conditions on the surface, especially the temperature hotter than a kitchen oven and the lack of water, are overwhelming for all known life forms. With the even hotter interior, the conditions underground do not offer any better shelter for life.

Although the atmospheric pressure at ground level is high, if we go further up the pressure and the temperature decrease. From Pioneer, Venera, and Magellan missions, at a height from about 45 to 70 km there is a layer of cloud droplets made of high 75%–95% sulfuric acid content, with the remaining being water. Inside this cloud at heights around 50 km the temperature and pressure are just what terrestrial life could find comfortable (50–0°C, 1.3–0.37 bar).[6]

---

[6] The high acidity could appear as a problem, but extremophiles are known to live in pH 1. The suspension time of aerosols is longer than in terrestrial clouds. Finally higher clouds shield this zone from the harshest UV radiation.

Venus may have not been always like today. The young Venus was probably much like the Earth. It is about the same size and most likely got a similar early atmosphere. As the young Venus cooled, the water in the atmosphere possibly formed oceans. The Sun had a luminosity of about 75% of present value, and Venus may have sustained a reasonable greenhouse, and conditions may have allowed life to form. Compared to Earth, something went wrong. Maybe continents and tectonic plates never formed, and CO_2 was never bound into the minerals through the weathering cycle as it does on Earth. Or maybe Venus had continents and oceans and the temperatures rose slowly and reached a critical value only a billion years ago. Or, Venus was hit by an asteroid, which evaporated the oceans. What we know is that at some point the temperature on the surface rose enough to evaporate the oceans. Once the water vapor, a strong greenhouse gas, entered into the atmosphere in large quantities, temperatures began to rise: Venus experienced a runaway greenhouse.

Ironically, the large amounts of water in the atmosphere apparently led to its permanent loss from Venus. Water vapor has since mostly been lost by breakup by solar UV light of water into oxygen and hydrogen with the latter low mass atoms escaping via thermal evaporation. While Earth has water oceans thousands of feet deep, all the water vapor currently in Venus' hot atmosphere if condensed would result in a layer only one foot deep. From the impact craters, the age of the present Venusian surface has been estimated to be only 250 million years. This means that if life arose in Venus before the oceans evaporated, then there is really no way fossils from that time could have survived. Furthermore, the greenhouse effect, although now strong, was even stronger when there was water vapor in the atmosphere. It could have melted the surface, destroying fossils. There are still at least two possibilities to search for signs of Venusian life. One is to make a sample return mission to collect particles from the sulfuric acid clouds. We may also search for signs of fossilized life in meteorites ejected from Venus in early impacts. Understanding what happened to Venus helps us to predict the behavior of Earth's greenhouse effect (Table 31.2).

Table 31.2 Atmospheric and surface properties of the inner planets

	Mercury	Venus	Earth	Mars
Min. Temperature	+70°C	Narrow	−89°C	−140°C
Max. Range	+430°C	Range	+58°C	+20°C
Mean Temperature[a]	+170°C	+460°C	+15°C	−60°C
Surface[b] pressure	Very low	92 bar	1.0 bar	8 mbar
Atmosphere[c]	Unstable, very thin, various elements	CO_2 97% N_2 3%	N_2 77% O_2 21%	CO_2 95% Ar 3%, N_2 2%

[a] Average temperature on the surface (e.g., in Mercury a wide range from about 430°C near noon to below −170°C late at night)
[b] The Earth-like planets have a solid rocky surface
[c] The outer planets (Jupiter, Saturn, Uranus, Neptune) have atmospheres without a clear-cut lower border, but a gradual transition from gas to liquid state. These "gas planets" are mainly composed of the same elements as the Sun and the ancient protoplanetary cloud. Their cold atmospheres contain molecular hydrogen and helium, with traces of methane, ammonia, etc.

A Brief Look at Earth

Life on Earth has entered nearly all the niches that have been studied. Alkaline and acid lakes, hot springs, deep subsurface crevices, the deepest sea bottoms are all populated. It appears that there are some limits for terrestrial life though. The upper limit in temperature is about 122°C. Life has been searched for, but not detected in volcanic springs at 250°C. The cold limit is more difficult to define. Many life forms can survive deep freezing to liquid nitrogen temperatures and recover. Life activity generally decreases as temperatures fall below freezing point, but in some cases some activity, e.g., DNA repair processes are still detected at −40°C. For water-based life the complete lack of water is clearly a limitation. For DNA/RNA-based life UV radiation is lethal, as it destroys the DNA and RNA. Life appears to adapt to many other obnoxious things, such as long cold, droughts, and the advent of an oxygen atmosphere.

In the future, astronomers hope to examine spectra of planets in other solar systems. How would we detect life on an Earth-like planet from space? We can actually do such an experiment and point our telescopes to the sky, and check what our Earth looks like from space. Soon after the new Moon when the lit portion is a thin crescent, one can usually see a faint glow covering the rest of the Moon. This fainter part, the earthshine, is light from the Earth reflected by the Moon (cf. Fig. 4.2). A near-infrared spectrum of the earthshine tells that the atmosphere has carbon dioxide, water, oxygen, and ozone. This is a signature of a photosynthesizing planet with water-based life. The strong water, oxygen, and ozone mark a difference to the spectra of Mars and Venus. If all photosynthesis ceased on Earth, oxygen would remain here for only 6,000 years, so when life on Earth will die, oxygen will disappear almost instantly. Its existence is a definite sign of life.

Jupiter – a Gas Giant

When Galileo Galilei observed the bright Jupiter with his telescope, he noticed four "stars" going around it. This was a very profound observation as now there was another point in the universe around which something went around, clearly challenging the view that the Earth was the cosmic focal point. Using Kepler's Third Law one could calculate the distance of Jupiter from the Sun and then from the Earth. Measuring the apparent size of the planet one can then calculate its real size. This can be done for any planet. The diameter of Jupiter is about 11 times the size of Earth.

In 1687 Isaac Newton set up the framework for calculating other important physical properties of Jupiter in his *Principia*. From the orbital periods of the moons and the known size of the orbits of the planets it was possible to calculate the mass of the Jupiter. Doing this simple calculation it is clear that Jupiter has a mass of about 330 Earth masses. The density is then about 1.34 of water. This is clearly a planet with plenty of light matter. This turns out to be hydrogen and helium gas.

Jupiter is a very active planet with complete wind patterns. It is a strong radio emitter with a strong extensive magnetic field. Its rotation time is less than 10 h.

Going down into the interior one would encounter first gaseous hydrogen gradually changing to fluid molecular hydrogen, then liquid metallic hydrogen. The interior of Jupiter is very hot with a temperature of tens of thousands of degrees. A molten rock/iron core is thought to exist in the center with a mass of about 20 Earth masses. Convective motions in the electrically conducting interior organized by the rapid rotation are thought to generate the magnetic field. Some theories even connect the surface winds with the deep core motions.

Jupiter has an extensive system of satellites. The four inner moons (the Galilean moons) Io, Europa, Ganymede, and Callisto are spherical. From their sizes and brightnesses, even before the space missions it was already clear that Europa had a high albedo (reflectivity) and Callisto a low albedo.

Six space probes have been sent toward Jupiter. *Pioneer 10* was launched in 1972 and passed Jupiter in December 1973. The sister probe *Pioneer 11* followed in 1973, and flew by Jupiter in 1974 on its way to Saturn and beyond. Many nice pictures were returned. *Voyager 1*, launched in 1977, made a flyby in 1980 continuing then to Saturn. It sent back pictures of Jupiter and the Galilean moons. *Voyager 2* launched a few days earlier made a flyby in 1981. After these, a decade-long break followed. In 1989 ESA and NASA launched the *Galileo* probe, and it arrived at the vicinity of Jupiter in December 7, 1995, the same day its probe entered Jupiter's atmosphere. On its orbit around Jupiter Galileo made about 10 flybys of each of the Galilean moons. Galileo was destroyed in 2003 by sending it into Jupiter to avoid a crash with the moons, which could have caused their bacterial contamination. The *Cassini-Huygens* probe made a flyby in 2000 and took 26,000 images of Jupiter and its moons. With the space missions, we know the composition and the conditions in the Jovian upper atmosphere. Unlike Venus, there appears to be no zones where life could exist. Even though an upper level of the atmosphere is at "room temperature" with liquid water droplets in the clouds, these are probably carried deep into the hot interior in a cyclic motion. In the face of having no solid surface and the cyclic heating, Jupiter itself does not seem to be suited for life. Let us turn to the moons.

The four Galilean moons have radii roughly in the same league as our Moon. They all have a typical surface temperature of about $-160°C$. They also have very thin atmospheres with pressures less than a microbar. In these cold and near-vacuum conditions there can be no (liquid) water on the surface. These moons are exposed to UV rays from the Sun. The inner moons, Io and Europa, are also bombarded by energetic particles accelerated by Jupiter.

The Active Io

Io, the Galilean moon closest to Jupiter, is geologically the most active body in the Solar System. It has several active volcanoes with plumes rising to 300 km. Heat pumping from tidal forces of Jupiter keep the lower crust of Io molten. The tides of Io's solid surface are up to 100 m. The surface is likely made of sulfur and its compounds, or silicate rocks. Though it is unlikely that Io could harbor life, there is one environment that may warrant further study: the hot vents and their

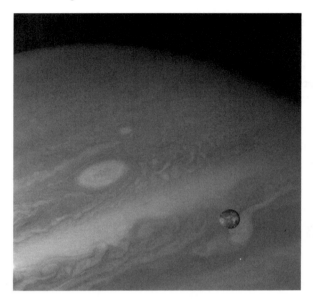

Fig. 31.6 Jupiter's innermost moon Io as seen against Jupiter's turbulent atmosphere by Voyager probe (credit: NASA)

surroundings. There could be a temperature regime suitable for life, but the other conditions may be too rough unless the life could shelter in a porous underground. The lack of water could still be a serious problem (Fig. 31.6).

Europa – Ice World with Prospects for Life

Europa, the second closest Galilean moon from Jupiter, is also the smallest of them. Its radius is 1,570 km, about 90% of our Moon's radius. Because of its high albedo, it was clear early on that the surface was made of something much more reflective than our own Moon.

The satellite missions have revealed an astonishing ice world. The rather whitish surface is very smooth, with only a few craters larger than 5 km. The surface shows fault lines and bears striking resemblance to frozen ice field that has experienced cracking at various times. Under the ice there is a saltwater ocean. Galileo flew by Europa at a distance of 315 km. Its magnetometers observed changes consistent with a shell of conducting matter such as an ocean with salty water. The combined depth of the ice and the ocean is about 80–170 km.[7] The thickness of the ice is not known.

---

[7] A second oddity in Europa is its rotation: it is not synchronized to its orbital motion around Jupiter. This means that it rotates slowly in respect to the Jupiter–Europa line. This slow rotation is not well known, but one rotation takes at least 12,000 years. As the rocky core is most likely synchronized it means that the surface and the core are detached from each other. This fits well into the idea of an ocean between the core and the icy outer layers.

Current estimates on theoretical grounds from surface features such as cracks, and the larger craters, range from a thin ice layer of a couple of kilometers to a thick ice of tens of kilometers.

It appears that here we have an ocean world. Could there be life? The direct way to answer this is to send a probe and drill though the ice. Provided funding for this is guaranteed, there is a major challenge in undertaking the effort. How can we search for life without affecting the environment too much or contaminating it with terrestrial life forms? There are also technical issues such as how to drill through ice that is kilometers or tens of kilometers thick. Scientifically the critical question regarding life is what is its energy source. Many of these questions can be tested under conditions we think are similar to what we expect to find on Europa and that we find on Earth (see Box 31.2).

Box 31.2 Lake Vostok in Antarctica: the test bed for Europa

In Antarctica, there are over a hundred lakes that are covered by ice all year round. The one that is considered as the best test bed for Europa is Lake Vostok whose size is 250 km × 50 km. The ice-covered lake, beneath Russia's Vostok Station, is nearly 4-km thick. The bottom of the solid ice is about 420,000 years old. Under the solid ice there is a 200-m slush zone, and then a lake with an average depth of 344 m. The two deep basins, Southern and Northern, are 400-m and 800-m deep, respectively. The residence time of the fresh water is thought to be about a million years. The temperature of the water is about −3°C, but it stays liquid because of the high pressure. Here we have an example of an ecosystem, or possibly two – one for each basin, in which no light is available from the top surface. It is cold and covered and protected from the hostile outside world by a thick ice layer. In the 1990s, a drilling experiment was carried out with the aim of obtaining samples from the slush and the lake. However it was halted at 130 m above the surface of the lake. The problem with the present hole is that it was made with tonnes of kerosene as antifreeze, and this could seriously contaminate the lake.

It has been pointed out that the lake itself is very hostile due to the very high concentration of dissolved oxygen. When samples will be obtained, it will show up as the first sterile lake on Earth or if life is found, it would prove to be a new type of an extreme environment where the microbes have adapted to high oxygen levels. In either case the result will prove interesting. To get samples in a clean way from Lake Vostok, the Jet Propulsion Laboratory is constructing a small robot probe that could drill into the lake with the trail behind it closing. This could be the only means for getting a clean sample from Lake Vostok and, in the future, from Europa's ice-covered ocean.

Under a several kilometer ice layer there is practically no light, so for life to exist there, one would need an alternative energy source. Since Europa is relatively close to Jupiter, tidal interaction creates heat in the core of the moon. It is conceivable that at the bottom of the ocean there are hydrothermal vents. On Earth, similar systems are found on the Mid-Atlantic ridge. These are full of life. They are rich in reduced compounds, which provide a source of energy for life independent of the Sun. The temperatures vary on a short distance of a few tens of meters from 200–400°C at the vent mouth to about +3°C of the sea bottom. These kinds of thermal vents have been suggested as cradles of life, and similar systems could be present in Europa. Besides Mars, Europa is considered to be the most likely site in our Solar System where life beyond Earth could be found.

Finally, a few words about Ganymede and Callisto. Ganymede is the largest moon in the Solar System, about 50% larger than our Moon. Callisto is slightly smaller. The surfaces look darker and older than Europa's, but models of these bodies still suggest that they have a thick ice cover under which there is a sea or more probably a zone of mixed water and ice slush. Being more distant from Jupiter, internal tidal heat will be less. These moons do not score very high in the possibility of life.

Saturn: The Gas Giant with Prominent Rings

Saturn is the most distant planet visible with plain eyes. When Galileo looked at it with a telescope, he was surprised to see Saturn as "three stars." Two years later the two companions disappeared and four years later in 1616 he drew the rings as half ellipses. It may have been a bit confusing to see so profound changes in the appearance of a "star." In 1655 Christian Huygens proposed that Saturn was surrounded by a continuous ring. The visibility, disappearance, then reappearance of the rings can be understood in terms of their tilt relative to Earth's orbit around the Sun. Occasionally, they are edge-on to the Earth and appear to vanish. Huygens also discovered Saturn's largest moon Titan in the same year. After Huygens the rings were frequently observed, and many astronomers suggested that they were solid bodies. Giovanni Cassini and Jean Chapelain thought that the rings are made of many small bodies, a view that took over 200 years to become commonly accepted.

Saturn is known to have the lowest density of planets in the Solar System, only 0.7 times that of water. This gas planet is made mostly of hydrogen and helium, and is quite similar to Jupiter with a mass about 100 times that of the Earth. Its beautiful set of rings is much more prominent than the rings around the three other giant planets. The satellite system of dozens of moons differs from that of Jupiter. It has one large moon, Titan. Nine other much smaller satellites have radii larger than 100 km.

Four space missions have visited the Saturnian system. The three first ones visited also Jupiter. Pioneer 11 flew by Saturn in September 1979, Voyager 1 in 1980, and Voyager 2 in 1981. Two decades of silence followed until the Cassini/Huygens mission was launched in October 1997. The Huygens lander was released on the Christmas Day 2004 and reached Titan's surface on January 14, 2005. The Cassini orbiter is expected to continue its operation until July 2010.

Titan – the Moon with Its Own Atmosphere

In 1944 Gerard Kuiper detected spectroscopically Titan's atmosphere. He concluded that it was methane. This was the state of understanding until the Voyager 1 mission, which revealed as late as in 1980 that the main component was nitrogen, and that the surface pressure is about 1.5 times that on the Earth's surface. The HST telescope with its sharp images has also been important for studies of Titan. Voyager 2 did not make a close approach to Titan during its flyby of Saturn.

The Huygens probe and Cassini have opened a new book in the study of Titan. Previously very little was known about the surface or the climate. The atmosphere turns out to be mostly nitrogen, with 1% methane in the stratosphere and 5% on the surface, and only trace amounts of other gases. The temperature on the surface is about $-180°C$. This is $10°$ less than expected from calculations. The effect is because the haze in Titan's atmosphere is effective in absorbing light, but transparent to infrared. During the 2 h 27 min descent Huygens observed a rather uniform layer of haze from an altitude of 150 km down to the surface, but no clouds. The winds on surface were very weak (< 1 m s^{-1}), whereas at an altitude of 120 km strong winds of 120 m s^{-1} were blowing.

After landing Huygens transmitted data for over one hour. It landed on a soft soil surface, similar to wet clay, lightly packed snow, and wet or dry sand. A stepwise jump in gas composition in landing indicated that the moisturizer in the soil was methane. The onboard cameras showed what appeared to be rocks, which could be silicate, but are more likely hydrocarbon-coated water ice. The pictures Huygens took during its descent reveal the appearance to be Earth-like. Dendritic structures similar to rivers and associated drainage structures are seen. These turn out to be of the right shape and structure for drainage from rain. They look dark in comparison to the icy surroundings.

Cassini took some radar images of 75 lakes in 2006. They are located in polar areas where methane and ethane are stable in liquid form. The lakes have low radar reflectivity, are located in topographic depressions, and have associated channels, similar to those seen by Huygens. The liquid in the lakes is methane, with some amount of ethane and possibly nitrogen. On a flyby April 2007, Cassini observed a lake so big that it has been called a "sea." It covers at least $100,000$ km^2, or 0.12% of Titan's surface (the Black Sea covers 0.085% of Earth's surface area). A second lake possibly a few times bigger has also been suggested from Cassini data. These would be enough to maintain a methane weather cycle (Fig. 31.7).[8]

What about life in this kind of a world? Titan is a cold place, and chemical processes proceed at a slow pace. Furthermore Titan never had a warm past (it may have a warmer future when the Sun becomes a red giant). At present there is no liquid water on the surface, nor any water vapor in the air. However, Titan's density

---

[8] The details of the formation of the clouds on Titan, the apparent rapid dissipation, and the torrential methane rains ($\sim$100 kg m^{-2}) are not well understood. It is however evident that Titan has a weather cycle in some respects similar to the one on Earth, except that water is replaced by methane. The composition of the lakes on Titan was confirmed spectroscopically in 2008.

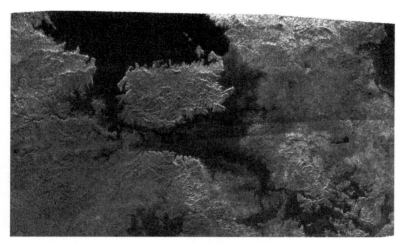

Fig. 31.7 Lakes and shorelines in Titan as revealed by a radar image taken by the Cassini orbiter (credit: NASA/JPL)

suggests that it is half rock and half (water) ice. According to models there could be, below a thick surface ice, an ocean of water and ammonia mixture. In the high pressure hydrocarbon clathrates could also exist, below which there would be a zone of high-pressure ices sitting on a rocky core. In some sense this resembles the Earth, except that water in Titan takes a similar place as silicates here. Titan appears to have volcanoes spewing out water, methane, and ammonia. "Water lava" lakes could survive for a hundred years in liquid form, a potential habitat for life to evolve.

The cold regions outside the volcanic or tectonic areas of Titan are harsh for life. If there is any it is bound to be quite different from ours. In any case, Titan may prove to be the best test bed we have for prebiotic chemistry. In the atmosphere, there is a zone of photoactive haze at 200–300 km from the surface. In the zone continuing up to a height of 1,000 km methane and nitrogen are bombarded by solar UV radiation and energetic particles creating ions. They are quite reactive and start forming longer molecules whose sizes and composition vary. This tar-like matter is called collectively tholins, and it is likely that various amino acids and long hydrocarbon chains can be formed. As the tholins grow heavy they start to precipitate and return slowly to Titan's surface. This all may be reminiscent of what happened in Earth's atmosphere around the time when life was about to form.

Enceladus, Thethys, and Dione are Saturnian moons smaller than Titan. On the July 14, 2005 Cassini flyby of Enceladus massive water vapor and ice-crystal plumes were seen near the south pole. These geysers have been modeled as emerging from a warm sea, possibly rich in organics. Some models suggest that the sea is 50-km deep and located under a 10-km ice layer. The near proximity of cold and warm may be important for the formation of ingredients for life. This should be borne in mind also in case of Europa and Titan. Observations of the plasma near Saturn suggest that Thethys and Dione inject particles into space. Thus these icy moons may also be geologically active.

The Outer Realms of the Solar System – Cold and Lonely

The outer parts of our planetary system are cold. The planets Uranus and Neptune are smaller versions of Jupiter and Saturn. They are too small to have metallic hydrogen in their interiors and probably have a larger proportion of compounds of hydrogen with other elements. Their interiors are thought be unsuitable for life for the similar reasons as for Jupiter and Saturn.

So far away from the Sun, water is in solid form. Methane and nitrogen turn liquid and then to snow. Triton, Neptune's largest moon, has volcano-like structures with possibly liquid nitrogen or methane erupting. Here the proximity of a volcano does not guarantee any warmth needed by life. The transneptunian objects, Pluto and half a dozen others, are large enough (in radius about 400 km or larger) to have a nearly spherical shape. These are nowadays classified as dwarf planets along with some objects in the asteroid belt (Box 31.1). We do not know much about these distant worlds. The first mass estimates of Pluto were indirect and involved a large range of values, up to about the mass of the Earth. In 1978 Pluto's small companion Charon was discovered, and soon the mass of Pluto was measured for the first time, and yielded quite a small mass, only one-fifth of the mass of the Moon or about 1/400 of the mass of Earth (Fig. 31.8).

Pluto's surface contains predominantly nitrogen ice. Interestingly, as we move outward in the Solar System, gases like nitrogen that is the main atmospheric constituent in a close-in object like Earth turn out to be solid ice in the farther out systems. Pluto has a thin atmosphere. Its interior is likely to be cold. If there are any liquid zones in the crust, they would seem to be places where life or even advanced prebiotic chemistry would have hard time proceeding at any significant speed. Other large transneptunian objects are even more distant than Pluto, and the conditions on

Fig. 31.8 The dwarf planet Pluto has two small satellites, Nix and Hydra, detected by the Hubble Space Telescope, in addition to the bigger Charon which was discovered in 1978 (credit: NASA, ESA, H. Weaver (JHU/APL), A. Stern (SwRI), and the HST Pluto Companion Search Team)

or in them are difficult to estimate. Should there be an Earth-sized object found in an orbit beyond Neptune, it could possibly have a subsurface sea of some liquid, but these speculations may be better suited for science fiction.

Comets and Asteroids

From the depths of the outer Solar System visitors drop by to the inner Solar System. These comets are relatively small and composed of ice and dust. On their visit they follow very elliptical orbits. The most distant part of the orbit, the aphelion, may be hundreds or thousands of astronomical units from the Sun, whereas they may just skim the Sun on their closest point of the orbit, the perihelion. Some come closer, and even plunge into the Sun and some get only as close as Jupiter.

On their long orbits, comets spend most of their time close to the aphelion. Their visit to the inner Solar System is quick and dramatic. As the comet approaches, the Sun starts to heat it. Somewhere near the same distance as Jupiter the heat usually becomes sufficient to wake up the comet. Volatile gases start to sublime. The coma usually brightens and a tail appears. The increased solar radiation heats up the comet and the solar wind acts on the gas and the particles creating one or two tails. After perihelion the reverse takes place and the comet ends up back in the loneliness of space. Some day it may return.

On the way to the inner parts of the Solar System, the comet feels the gravity of the giant planets. This will usually change its orbit a little bit. If it wanders closer, the planet may capture the comet, say, on a much smaller orbit. Rarely, it may even collide with the planet as happened when the broken up comet Shoemaker–Levy spectacularly plunged into Jupiter on July 16–22, 1994 (see the color supplement).

Comets are interesting in many ways. As mentioned, they provided a delivery vehicle for important ingredients to the planets early in our young system. Dozens of molecules in comets have been detected by radio astronomical spectroscopy. Generally, they are the same ones as observed in the cold interstellar clouds. Molecules such as water, hydrogen cyanide, formaldehyde, the molecules thought to be the first building blocks of life, have been found in comets.

Several space missions have aimed at comets and have provided a wealth of information. The successful ones include *Stardust* (Comet Wild), *Deep Impact* (Temple 1), *Deep Space 1* (Borrely), *ISEE-3* (Giacobini-Zinner), and five missions to Comet Halley. The Stardust was the first ever mission that returned samples from beyond the Moon. Now let us forget the tail and the coma and consider the comet itself. All the comets visited so far by the space probes are individuals. Comet Wild is quite spherical, whereas Borrely and Halley are rather elongated and look a bit like sweet potatoes. We take a closer look at Temple 1.

During the impact mission Comet Temple 1 was at a distance of 1.5 AU from the Sun. Its size, 8×5 km, is typical. From the impact it became clear that the comet has a dust layer of several tens of meters on the surface and appears layered further deep.

It also has low-lying areas with more craters and higher areas that appear younger. The internal structure is highly porous as the mean density is only around 0.1–0.5 that of water. The side facing the Sun has a temperature of about 70°C and the shadowy side is cooler at about −3°C. The surface is too warm for ices. The infrared Spitzer telescope saw signs of clays and carbonates in the impact plume. These suggest that somewhere in the comet liquid water is present at least transiently. This is important for prebiotic chemistry and even possible formation of life because this implies that "concentration dilution" cycles could be possible in comets. Furthermore, surface chemistry involving minerals, clays, dust, ices, energetic particles, and solar radiation makes the formation of complex molecules possible in principle. Something similar to tholins could form. The return mission from Comet Wild suggested that in it water had played little role in mineralogy. On the other hand the samples included a large set of rather complex molecules.

The difference between comets and asteroids is not always straightforward. An important difference appears to be the "fluffiness." Asteroids and meteorites can have similar, though maybe not identical, chemistries on their surface. This was shown by the Murchison meteorite that fell in 1969 to Australia. Several dozen amino acids and other complex organic molecules have found from it.

Comets are not likely to survive an impact on a planet. Asteroids on the other hand on impact could break up and eject some material from the planet into space. This ejected matter could in principle carry life, such as bacteria in it. If the rock is of the order of 1 m, it could provide a means of transport for life between the planets in our Solar System. Quite possibly this kind of transport has happened several times. It is also an interesting curiosity that this could provide the only means for detecting fossilized life from the ancient Venus via impacts that occurred there long ago!

As a result of improved understanding of life on Earth and new discoveries of internal processes in Mars and smaller Solar System objects, the list of sites suitable for life or at least chemistry of life in our Solar System has expanded from the one we have known so far. As our own system has several places where some kind of life could exist, the number of life-bearing candidates could rocket, when we look at the stars in the whole Milky Way. But are planetary systems common around stars? And are they good for life? We will discuss such questions in the next chapter.

Chapter 32
Extrasolar Planetary Systems and Life in other Solar Systems

The family of planets grew for the first time in written history when, in 1781, William Herschel discovered a new planet, Uranus, which he first took for a comet (Chap. 11). Once an additional planet was found, even though accidentally, the possibility of others was more readily considered. In the late 1700s, this was also inspired by the empirical Titius–Bode law which at that time seemed to predict the distances of the known planets through Uranus nicely except for a prediction of a nonexistent planet at 2.8 AU (see Box 11.1).

The Increasing Number of Planets

In 1801, the Italian astronomer Giuseppe Piazzi (1746–1826) discovered an object, which he named Ceres, almost exactly at 2.8 AU from the Sun. However, Ceres turned out to be much smaller than any planet, less than 1,000 km in size. This, with the subsequent discovery of other similar, smaller objects between Mars and Jupiter, ultimately resulted in the demotion of these small objects to a new category, asteroids (they appeared star-like through the telescopes of the time). Thousands of asteroids are now known, some of which orbit through the inner Solar System. There are icy asteroids outside the orbit of Neptune. One of these, the former planet, Pluto, along with the largest asteroid, Ceres, have recently been demoted and elevated, respectively, to be dwarf planets. These pull themselves into a round shape but otherwise have no dramatic gravitational effects even on objects in nearby orbits.

Besides the dominating force of the Sun, the planets gravitationally tug on each other in varying degrees dependent on their masses and distances. Even including the effects of planets out through Jupiter and Saturn, the calculated orbit of Uranus did not quite match its true positions. From these small differences, the location of a new unknown planet tugging at Uranus was calculated by astronomers. Soon thereafter Neptune was found close to the predicted position. We described the checkered history of this discovery in Chap. 11. The same perturbation method was used in an attempt to find a ninth planet. The discovery of Pluto in this search was a peculiar

P. Teerikorpi et al., *The Evolving Universe and the Origin of Life*
© Springer Science+Business Media, LLC 2009

case, as Pluto is much too small to have perturbed Neptune, which follows a path governed by the Sun along with the effects of the planets interior to its orbit. The story of new planets has not been a smooth progression, starting with millennia in which the list stopped with the naked eye planets through Saturn, then a burst of enthusiasm and success starting with the accidental discovery of Uranus, the methodical discovery of the asteroids and Neptune and ending with Pluto's discovery via persistence and good luck.

Astrometric and Velocity Attempts to Detect Extrasolar Planets

With the discovery of new planets in our Solar System, the notion of planets around other stars was also more acceptable. Astronomers assumed these other solar systems were probably like our own indicating that the discovery of these "extrasolar" planets would be difficult. The four planets closest to the Sun (Mercury, Venus, Earth, Mars) are small rocky planets with the most massive, Earth, being about 1/300,000 the mass of the Sun. The outer four (Jupiter, Saturn, Uranus, Neptune) are gas giants but even the most massive, Jupiter, is only about 1/1,000 the mass of the Sun.

According to Newton's third law of equal action and reaction, because the Sun causes the planets to orbit around it, the planets cause the Sun to move also. As shown in Fig. 32.1, looking at our Solar System from outside, Jupiter has an effect greater than all the other planets, which would cause the Sun to come toward an observer outside our Solar System then go away at $13\,\mathrm{m\,s^{-1}}$. This is a miniscule but measurable variation compared to the orbital speed of Jupiter. In terms of position, as Jupiter moves in its 5 AU orbit, the Sun moves in a much smaller 5/1,000 AU circle around their mutual center of mass. These two small effects in velocity and

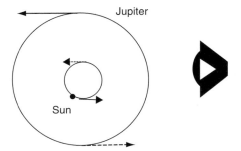

Fig. 32.1 Schematic of the Sun and Jupiter as they orbit around one another. As seen by the observer on the right, the small *circle* is the Sun which would wiggle up and down in position as well as moving toward and away (*solid* and *dashed arrows*) as Jupiter orbits in opposite positions and motions (*solid* and *dashed arrows*) in its much larger orbit. The size of the Sun's orbit relative to Jupiter's is much exaggerated. In reality the ratio of the two radii corresponds to the 1,000:1 ratio of the masses

position are two ways to detect planets around other stars (exoplanets). The velocity effect occurs only if the orbit is seen nearly edge-on.

Because of the tiny effects, a series of observations over many orbits of the planet were thought to be needed for either method. This was probably technically beyond reach before the advent of computers and new technologies. However, claims about detection of planets had been made as early as the nineteenth century. The first attempts were *astrometric*, in which the precise position of a star was measured on repeated occasions hoping to find small changes in the position as the planet and star orbit one another. As mentioned earlier, a Jupiter-sized planet orbiting a solar-type star at a distance of 5 AU would move the star only about 0.005 AU. Seen from a distance of 2 parsecs or over 400,000 AU this would amount to a tiny 0.0025 arcsecond (less than one millionth degree) angular wobble in the star's position over the planet's 11-year orbital period. This corresponds to about one-thousandth of the size of a stellar image blurred by the Earth's atmosphere from earth-based observatories. What kind of star allows one to observe this small a size? Clearly, the star should be close to the Sun to maximize the angle. The best chances for finding a planet could be around a red main sequence star which has a smaller mass than the Sun, and hence a larger wobble for a given mass planet. The faintness of the light from such a star provides an additional benefit because the image of the star is not saturated beyond measurability.[1]

Astronomers studied two such nearby stars in order to detect a possible wobble. Barnard's star, a red main sequence star at a distance of only 1.83 parsecs, is the second closest star to our Sun and therefore was a good candidate. In 1963 Peter van de Kamp (1901–1995) claimed first to have found one planet and later in the 1980s two planets around Bernard's star. He had inspected the star's positions on photographs taken at Sproul Observatory in Pennsylvania between 1938 and 1981. It was not until several decades had passed that a consensus formed that these had been spurious detections caused by changes in the telescope after its objective lens had been removed for cleaning then reattached.

The second candidate in the position wobble search was Lalande 21185, another red main sequence star at a distance of 2.54 pc, the fourth closest star to our Sun. The first claim of a planet around it came from Susan Lippincott at Sproul Observatory in 1960. Later George G. Gatewood (Allegheny Observatory, Pennsylvania) reported in 1996 the detection of companions that were much smaller than Lippincott's planet. Van de Kamp is also said to have detected a planet of five Jupiter masses around the sun-like epsilon Eridani. None of these claims has been confirmed yet, but as with other close-by stars these stars are on the task lists for space telescopes. All these old observations were difficult to analyze, and were hampered by the limited capabilities of the techniques and instruments.

---

[1] Some early claims involved the binary star 70 Ophiuchi. In 1855, Capt. W. S. Jacob at the East India Company's Madras Observatory reported that anomalies in the motion of the pair around each other made a planet "highly probable" in this system. In the 1890s, Thomas See (US Naval Observatory) stated that the orbital anomalies proved the presence of a dark body with a 36-year period around one of the stars in 70 Ophiuchi. These are now viewed as erroneous detections.

Ultimately, use of the variation of velocity was more successful in the detection of planets. The first published discovery to receive subsequent confirmation was made in 1988 by the Canadians Bruce Campbell, G. A. H. Walker, and S. Yang. Their radial velocity observations suggested that a planet orbited the star gamma Cephei. It was not until 2003 that this detection was confirmed.

Extrasolar planetary astronomy really began for certain in 1992 when Alexander Wolszczan and Dale Frail announced their discovery of two planets and possibly a third one orbiting the pulsar PSR B1257 + 12. This caught astronomers by surprise because pulsars were really the last stars around which one would expect to find planets – after all they are thought to be stars that have undergone a supernova explosion, which ought to destroy any planets. The regular radio pulses of the pulsar permitted extremely accurate measurements of perturbations caused by planets and thus their detection. Pulsars are accurate clocks, and if a planet moves such a clock about, then there is a small delay or advancement of this clock. The method is related to the speed (Doppler) method, but rather than measuring the velocity of the star at any given moment, an accumulated delay in pulse times integrated over time is measured. This corresponds essentially to a measure of displacement in distance. This method can provide accurate timing to within an order of (tens of) milliseconds or detect a small shift in the pulsar's position of about 0.00002 AU.[2]

Because of the expected velocity variations were so small, in 1995, astonishment was great when the first planet around a *sun-like* star was discovered using the velocity method. Michael Mayor from Geneva Observatory and his student Didier Queloz announced a planet orbiting the star 51 Pegasi with a period of 4.23 days. The planet has a mass of at least 0.47 Jupiter masses, and its orbit is only at a distance of 0.05 AU from the star (about 0.01 Jupiter's distance or eight times closer than Mercury is to the Sun). Because of this closeness, it produced velocity variations of $60\,\mathrm{m\,s^{-1}}$, much bigger than the $13\,\mathrm{m\,s^{-1}}$ Jupiter produces for our Sun. Consequently, the velocity variations of 51 Peg could be detected over a much shorter series of observations than expected.

In our Solar System, giant gas planets are further out and rocky planets are in the inner parts. However, the first new planet around another sun-like star turned out to be a giant planet very close to the star. It did not fit the picture based on our planetary system, but it did result in detectable perturbations. The number of known planets is now in the hundreds with the velocity wobble dominating as the method of detection. It appears that close-in giant planets seem to be typical for these systems. These giants would have such strong disturbing effects that Earth-like planets at the proper distance for life would not have stable orbits in such systems. Does this mean that our Solar System is a rare exception to the rule? Do we expect there to be *any* planets like the Earth circling other stars?

---

[2] Planets around pulsars probably suffered much from the supernova. They may represent the surviving cores of Jupiter-like planets. Stars that explode as supernovae do so quickly compared to the time required for life to evolve on Earth. They leave behind a neutron star which continues to threaten life on any nearby planet. All three planets around the mentioned pulsar seem to share roughly the same orbit plane (like the Solar System), but the orbits are smaller than Mercury's orbit.

Optical spectrographs used today can detect very small shifts of spectral lines. With a single observation, they are able to measure a star's velocity to a remarkable accuracy of $0.6\,\mathrm{m\,s^{-1}}$. As stated earlier, Jupiter in its orbit causes the Sun to approach then recede from an external observer with speeds of $13\,\mathrm{m\,s^{-1}}$. To detect a Jupiter-like object at about 5 AU from a sun-like star would require many independent observations over many orbit cycles. At almost 12 years per orbit, a long period of time would be required. Even though the Earth-like planet would have a period of one year, its velocity perturbation of $0.1\,\mathrm{m\,s^{-1}}$ is pretty hopeless with present observational technology.

Increasing the size of the planet or bringing it closer to the host star all increases the velocity variation and improves the chances of detecting the planet. Thus this method is biased toward finding giant close-in planets. Furthermore, one has to observe the planetary influence over many orbits before claiming to discover a planet, so there is again a "close-in" bias. Note also that planets with orbits perpendicular to the line of sight cannot be detected with this method. Since most of the extrasolar planets have been found in this way, we thus would expect that there may be many systems like our own but the present, most successful technique, cannot easily find them. Indeed, in astronomy we often run into the problem of selection effects, as we have to make observations from far away and cannot travel among the stars. The Malmquist bias mentioned in Chap. 21 is another example – at large distances we can see only the brightest stars and galaxies, the tip of an iceberg.

Other Detection Methods

Several methods are used now to find exoplanets, and in some respects they are all complementary. They each have also their own limitations. Another method that is sensitive to the orientation of the orbit is the search of eclipses of a star by a planet. Here the benefit is that one can search for eclipses caused by planets from a huge number of stars – essentially all the stars in the field of view of a camera. As shown schematically in Fig. 32.2, a Jupiter-sized planet passing in front of the Sun as seen from a distant solar system would cause a 1% flat-bottomed dip in the brightness of the Sun lasting for about 30 hours. For making a reliable detection one should observe at least three eclipses, which in the case of Jupiter would happen on a single day and about 12 years apart. This method is clearly suitable for short period orbits. Combined with the spectroscopy method, this is at present the only means to measure accurately the size and thus the density of the planet. Most of the planets measured this way have densities comparable to the density of water, but some have turned out to be puffed up planets with only one-quarter of water's density.

We discussed earlier gravitational lensing. Consider a light ray from a distant star on the way to our telescope. If an intervening object lies very close to the line of sight to the distant star, then light from the star gets slightly bent as it passes

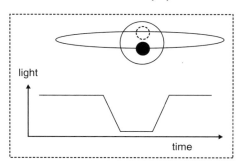

Fig. 32.2 Discovery of exoplanets by eclipses. The planet (*black circle*) crosses in front of the star's disk (*white circle*) causing a dip in the star's light (see light vs. time graph). When the planet (*dashed circle*) is behind the star, it has no effect on the star's light (higher *horizontal line* in graph)

the intruder and eventually gets focused onto our telescope. The star gets brighter. In practice, the influence of the intervening star and the planet is seen as a double event. The planet is a small brightening on the side of the larger stellar brightening. A gravitation lensing event for a particular planet is unique. The likelihood for it to be seen ever again is small. If the detected planet's orbital plane happens to be seen edge-on, then in principle it could be studied further using eclipses. This method can detect distant planets, and it may be the best way to find Earth size planets.

Why do not we just look at a star with a telescope and see if there are planets close to it? This method, direct imaging, sounds straightforward, but in practice it is very challenging, because of the huge contrast in brightness. Our Sun outshines the light reflected from Jupiter by a factor of one to hundred million. Clever means have been developed for reducing the effect of the bright starlight. One that has proven fruitful is to have a telescope in orbit, above the air which would smear the image. The ability of a space telescope to detect small angles is limited mainly by diffraction due to light waves. For the Hubble Space Telescope this so-called Rayleigh limit for visible light is about 0.055 arcseconds. This equates to seeing a planet at a Jupiter's distance from a star 95 pc (310 light years) away. In practice, the hugely bright star still poses a real problem, because its brightness overwhelms the planet by 1,000,000 to 1 even at the first diffraction minimum, the most favorable separation to see a planet. It would take one week of the precious HST telescope time for the planet to be detected. In the few cases when planets have been found with direct imaging, they have been rather far from the central star.[3]

The European Southern Observatory announced in 2007 a new tool for planet hunters, the integral field spectrograph (developed by Niranjan Thatte and his team).

---

[3] Based on resolution alone a planet on an Earth-like orbit could be detected from 18 pc. Planets close to the star will be hidden in its glare. Planets further out from the host star are easier to see, especially if they are big and reflect well light.

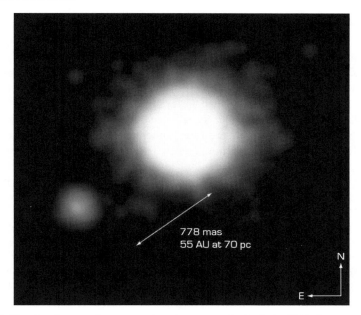

778 mas
55 AU at 70 pc

N

E

Fig. 32.3 The first image of an extrasolar planet was obtained, in 2004, by a team led by Gael Chauvin, using the 8.2 m VLT Yepun telescope (ESO), with a special adaptive optics facility in infrared (see the color supplement for a picture of the VLT). The host star 2M1207 is a faint brown dwarf star in the constellation Centaurus, not visible by naked eye. The planet is about five times more massive than Jupiter (credit: ESO)

It uses pictures taken in rapid succession in different wavelengths. In such images, the various disturbing effects will change as the wavelength changes, while the star and the planet should stay at the same position independent of the wavelength. It will be used in the Very Large Telescope of ESO in Chile. The VLT is presently the leading optical ground telescope with its four 8.2 m units which can be used separately or in concert.

To date, almost all exoplanets have been found by ground-based telescopes (Fig. 32.3). The HST has confirmed a few of them. The situation will change in future. The *COROT* space telescope, led by CNES (France), was launched in 2006. One of its main tasks is to detect planets by observing the small brightness variations caused by eclipses occurring when the planets transit in front of their central star. There are also a whole row of space missions in the planning stage set to search for planets (e.g., *KEPLER*, *New Worlds Imager*, *Darwin*, *Space Interferometry Mission*, *Terrestrial Planet Finder*, and *PEGASE*).

Finally, observations of dust disks around young stars can provide indirect means for detecting the presence of a planet. Such disks sometimes show circular regions void of matter. The gaps are understood as regions where a forming or newborn planet is cleaning up its orbital neighborhood.

Characteristics of Extrasolar Planets

Because of heavy observational selection effects most of the planets found so far are massive gaseous planets in rather small orbits (about 40% of orbits have a semimajor axis <0.4 AU). Ironically, the first planets to be detected *en masse* were the types that were least expected on theoretical grounds.

The planetary orbits in our Solar System can be characterized as nearly circular, whereas highly elongated orbits are found among comets. The extrasolar planets have confused the picture also in this respect. Their orbits are typically rather elongated, and only 10% have roundish orbits. Furthermore, only about 10% of known planetary systems have several known planets. In reality this number is bound to increase, as new planets will be found in systems where only one planet is now known.

Systems which feature eclipses and have measured velocities are a good source of information. From Doppler shifts, we can calculate the orbital parameters and the velocity of the planet when it causes the shallow eclipse. We can also infer the mass of the planet. From timing of the four contacts in the eclipse we can then calculate the size of the planet and get a lower limit for the size of the star. Since we know the mass and the size of the planet we can measure its density. Only a dozen have been measured, and they have turned out to be gas giants.

The masses of detected planets go down as methods improve and more data are accumulated. In 2000, Saturn-sized planets were found, and later Uranus- and Neptune-sized planets were reported. The first indication of a rocky planet was found in 2007. Stéphane Udry and his coworkers at Geneva observatory reported the two least massive planets found so far orbiting the same star Gl 581. The larger, an 8 Earth mass planet, has an orbital radius of 0.25 AU, whereas the 5 Earth mass planet is orbiting the star at only 0.073 AU. It has a 13-day orbital period. The latter one is particularly interesting as its distance from the star is such that water could be in liquid form. This means that an ice world is practically ruled out, a gaseous planet is not likely because of the small mass, so that leaves us with a rocky planet, which may or may not have water in liquid form on it.[4]

Debra Fischer from San Francisco State University and Jeff Valenti (Space Telescope Science Institute) found in 2005 that the presence of planets strongly depends on the *metallicity* (amount of iron with respect to hydrogen) of the central star. The fraction of stars with planets increases as the iron abundance increases: stars with a metal abundance half of the Sun have planets in about 2% of cases studied, whereas stars with twice the metals of the Sun had planets in about 10% of the cases. This can be understood in the context of planet formation. To build up the mass in

---

[4] Eclipses provide another important bit of information. If one compares the spectrum of the star during the eclipse to a spectrum obtained outside the eclipse one can see two differences. First there is a small decrease in the total flux. In addition, if absorption of light takes place in the planet's atmosphere, then some specific lines can have extra absorption. The effect is minuscule, but when detected, will tell about the temperature and density in the atmosphere of the planet.

gaseous planets within the accretion disk, ices may be needed. One would expect ices to be more common in high metal environment where oxygen is also more abundant.

On the other hand, low metallicity does not rule out planets. They have been detected around metal poor stars. Extreme cases are a giant star HD 47536 and a main sequence star HD155358. Their metal abundances are only one-fifth of our Sun. Both stars have two known planets.

Binary Stars and Planets

If a third body, say, a planet, is inserted in random into a binary system, it is likely that it will be sooner or later ejected from the system. There are, however, some families of orbits that are dynamically stable where planets could exist. A tight binary, could have a common planetary system, where the planets orbit the system regarding the binary as a single "core." In a very wide binary system each of the stars could in principle have their own planetary systems. Totally different types of planetary orbits are also possible in a binary system. Some of these require specific mass limits for the binary components. If the lighter binary component is less than 1/26 of the heavier member then *Trojan* orbits are possible. These kinds of orbits are found in our Solar System in association with each giant planet. The Trojan asteroids stay close to an equilibrium point, which forms an equilateral triangle with the two heavier components, the Sun and the giant planet. Other stable orbits also exist, but we will not go into detail about them.

Understanding Planetary Formation

The standard picture for planet formation (Chap. 29) explained our Solar System and is likely to be correct in general. However, the separation of inner rocky planets and outer gaseous planets reflects the temperature of the protoplanetary disk, and whether it was too high or too low for water ice to form. In this picture, there was no room for a giant planet to form very close to the star. Thus the exoplanets with an orbital radius a less than 0.4 AU pose a serious problem. Hot Jupiters (with $a < 0.05$ AU) are an additional problem (they compose about 10% of the known planetary population). The solution to this dilemma was suggested by Peter Goldreich and Scott Tremaine as early as 1980. They proposed that once a planet is formed in a protoplanetary disk, it might experience migration, due to angular momentum exchange between the planet and the gas disk. Computer studies indicate that this migration can be rapid. The planet moves inward because the net torque it feels from outer part is larger than the torque from the inner part. This very rapid migration (Type I) has a time scale of about one-tenth of the accretion disk's lifetime or

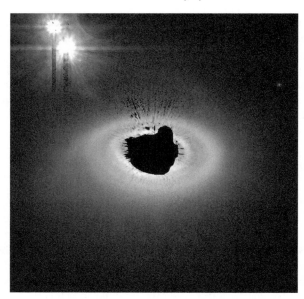

Fig. 32.4 The dust disk around a young star 320 light years away in the constellation Libra. The disk may be a birthplace of planets, but its structure is also influenced by two nearby stars in this triple system. The black patch marks the region where light from the star has been masked out in this coronagraphic image by the Hubble Space Telescope (credit: NASA, M. Clampin (STScI), H. Ford (JHU), G. Illingworth (UCO/Lick), J. Krist (STScI), D. Ardila (JHU), D. Golimowski (JHU), the ACS Science Team, and ESA

shorter. A different type of migration (Type II) sets in when the planet has grown large enough to form a gap in the accretion disk. After this the planet drifts slowly (in a low viscosity disk it may stop altogether). These processes could explain how the hot Jupiters got close to solar-type stars. A stopping mechanism is also needed (e.g., tidal or magnetic torques from the star, reaching the inner edge of the accretion disk or the dissipation of the disk) (Fig. 32.4).

The planetary migration or the standard accretion models would predict rather circular orbits, as seen in our Solar System. However, in extrasolar systems we see elongated orbits. The most natural explanation to this are strong gravitational interactions between two planets, when they get onto suitably resonant orbits. Then the effect from interactions may grow nonlinearly, at which point the orbits of the planets change. The changes may be modest, as has possibly happened among giant planets in our Solar System, or they can be more dramatic. One of the planets can be ejected from the system or end up on an eccentric orbit.

So as you can see, the study of extrasolar planets is an active field, changing rapidly. It is, as yet, too difficult to draw firm conclusions about what types of planets form under various conditions. As of August 2008, the count of exoplanets stands at 300. Most of them are Jupiter-like.

Are Any Exoplanets Suitable for Life? Habitable Zones

The frequently detected gas giants close to the central star appear to be totally unsuitable for life. Even if water, oxygen, or other important atoms or simple molecules are found in their atmospheres life would not find it easy there. There are, however, two recently reported planets that for the first time could in principle be suitable for life, but with a narrow zone of comfort.

In 2005 one planet was reported orbiting Gliese 581. This was a Uranus-sized planet with an orbital period of about 5.3 days. As we already mentioned, in 2007 two additional planets were found in this system with masses of five and eight times that of Earth. What makes this discovery more interesting is that these two new planets are in the *habitable zone* of the red dwarf Gliese 581. The first planet may be tidally locked and the second one is barely in the habitable zone.

When considering life, it is convenient to restrict ourselves to some simple requirements. In particular, the conditions have to be such that water remains in liquid form for a reasonable time. It may freeze for a while in winter – we know life can get along with this – but it is not allowed to boil. In a normal atmosphere the temperature limits for liquid water are 0 and 100°C. The freezing point is not very sensitive to a change in pressure, but the boiling point is somewhat more so. If the air pressure were twice the present value, the boiling temperature would be 121°C. A temperature range between 0° and 50°C would be more suitable not only for life as we know it but also for a stable water world.

If we know the host star's luminosity, the distance of the planet, and its radius then we can infer the temperature of the planet assuming thermal equilibrium. The albedo (reflectivity) and the rotation of the planet have to be factored in also. The greenhouse effect plays a role in the final temperature, but it is hard to estimate without more knowledge of the planet. In our Solar System assuming a reflectivity equal to 0.5 (between the values for Venus and Earth) and a slowly rotating planet such as Earth or Mars and zero greenhouse we get a range for the habitable zone with the present conditions from 0.75 to 1.05 AU. If the albedo is 0.2, closer to Martian value we would get a range from 0.95 to 1.32 AU. The Earth sits nicely inside these limits. The limits can be pushed inward by increasing the albedo, and outward by reducing it. Also one should keep in mind the possible greenhouse effect.

As a star evolves, its luminosity changes. The Sun has increased in brightness by about 30% during the age of the Solar System. When the Sun was fainter in the past, the habitable zone was closer (by the square root of the luminosity). For an albedo of 0.5 this would mean a lower limit of 0.66 AU and for an albedo of 0.2 an upper limit of 1.16 AU, so Earth still remains in principle within these limits. It is interesting to note that earlier Venus may have been a good place for life, and for Mars to be in the habitable zone it should have at any time had a relatively strong greenhouse effect. In the future, as the solar luminosity increases, the habitable zone will move out, eventually reaching Jupiter and Saturn. For a new exoplanet the estimation of a habitable zone can be done with the numbers given above scaled by the square root of the luminosity of the central star. What does this mean? If the luminosity of the star is greater, then the habitable zone will be at a larger distance. For a star nine times brighter than the Sun, the habitable zone would be at a radius of 3 AU rather than one AU.

This definition of the habitable zone appears clear, but it excludes some potential sites of life in our Solar System, such as the Jovian moon Europa, and the Saturnian moons Titan and Enceladus. These would be liquid water ocean habitats like the undersea "black smoker" sites, which would be quite independent of the Sun as long as some sort of internal heat source is available. Also, in the cold frontiers of the planetary systems outside the usual zone of habitability, life may end up completely chemotrophic (deriving energy from chemical reactions rather than solar radiation). These prospects have to be held in mind when considering life in other planetary systems.

Another thing that life needs is a shield from the space vacuum and high energy particle and cosmic-ray bombardment. Life could have a "hard cover" shield such as ice (e.g., Europa). Or, the shield could be an atmosphere or a magnetosphere (e.g., Earth). This in turn creates an interesting problem in planets around M spectral class dwarfs (like in the smallest planet in the Gl 581 system). For a planet to be within the habitable zone of a low luminosity red dwarf it has to orbit at a small distance. Because of the small distance the rotation about its axis gets synchronized with the orbital motion, so that the same side of the planet faces the star. At the same time the opposite side points away from the star. On this very cold night side the atmosphere will snow down! Only a rather thick atmosphere with an effective circulation could save such a planet from an unshielded demise.

The spectral type of the star has important attributes to the development of life. Three properties rise above others. The first is the time the star stays in the main sequence. O to A type stars that evolve in less than about 2 billion years may not provide sufficient time for life to develop photosynthesis. The second is the UV flux, which is detrimental for life. It is relatively strong in these same spectral classes. On the other hand, planets around M spectral class dwarfs have plenty of time at their disposal. If life were to form on such a planet, then in addition to the problem of the tidal locking, the third problem may arise, which is the variability of the star. M dwarfs tend to have active chromospheres flaring quite frequency. That leaves F, G, and K stars to consider.

In our Milky Way galaxy all locations may not be equal for life. Out in the stellar halo or in the outer edges of the disk the metal abundances are low; thus, the conditions may be less suitable for the formation of giant and rocky planets, and thus emergence of life. In the inner parts of the Milky Way more energetic young stars are found. Supernovae and other cataclysmic events take place more often. These will not hinder the formation of planets, but the frequent near or full extinctions from these disastrous events may be a hindrance for the establishment of emerging lives.

To recap the astronomical conditions, life as we know it needs temperatures that permit liquid water, a shield from both space and the harmful radiation, an appropriate spectral-type star, and a good location in the host galaxy where the metal abundances are not low and cataclysmic events are at a minimum.

Survivability of Earths and How to Detect a Life-bearing Planet

How would a small rocky planet survive in a merry-go-round of the planet formation? A giant planet moving relatively slowly in type II migration would quite certainly "sweep" all smaller planets. They can be accreted by the giant planet or ejected onto new orbits early in the planet formation. It is thought, however, that some smaller planets could survive these stages. Of course all planetary systems may not need to have giant planets.

The very elongated orbits of giants however are really more of a problem for Earth-like planets, as it is likely that all planets within the extremities of the orbit of a giant planet will at some point come into close gravitational interaction with the larger planet. The fate of an Earth-like planet could be to settle onto a new orbit, or possibly even to be ejected out of the planetary system. These kinds of orbital changes are not beneficial for life at any stage, so it seems quite unlikely that a life-carrying planet could be found in a system which has a giant planet on a very eccentric orbit.

Even though the gas giants themselves are not suitable for life, we should bear the possibility that the moons of these planets may be close in size to Earth and could have conditions suitable for life, provided that the planet and the moons are in the habitable zone of a star.

A life-bearing planet may have some common properties. Possibly it will have beings with DNA and proteins. Life might be water based and so on. Possibly some of these assumptions are not necessary. If we have life on a planet then we should see in any case signs of an atmosphere in a nonequilibrium state. On Earth it means oxygen and ozone. Somewhere else it could mean other combinations of gases, but without better knowledge this is what will be looked for. The other ingredient that will be searched for is water. These are all seen in our atmosphere and not for example in Venus and Mars.

Oxygen, ozone, and water can be measured with infrared spectroscopy. One could also look for signs of chlorophyll. Its spectrum features what is called a red edge. At a wavelength from 700 to 750 μm the reflectance of chlorophyll increases rapidly, so in near infrared plants look very bright. A sharp jump in the reflectance spectrum could be searched for. The exact wavelength of this edge may depend on the host star and the available pigments used for harvesting light of the central star.

Recently new prospects were opened for exploring the atmospheres of exoplanets, when S. Berdyugina and D. Fluri (Zurich's Institute of Astronomy) and A. Berdyugin and V. Piirola (Tuorla Observatory, Finland) for the first time detected light reflected from the atmosphere of an extrasolar planet. In the method the researchers measured the variations of the polarization of the light from the star plus planet system while the planet orbits the star and about every two days crosses in front of its disk. From these variations they could infer the size and some properties of the atmosphere. Light becomes polarized when it is scattered by atoms or molecules, the same process that makes our own sky blue. It is interesting that this pioneering polarimetry study of the "hot Jupiter" over 60 light years from the Earth was made with quite a small instrument, the 60-cm KVA telescope located at the island of La Palma and which was operated by remote control thousands of kilometers away from the observatory.

We are Here

For two-way communication, it is really meaningful to consider only our own Galaxy. The next large galaxy, the Andromeda, is so far away that one single dialog would take about 5 million years.

Several attempts to inform "others" about our civilization have been made. The oldest and possibly the most effective one is the trail we leave without even noticing it. During the past 60 years we have been broadcasting in radio, and this is spreading into the surrounding space at 1 light year each year. Now the bubble of Earth's radio signals is about 60 light years (18 parsecs) in radius. About a thousand stars are found inside this radius and in principle could be listening to our radio programs.

The spacecraft Pioneer 10, launched on March 2, 1972, and Pioneer 11, launched about a year later, carried on board a pair of aluminum plaques. The plaque includes information about our location in respect to pulsars in our galaxy, our position in the Solar System, a picture of a human male and a female scaled to the size of

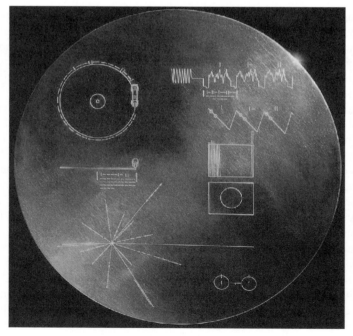

Fig. 32.5 The Voyager Golden Record cover with instructions helping possible extraterrestrials to read our message (credit: NASA)

the spacecraft. The two Voyagers launched in 1977 included a "Voyager Golden Record." It included pictures and sounds from Earth and the human culture. On the cover Earth's position is shown as are instructions for reading the record. Voyager 1 passed the 100 AU distance mark in July 2006 and is the most distant man made spacecraft. It will enter the interstellar space possibly by 2020. The other three space probes are on their own trajectories going out of the Solar System. It will take tens of thousands of years before the spacecraft will approach any star (Fig. 32.5).

Intentional signals at radio waves were sent in 1974 from the 300-m Arecibo radio telescope toward the globular cluster M13. The stream consisted of 1679 bits of 0s and 1s. If the bit stream is plotted into a 73×23 rectangle, then the "1" bits reveal the information, which tells what we are, what we are made of, and where to find us. It also tells our number system and lists the most important elements. This message will reach M13 in about 25,000 years. The globular cluster is a relatively metal poor environment, so the possibility for a rocky Earth-like planet to form is slim, meaning that the chance for a civilization to receive our message is small.

Radio SETI

The search for extraterrestrial intelligence (SETI) was pioneered by Frank Drake, when he made a microwave search for signals from other solar systems in 1960. Independently, the Cornell University physicists Giuseppe Cocconi and Philip Morrison suggested a year earlier that microwaves could be used for interstellar communication. In the early 1960s the Soviets were very active in SETI. Since then SETI work has continued at various levels of enthusiasm. In the early days several difficult selections had to be made, such as the wavelength, bandwidth, integration time, measuring mode, and which stars to look at. These have been overcome by modern receivers, which can record tens of millions of frequencies at high time resolution, and then combine and analyze in various ways. The virtual supercomputer seti@home has been a very powerful idea and tool for this. The SERENDIP (Search for Extraterrestrial Radio Emissions from Nearby Developed Intelligent Populations) project is run as a piggyback program on the Arecibo telescope by University of California, Berkeley. Its sister project at Parkes radio telescope is run by the SETI Australia Center of the University of West Sydney. A further sophistication would be the use of small telescopes in an interferometric mode, where the data from each telescope are correlated. This would enable searches over wide fields and in large number of frequencies. This concept is now in use in the Allen telescope Array in California, which is being built at present. In its final constellation it will have 350 telescopes, each 6.1 m in diameter.

Light in highly directed nanosecond pulsed laser beams might be a more effective means for interplanetary and interstellar transmission. These kinds of pulses do not commonly occur in natural sources. Optical SETI projects at University of California, Berkeley, and Harvard University have been looking for such pulses. Several thousand stars have been searched.

The Drake Equation Or "Is There Really Anybody out There?"

Looking at the starry sky in a dark night will easily let one's imagination fly and wonder if anybody on a planet of that star is looking at our Sun and wondering if anybody here is looking at her star. In 1961 Frank Drake held a SETI meeting at West Virginia. For this meeting he itemized an agenda containing the different steps to calculate the number of civilizations in our galaxy. In doing so he had formulated the "Drake equation." In this formula he multiplied various numbers together and got out the number of civilizations in the Milky Way. The factors include:

- The number of stars in our galaxy, the average rate of star formation
- The frequency of stars with planets, the number of planets in these systems
- The likelihood of planets being suitable for life
- Various probabilities from emergence of life to the emergence of civilization
- Finally to a state where technical communication is possible

Most of these numbers and frequencies that are based on astronomy are now known to a reasonable accuracy, but the last few biological and technological figures are still only well-educated guesses. However, although the equation does not give us solid numbers within a narrow range, it gives us interesting lines of thought. Various scientists have come up with answers ranging from one to a billion civilizations in our galaxy. In fact, one may speak about "pessimists" and "optimists" and use the following extreme approximations for the Drake equation.

An optimist thinks that the probability for the civilization to emerge on a planet suitable for life (which has also emerged with a high probability) is high, almost one. Then their present number in the Galaxy, very roughly, is similar to the lifetime of a typical civilization in years. So, if civilizations exist 1 million years, the optimist would not be surprised to find one million of them in our galaxy! On the other hand, the pessimist views the spontaneous origin of life and/or the subsequent evolution to the civilized level as very improbable on the scale of the Galaxy. So the number of civilizations is much less than their lifetime and in practice we would be the only one here, perhaps along with some dead remains of ancient cultures on their planets scattered across the desolate Milky Way...

The Fermi Paradox

Physicist Enrico Fermi, and even before him Konstantin Tsiolkovski, "the father of space travel," cited the fact that in the long history of our galaxy and considering the spread of humans to every part of the Earth, a typical planet like the Earth should already have visitors. The failure of detecting radio messages accentuates this riddle. Specifically, the "Fermi paradox" is the apparent contradiction between the "optimistic" estimates of the probability of extraterrestrial civilizations and the lack of evidence for such civilizations. Possibly the "pessimists" are right and there is nobody out there, and we are the only technically advanced culture in the Milky Way.

Perhaps other societies could not care less about us. Maybe the radio transmission phase lasts only for a fraction in the life of a civilization, as might be suggested from what is happening on Earth. All transmissions are now going into optical cables, and even satellites use less and less power. It appears that Earth is slowly turning radio silent. There are thus dozens of possibilities in answering Fermi's question "Where are they?" All this may be shrouded by an enigma for a long time: we will not know the correct answer if we never hear from another civilization. But if we get in touch with someone out there the paradox ceases to exist and other fascinating questions about the cosmic life and cultures will fill our minds.

If we are the only civilization in the Milky Way, then it is unlikely that we will ever get in touch with another civilization in another galaxy. If we end up self-destructing, in any of the many available ways, then we understand why – a technically capable civilization does not live long enough even to ask "for whom the bell tolls." If, on the other hand, civilizations exist, and do live longer, then it could be possible to get in touch with one. To get a contact (or even hear from another civilization) would have very profound consequences on humans. Here we have to bear in mind that then, on statistical grounds, that civilization would also be very likely vastly ahead of our 70-year radio broadcasting culture. One is left to wonder if a real communication is possible at all between such very different levels – again a source of inspiration for science, philosophy, and science fiction.

Chapter 33
Human's Role in the Universe

We have discussed how understanding of the structure of the universe, and of its dimensions, age, and formation, has greatly changed our idea of the cosmic role of mankind. The discovery of astronomical cycles and the prediction of future celestial events were important early activities on the road to science. Initially, these cycles were used to determine the seasons for agriculture, and also to entertain other ideas which seemed important at the time, but were later found to be blind alleys (such as astrology). The available observations made it natural to view the Earth as the center of the universe. The starry sky, with the moving Sun, Moon, and planets, all rotated around Earth once a day. Eventually the celestial bodies were taken to be material objects, perhaps created by God but in any case not being gods themselves. Attempts to understand their motions in the sky led to today's science.

Immense Space, Deep Time, and Common Life

Despite early proposals by Greek philosophers such as Anaxagoras that celestial objects were made of the same substance as the Earth, and by Aristarchus that the Earth circled the Sun, the Earth-centered view persisted through medieval times. The universe was considered to be of a finite comprehendible size, enclosed by the revolving celestial sphere of stars. God served as the prime mover of celestial objects in a perfect unchanging realm. Other living things, although also God's creations, were clearly inferior to humans, who, endowed with the capacity to reason, comfortably wore the crown of Earthly creation.

This great concept began to crumble as Copernicus introduced the idea of the heliocentric universe, leaving Earth as just one planet among the others orbiting the Sun. The seventeenth century realization that the stars were heavenly bodies like our own Sun, located at incomprehensibly far distances, changed completely the position of our Sun and of ourselves in the cosmos. The Sun was just another star like the others, endlessly traveling its lonely path through space. Starting with estimates that Earth–Sun distance was 150 million km, and that even the nearest stars

were over 200,000 times more distant, the size of the observed universe continued
to grow to unexpectedly large values.

During the late nineteenth and early twentieth centuries, astronomers found that
our Sun was a member of a huge star system, the Milky Way. In an interesting
variation of the Copernican controversy, our Sun was found to be far from the center of the Milky Way galaxy. With improved measurements of distances outside
of our Galaxy, this huge 100,000 light-year wide system turned out to be just one
among multitudes of others, such as the nearby (over two million light-year distant!)
Andromeda galaxy. In modern cosmology, Copernicus' ideas have been extended to
the ultimate in that our home location in the universe does not appear to be special
in any way.

Beyond this immensity in space, we have recounted the discovery of the vast extent in time, with the toppling of the Biblically estimated age of about 6,000 years.
Biological evolution and the deposit of the huge thickness of geological sediments
suggested longer time spans, and after the discovery of radioactive decay, an immense age of 4.6 billion years has been determined for the Earth. These huge times
might suggest a universe of indefinitely long age. However, the dark night "paradox" and the discovery of the recession of galaxies from one another supported the
ideas that the world cannot be infinite in time or space. Finally, with the discovery
of the cooled-down cosmic background radiation, there was an acceptance of the
basic idea of many creation stories, a definite origin of the universe. However, this
origin has been pushed a 14 billion years into the past.

Fig. 33.1 There may be a diversity of lives in the Milky Way, or our own example may be the only
one – we just do not know. In this encounter, imagined by Georges Paturel, one might as well put
the words to the mouth of our delegate (all is relative . . .), but we wonder if there is anything even
remotely similar to us out there

Considering that the observable universe is billions of light-years in extent, billions of years in age, including hundreds of billions of galaxies, many with hundreds of billions of stars, we certainly realize our meager role in the development of this vast world. Indeed, at first glance it might seem that humans, or even life itself here on the Earth, make no difference to the universe.

Starting from old ideas of the Divine creation of life, which leave many details without answer, we have described how biologists now view the origin and development of life via special but ordinary processes here on Earth. We have described the current understanding of how in suitable conditions life may get started spontaneously, via chemical reactions between some abundant elements and their compounds. We do not know yet how exactly this can happen, or whether it happens easily, or requires unique conditions. In any case, the cosmos is so vast, and it has had so much time to produce all kinds of chemistry, at so many different locations, that it is also possible that the "chemistry of life" has started many times in many places in the universe. It would also seem that if life starts somewhere, and if the conditions stay favorable for a long time, then evolution may produce complex life forms (Fig. 33.1).

On the Other Hand, a Fine-Tuned Universe with Unique Life?

There are many unknown critical factors: We do not know whether the origin is very difficult, or whether the favorable conditions would be quite rare or even unique. Depending on these factors we may be just one representative of multiple different biospheres existing on planets scattered through our Milky Way and the universe, or, at the other end of the odds, we are just one lone phenomenon, an unlikely and odd product of the cosmic chemistry. Thus, even if we already know earthly life to be a very small factor on a cosmic scale, we are still left with a rather uncertain view of the role or possibility of life, as a whole.

However, there is one perspective that seems to give sentient life strange cosmic significance. Only a sentient, intellectual species can become aware of the universe, and understand its natural rules and regularities. It seems that through thinking, self-aware species the universe becomes aware of itself!

A sentient species can also recognize the links between the physical laws and its own existence, i.e., the preconditions that have allowed its presence in the universe. In this respect, we have realized that our kind of life is strictly dependent on many specific physical parameters, both on a large, cosmic scale, and also locally here in our planetary system. Even if life is a result of cosmic chemistry, this chemistry and physics function exactly so that it can produce and maintain life. It almost seems that life is a specific product of the cosmic evolution of the universe, not just a local accidental event.

If there are other universes, maybe existing in parallel to ours, with different cosmic constants and parameters, those would not have produced life, or not at least life functioning the same way as on the Earth. We do not know whether any possible different lives in such unseen worlds could also be sentient. But, it has become more and more apparent because we are what we are, then we *must* be living exactly in *the kind of universe where we are*. Any other, slightly different world would not have produced us. But does this mean that the universe is just as it is for the purpose of producing sentient life like us? This is quite too ego-centric for us, who as a result of the Copernican Revolution have learnt to view ourselves as just specks of dust in a vast cosmos. If the laws and structures of the universe reflect our presence, this is just because otherwise we could not be here wondering about this state of affairs. This so-called *anthropic principle* is a rather common attitude among scientists.

So, what are the properties of our universe that have made life possible, and what special was required for sentient life? As we examined our Solar System, it was interesting to see that Venus, a "sister planet" to the Earth, by being a bit closer to the Sun than Earth, wound up being totally unsuitable for life as we know it. Similarly, Mars, although possibly suitable for life in the distant past, is now a rather hostile place. On the other hand, our better understanding of the life on Earth has revealed ocean floor "black smokers" and even extreme deep rocky subsurface sites where the most archaic micro-organisms flourish. These results have extended the range of suitable environments for the origin of life, essentially where an energy source and liquid water exist, such as satellites of Jupiter-like planets.

The immense age deduced for the Earth indicates that a huge time was required before sentient technological life appeared here. Generalizing, habitable planets need to revolve about their parent stars in very stable orbits for long enough times to allow life to evolve. Although many other planetary systems have been found, their close-in Jupiter-like planets hamper the existence of Earth-mass planets on stable orbits, at suitable distance from the central stars. However, since it is very difficult to discover other than such massive close-in planets, it is reasonable that many, or perhaps even most, planetary systems may still be like our own and possibly contain Earth-like planets.

As mentioned in our discussion of the number of dimensions (Chap. 18), stable planetary orbits would not be possible in a universe with more than three actual spatial dimensions. Also the electron orbits in atoms would be unstable there, making very problematic the chemical bonding, required to form the complex molecules for life. Some theories propose that there would be many dimensions of space, and most of these would have wrapped up during the early stages of our universe, leaving us with three spatial and one time dimensions. We can imagine other universes maintaining four, five, or even more of spatial dimensions, but life as we know would not be possible in such exotic places.

Natural Laws and Universal Constants

Also natural laws and constants are exactly such that they have allowed chemical evolution from the Big Bang to humans. For instance, if all the hydrogen had been converted to helium in the Big Bang, then we would have neither any water, nor life in the universe today. This would have happened if the nuclear force binding together protons and neutrons had been slightly stronger than it is. In our actual universe, the nuclear force is just strong enough to bind a proton and a neutron into a deuterium nucleus, but it is not strong enough to hold two protons together (overcoming their strong electrostatic repulsion). In an alternative universe, an addition of only 3.4% to the nuclear force would make it able to stabilize a nucleus with two protons to form ^{2}He-nuclei. Such helium nuclei would have formed so easily in the Big Bang, that nearly all the hydrogen would have been converted to helium. There would be no hydrogen compounds in such a universe, and no long-lived stars that use hydrogen as fuel.

On the other hand, had the nuclear force been only 9% weaker, it would not have held together the deuterium nuclei, an essential link in the chain for making the heavier elements. Without deuterium we would not have carbon, or the compounds like the proteins and nucleic acids. The nuclear force needs to be, with an accuracy of a few percents, just what it is in order for life to form.

The importance of the precise value of the nuclear force was realized by Fred Hoyle. In the 1950s he pointed out that the nuclear reaction that produces carbon (fusion of three helium nuclei) happens efficiently only if the nuclear force is exactly of a particular strength. He thus proposed on the basis of our existence as carbon-based life forms, the value of this particular nuclear parameter. Experimental nuclear physics over the following years proved that Hoyle was right: formation of carbon in stars is really strongly dependent on the value of the nuclear force. Hoyle also pointed out another lucky coincidence, the conversion of carbon to oxygen in stars does not happen as efficiently as the formation of carbon, and allows carbon to accumulate. Life would not have had much chance to succeed in a world with an abundance of oxygen over carbon. If multiple universes exist, then these rather unique parameters make our universe a good place for life, as compared to most others.

We have seen that one of the most abundant elements, carbon, had just the right chemical affinities to form four covalent bonds and make large complex molecules. It also turned out that the most common compound in the universe, H_2O, functions as the optimal solvent to support biochemical reactions. It seems that the potential for the full chemical repertoire of life was included in the special properties of carbon and water. These in principle are available in planets throughout the Milky Way.

One physical process essential to the formation of the first stars was the slightly uneven density distribution in the initial radiation and nuclear plasma that formed in the Big Bang. These allowed the uneven distribution of the early hydrogen and helium clouds, which then allowed contraction to form the first stars, the factories producing the very first heavier elements necessary for life.

Beyond the important subatomic parameters for processes within stars, there is also the weak force of gravity, the builder of cosmic structures. If this force had been any weaker, or on the other hand, any stronger, then the formation of stars would not happen as it does now. If the force is weaker, the heavier elements would not be produced, or alternatively, if stronger, the stars would develop so fast that their planetary systems would not last long enough to allow life to evolve. Again, other universes with different gravitational parameters would be unsuitable for life.

The age of the universe and stars is crucial for the existence of life. If the universe was very short lived (say, would last only for a million year) life could not exist. The elements of life, the carbon and others, have been formed in the nuclear reactions inside stars and have been blown out to the interstellar clouds as these stars have exploded. Likewise, formation of new stars and their planets takes time. Earth-like planets would not be possible around stars without significant amounts of the more massive atoms in their protoplanetary disks. Stars during the early history of our Galaxy would not have had planets suitable for life. Accumulation of the required elements in the gas clouds from which later stars and planets would form would take awhile, but the required time is uncertain.

Then, after the synthesis of the heavier elements, and formation of new stars and planetary systems containing these elements, life may have started on some of those planets. Still after that, its evolution into complex life forms would have taken billions of years, as it has taken in our case. We can understand that this gradual, long-lasting process has been the precondition to our existence (Fig. 33.2).

Fig. 33.2 The constants and laws of Nature are such that a star like the Sun keeps shining about 10 billion years, allowing the birth and evolution of life on suitable planets orbiting the star. In this picture our Sun is seen from an unusual perspective, setting below the rim of a crater on Mars in 2005 as imaged by the Mars Exploration Rover Spirit (credit: NASA/JPL/Texas A&M/Cornell)

Focus on the Solar System

However, if the evolution of life, from its initial elements to the cells and complex biochemistry, took a long time, it also did take some exceptional local conditions: indeed it appears that our blue planet offered just right things for life to survive and to develop further. As in the story of the three bears, this planet has been a very cozy home for Goldilocks. Indeed, the Earth is located at a very convenient distance from the Sun, such that, together with a protecting greenhouse atmosphere, it maintains suitable temperatures to keep water in liquid form, at least for most of the time. However, these temperate eras have been interrupted by times when the greenhouse atmospheric gases have been lost, and the temperatures have plunged below freezing for millions of years. During a few of these glaciations, temperatures have remained so cold, for so long, that the whole planet has become covered by ice. The "snowball earth" might have remained so forever, had not the warm heart of our planet released enough gaseous CO_2, to restore the greenhouse to warm up the atmosphere. And, fortunately, we have noted how water has the property of being denser than ice. Therefore, the oceans froze over only from the surface, and life had a chance to survive underneath in the liquid water, protected from freezing and drying as seen today in lakes in Antarctica. Finally, the hot, molten outer core and the solid iron inner core of the Earth have provided the magnetic field to protect life from harmful cosmic radiation threatening fragile young life on any planet.

Early cosmic impacts of comets and asteroids were also important for making the Earth habitable: they brought here at least most of the water and gases that we have here now. Also, a very large impact, at the very early stage of the formation of the planet, gave us our heavenly companion, the Moon. These large impacts and the Moon itself have been very good for us. They tilted the Earth's axis so that the southern and northern hemispheres take turns in facing the Sun: this creates the seasons and helps to even up the temperatures among the different parts of the planet. The impact also gave Earth its spin. This spin was much faster early on, but has slowed down to its current 24 h per turn, creating for us this daily rhythm of light and dark, instead of one side being locked to be constantly dark, and the other one light. The presence of the Moon continues to stabilize the axis of our planet so that the global climates do not vary randomly. These factors are clearly important for the conditions of life here on Earth, but we do not know exactly how crucial a big satellite is to the origin and continued existence of life. It might be a restrictive factor on how common place life on other planets like Earth would be.

Collisions with comets and asteroids have had both physical and biological impacts on life on Earth. They may have been fortunate for life in the early Earth by transporting the seeds of life from one planet to the other in the near Solar System (say, from Mars to Earth, or vice versa). They have also been significant for the later evolution of life, by causing, at repeated times, massive extinctions in the existing biosphere giving a chance for new species to emerge. Although these events have been catastrophic to those species that have gone extinct (like the dinosaurs), they have been very useful for those species that got a chance to emerge (like the mammals). However, too frequent comet impacts might make the existence of any complex species very short. Therefore, our slow evolution and (so far) safe existence here has been greatly assisted by the large planet Jupiter. It has by now cleaned away the majority of rocky bodies, which were frequent travelers through the early Solar System. The effects of these impacts are plus or minus with the net effect being hard to judge.

Life Affecting Itself and Its Planet

The biosphere itself has greatly modified planetary conditions. This apparently happened in a complex interplay between the physical and biological systems, and certainly complicates details beyond the simple question of how common life is in the universe. For instance, the appearance of oxygen generating photosynthetic organisms produced oxygen into the atmosphere - and this has greatly affected the conditions for living species. The oxygen atmosphere allowed production of the ozone layer, which effectively blocks out hard UV and shorter wavelength radiation. This shield against radiation damage allowed life to move from water to dry land. The nitrogen-oxygen atmosphere is most transparent to visible light wavelengths, a fortunate match to the most abundant wavelengths produced by our Sun, thus allowing a major part of the solar energy to enter the Earth's surface. These are the wavelengths used by the photosynthetic biota for energy capture for carbon fixation and, not unexpectedly, also the wavelengths at which most animals see.

Over time, the evolution and generation of species has been driven by Darwinian process, via the genetic variation from mutations and establishment of the surviving progeny via natural selection. It is a common belief that the selection is ruled by simple "survival of the fittest" by the strongest competitor, but actually the criteria for "the fittest" are not so simple. The surviving species and individuals adapted to their immediate environments. In many cases they would have been those who could interact with the surrounding in a sustainable way, rather than those who would over-exploit it. Beyond simple competition, evolution was also driven by beneficial interaction between species, e.g., in diverse symbiotic microbial mats where the nutrients are passed from one layer to the next, or in food chains and ecosystems formed by higher organisms. Thus competition has induced specialization and differentiation of competing groups, and it has driven the diversification of new species. The growing complexity of ecosystems has apparently created more and more diverse niches, supporting the survival with increasingly different life strategies, diverse species, and more complex life forms.

The evolution of species has been also strongly affected by local and global events. Long-term geological changes, such as the drifting of the continents, have changed the climate over long periods, and the biosphere itself has modified the atmosphere. Jointly, these processes have made the climate change over long time spans, varying between global glacial and temperate periods. Major natural catastrophes via the impacts have repeatedly caused mass extinctions and wiped out large portions of the biosphere. The accidental catastrophes have often caused a major break in the existing ecosystems, wiping away most of the biota, and clearing up empty space for new colonization. At these times the renewed conditions have given rise to new biological lineages and new types of ecosystems and a new turn to the evolution of the biosphere. However, in between these major changes, the biosphere usually has been evolving steadily, with small gradual changes and adaptations. The paleontologist, Stephen Gould (1941–2002), called this variation of calm and stormy phases *punctuated equilibrium*.

Within this stormy evolution of the biosphere, fairly stable conditions are undoubtedly needed for the evolution of intelligent, technically capable species. From experience on Earth, it would seem that dry land is required for making a technical culture. The first step for the technical exploitation of the environmental energy sources is the efficient and adequate production of food. Since the invention of agriculture about 10,000 years ago, we have luckily lived in a stable climate without ice ages.

A Matter of Time

The 14-billion-year age of the universe, the 4.6-billion-year age of the Earth, and the 3.9 billion years of the biosphere, and even the durations of the various geological eras are difficult to comprehend. We can visualize the geological time spans by scaling the whole history of Earth to a time line of one year – within this same

Fig. 33.3 In December 1968 three men – Frank Borman, James Lovell, and William Anders – orbited the Moon ten times in the Apollo 8 mission. Three centuries after Newton's *Principia*, this picture of the Earth rising above the Moon's horizon has come to symbolize one result of science and technology: overcoming and making use of gravitation for spaceflight (credit: NASA)

scale, the age of the universe would be three years. If we think that the Earth was formed on the first day of January of the third year, then the oldest existing rocks were formed, and apparently, life started around the tenth of February. Then there was a very long period of evolution through the spring, the summer, and the autumn. The first primitive animals appeared in the mid-November, the first plants grew on dry land on December 10th, and the era of dinosaurs ends up in a catastrophe on the evening of December 26th. Homo sapiens separates into its own species on December 31st, at 6 p.m., and the last glaciation recedes from Scandinavia about one minute to midnight. The beginning of our western calculation of time happened only 14 s before the midnight or the present time.

So only for the tiniest fraction of the existence of our planet has a technical civilization existed. Its continued existence over a long time is in doubt. Considering this fraction it would indeed be fortunate to have a neighboring star in the Milky Way currently having a civilization which could visit or even communicate with us. We wonder if this might be the explanation of the Fermi Paradox which we encountered in Chap. 32, though other explanations are possible, too.

Given the vast numbers of stars in the universe, we have hope for the existence of life elsewhere, but the existence of complex life forms, intelligence, or even a technical civilization has a decreasing likelihood. The huge complexity of life, and even more, the existence of sentient, intellectual life that can comprehend the surrounding universe, is telling us a message of slow evolution, taking place over immense lengths of time. Long time scales and massive synthesis of elements require a cosmos of great age – so, only a large, long-lasting and slowly evolving universe could have produced complex life. However, in the vast, cold, and old universe ruled by dark matter and dark energy, some warm and safe local niches of ordinary matter are needed to make home for life. The Earth is one of them (Fig. 33.3).

We started with the words by biologist Huxley, expressing how we, living now on Earth, as well as all those human beings that have ever lived in the past, stand before the mystery of the universe. Another quote sums up the result of the long exploration in this book. We have seen how human understanding has expanded out to the universe, and our central position has evaporated in this process. But we have also seen that the universe, its history, and even the values of its physical constants are intimately connected with the emergence of life on our Earth, our own existence, and the possibility of other worlds like our own. This gives us hope that even if the nearest site of alien life is very far away or even if we are truly alone in our cosmic solitude, we may finally reach real understanding on the phenomenon of life in the universe, and consequently, on ourselves.

We shall not cease from exploration
And the end of all our exploring
Will be to arrive where we started
And know the place for the first time.
Four Quartets by T S Eliot.
Quartet number 4: Little Gidding.

Recommended Reading

Here we list some literature which we have found useful when preparing this book and which contain supplementary material for the reader who wishes to delve slightly deeper into some items. A useful list of many good Web sites, also grouped according to the parts of our book, may be found at http://bama.ua.edu/~byrd/Evolving_UniverseWeb.doc with "control click" links directly to the documents.

As a general astronomy book, with a clear exposition of basic concepts and theories, we recommend the much applauded

H. Karttunen, P. Kröger, H. Oja, M. Poutanen, K.J. Donner (eds.): *Fundamental Astronomy* (5th edition, Springer, Berlin, 2006); also available in German (*Astronomie – Eine Einführung*, Springer, Berlin, 1990) and in Finnish (Tähtitieteen perusteet, 3rd edition, Ursa, Helsinki, 2000)

and as a good introduction to biology

W.K. Purves, D. Sadava, G.H. Orians, H.G. Heller: *Life, the Science of Biology* (7th edition, Sinauer Associates, Sunderland, MA, 2004)

Part I

T. Heath: *Aristarchus of Samos* (Dover, New York, NY, 1981)
M. Hoskin: *Stellar Astronomy* (Science History Publications, Chalfont St Giles, UK, 1982)
T. Kuhn: *The Copernican Revolution* (Harvard University Press, Cambridge, MA, 1957)
S. Sambursky: *The Physical World of the Greeks* (Routledge & Kegan Paul, London, 1963)
S. Webb: *Measuring the Universe* (Springer & Praxis, Chichester, UK, 1999)

Part II

J. Banville: *Kepler*. Minerva paperback, 1990
P. Davies: *Superforce. The Search for a Grand Unified Theory of Nature* (Heinemann, London, 1984)
P. Davies (ed.): *The New Physics* (Cambridge University Press, Cambridge, UK, 1989)
B. Greene: *The Elegant Universe. Superstrings, Hidden Dimensions, and the Quest for the Ultimate Theory* (Vintage Books, New York, NY, 2000)
B. Greene: *The Fabric of the Cosmos. Space, Time and the Texture of Reality* (Vintage Books, New York, NY, 2004)

M. Kaku: *Einstein's Cosmos. How Albert Einstein's Vision Transformed Our Understanding of Space and Time* (Phoenix, London, UK, 2004)

B. Mahon: *The Man Who Changed Everything. The Life of James Clerk Maxwell* (Wiley, England, 2003)

Y. Ne'eman & Y. Kirsh: *The Particle Hunters* (Cambridge University Press, Cambridge, UK, 1986)

V.J. Ostdiek & D.J. Bord: *Inquiry into Physics* (West Publishing, St. Paul, MN, 1995)

E. Segrè: *From Falling Bodies to Radio Waves. Classical Physicists and Their Discoveries* (W.H. Freeman, New York, NY, 1984)

E. Segrè: *From X-Rays to Quarks. Modern Physicists and Their Discoveries* (W.H. Freeman, San Francisco, CA, 1980)

S. Weinberg: *The Discovery of Subatomic Particles*. Scientific American Library (W.H. Freeman, New York, NY, 1983)

Part III

A.D. Aczel: *God's Equation. Einstein, Relativity, and the Expanding Universe* (Random House, New York, NY, 2000)

Yu.V. Baryshev & P. Teerikorpi: *Discovery of Cosmic Fractals* (World Scientific, Singapore, 2002); also available in Polish (Wydawnictwo WAM, 2005) and in Italian (Collana "Saggi scienze," 2006)

R. Berendzen, R. Hart, D. Seeley: *Man Discovers the Galaxies*. (Science History Publications (a div. of Neale Watson Academic Publications, Inc), New York, NY, 1976)

G.G. Byrd, A.D. Chernin, M.J. Valtonen: *Cosmology: Foundations and Frontiers* (Editorial URSS, 2007)

A. Fairall: *Large-Scale Structure in the Universe* (Wiley, New York, NY, 1998)

E.R. Harrison: *Darkness at Night* (Harvard University Press, Cambridge, MA, 1987)

E.R. Harrison: *Cosmology – The Science of the Universe* (2nd edition, Cambridge University Press, Cambridge, UK, 2000)

D. Layzer: *Constructing the Universe*. Scientific American Library. (W.H. Freeman, New York, NY, 1984)

J.-P. Luminet: *The Wraparound Universe* (A K Peters, New York, NY, 2008)

S. Mitton: *Fred Hoyle. A Life in Science* (Aurum, London, UK, 2005)

A. Sandage: *The Mount Wilson Observatory. Breaking the Code of Cosmic Evolution* (Cambridge University Press, Cambridge, UK, 2004)

E.A. Tropp, V.Ya. Frenkel, A.D. Chernin: *Alexander A. Friedmann. The Man who Made the Universe Expand* (Cambridge University Press, Cambridge, UK, 1993)

S. Weinberg: The First Three Minutes (2nd edition, Basic Books, New York, NY, 1993)

C.A. Whitney: *The Discovery of Our Galaxy* (Angus & Robertson, London, UK, 1972)

Part IV

T. H. van Andel: *New Views on an Old Planet. Continental Drift and the History of the Earth* (Cambridge University Press, Cambridge, UK, 1985)

R. Baum & W. Sheehan: *In Search of Planet Vulcan – the Ghost in Newton's Clockwork Universe* (Plenumtrade, New York, NY, 1997)

A. Brack (ed.): *The Molecular Origins of Life. Assembling Pieces of Puzzle* (Cambridge University Press, Cambridge, UK, 1998)

M. Chown: *The Universe Next Door* (Headline Book Publishing, London, UK, 2003)

S. Jay Gould: *Ever Since Darwin. Reflections in Natural History*. (W.W. Norton, New York, NY, 1977)

J. Gribbin: *Deep Simplicity. Chaos, Complexity and the Emergence of Life* (Penguin Books, London, UK, 2005)

B. Jakosky: *The Search for Life on Other Planets* (Cambridge University Press, Cambridge, UK, 1998)

A. Hallam: *Great Geological Controversies* (Oxford University Press, Oxford, 1983)

G. Horneck & P. Rettberg (eds.): *Complete Course in Astrobiology* (Wiley, New York, NY, 2007)

C. Tombaugh & P. Moore: *Out of the Darkness. The Planet Pluto. A Twentieth Century Adventure of Discovery* (Stackpole Books, Harrisburg, 1980)

S. Webb: *If the Universe is Teeming with Aliens... Where is Everybody?* (Springer & Praxis, Chichester, UK, 2002)

Index

Printed in the United States of America